Green Energy and Technology

For further volumes:
http://www.springer.com/series/8059

Marius Paulescu · Eugenia Paulescu
Paul Gravila · Viorel Badescu

Weather Modeling and Forecasting of PV Systems Operation

Springer

Marius Paulescu
Department of Physics
West University of Timisora
Timisora
Romania

Paul Gravila
Department of Physics
West University of Timisora
Timisora
Romania

Eugenia Paulescu
Department of Physics
West University of Timisora
Timisora
Romania

Viorel Badescu
Candida Oancea Institute
Polytechnic University of Bucharest
Bucharest
Romania

and

Romanian Academy
Bucharest
Romania

ISSN 1865-3529 ISSN 1865-3537 (electronic)
ISBN 978-1-4471-4648-3 ISBN 978-1-4471-4649-0 (eBook)
DOI 10.1007/978-1-4471-4649-0
Springer London Heidelberg New York Dordrecht

Library of Congress Control Number: 2012949372

Printed on acid-free paper

Springer is part of Springer Science+Business Media (www.springer.com)

To those who know that prediction with no place for doubt is a superstition

Foreword

The world population is constantly increasing and the world electricity consumption will presumably double by 2050 with potential dramatic effects on our climate. It is expected that worldwide primary energy demand will increase by 45 %, and demand for electricity will grow by 80 % between 2006 and 2030.[1] Consequently, without decisive action, energy-related greenhouse gas (GHG) emissions will more than double by 2050, and increased oil demand will intensify concerns over the security of supply. There are different paths toward stabilizing GHG concentrations, but a key issue in all of them is the replacement of fossil fuels by renewable energy sources.

The EU's dependence on imports of fossil fuels (natural gas, coal and crude oil) from non-EU countries, as a share of total primary energy consumption, rose from 50.8 % in 2000 to 54.2 % in 2005.[2] In addition, baseline scenarios show a rising dependence on imports for most fossil fuels, although this is particularly relevant for gas, with imports (as a percentage of primary energy consumption) rising from around 59 % in 2005 to up to 84 % by 2030. In order to correct this situation, and considering that many countries have decided to lessen their dependence on nuclear energy, the European Union has adopted the goal of having 20 % of its electricity supply from renewable energy sources by 2020, along with a commitment to achieve at least a 20 % reduction of greenhouse gases by 2020, compared to 1990 (European Directives 2009/28/EC and 2009/29/EC).

Wind and solar power are presently considered as the sources of renewable energy with the best chance to compete with fossil-fuel energy production in the near future. However, for all the present and future wind turbines and solar power plants to be worthwhile, there must be sufficient wind and solar energy potential available. What happens when these conditions are not met? Wind and solar energy forecasting

---

[1] IEA (2009) World energy outlook. International Energy Agency, OECD Publication Service, OECD, Paris.

[2] EEA (2008) Energy and Environment Report 2008. European Environmental Agency Report EEA Report No 6/2008, Chap. 2. http://www.eea.europa.eu/.

techniques as well as electrical grid management developments aim to answer such questions and have the goal of helping developers of renewable energy power plants to decide where to install and how to operate them as well as to help the grid operators manage this per definition intermittent production input more efficiently. Indeed, for the operational management of electrical grids, integrating different power sources and dealing with the highly spatially distributed locations of the power plants together with the intermittent, weather-dependent production becomes a very important aspect and determines if the energy production will remain balanced with the demand. In other words, the increased penetration of renewable energies implies that the electrical grids will have to adapt to this new situation: the intermittent, difficult-to-predict character of this kind of energy production will be a growing challenge for the Transmission System Operators (TSOs) which have to cope with the dangerous risks of grid instabilities. New forecasting systems as well as enhanced grid management techniques are therefore needed to increase the predictability and integration of renewable energies for widespread penetration.

Furthermore, what is today needed is a European approach which would allow increasing electricity transfers between the countries. Meteorological conditions in Europe are such that the wind is likely to blow or the sun is likely to shine at some place in Europe: in order to increase the penetration of renewable energies, it is mandatory to consider the electricity exchanges on a more extensive scale.

In the near future, the number of relatively small decentralized production units will grow dramatically, which will be efficiently managed only by introducing "intelligent" technologies such as smart grids. Another aspect to guarantee electrical grid stability lies in the development of flexible storage capacities which will allow storage of excess energy and delivery of missing energy when necessary. Energy storage is therefore getting a strategic role and will have to be associated with the smart grids in order to adapt in real time and efficiently the energy production to the fluctuating demand. It will help to combine centralized and decentralized (intermittent) production systems. For this purpose, short-term weather—and production—forecasts will play a major role when considering the whole of Europe.

Accurate power forecasting, efficient and intelligent grid management, and increased flexible storage capacity are mandatory for the efficient development of the future energy policies in Europe and elsewhere, not to mention the benefits in terms of climate change.

June 2012

Dr. Alain Heimo
Chair COST Action ES1002 Weather
Intelligence for Renewable Energies WIRE

Preface

In the last years, the weight of solar electricity in the energy mix experienced an impressive augment and this trend is expected to continue. It means that a higher number of solar power systems, photovoltaic or solar-thermal, with inherent variable weather dependent energy production, are fed into the grid. As a result, forecasting the output power of solar systems, for the next minutes up to several days ahead, are of high importance for proper operation the grid. Accurate prediction of solar irradiance is of utmost importance, as this is a measure of available *fuel* of the solar power generator at a given future moment of time.

Apart from wind resources where the forecasting of wind speed is in a rather mature stage, forecasting of solar energy is just in an early stage. In the last years, a few projects dedicated to this matter, like European COST Action ES 1002—Weather Intelligence for Renewable Energies,[3] were deployed around the world. Many research groups started to put great efforts into enhancing the performance of the actual models or to devise more performing better.

The forecasting of the output power of a solar system involves modeling tools which generically should exhibit two functions: first, to predict the solar resource and second, to model its conversion into electricity. A large variety of models and approaches can be considered for implementing the first function. For nowcasting solar irradiance, statistical extrapolation of measurements seems to be an adequate approach, while for tens of hours ahead numerical weather prediction models represent the best solution. For fulfilling the second purpose, the model is chosen in respect to the application: solar-thermal or photovoltaics. All these demonstrate that the syntagma *forecasting the output power of solar systems* covers a very large area of research from atmospheric physics and meteorology to physics of solar cell and advanced electronics.

This book is focused on two subjects: (i) modeling and nowcasting of solar irradiance at the ground and (ii) modeling the output power of PV converters in specific operation conditions. Models developed by the authors along with other

[3] COST Action ES1002 Weather Intelligence for Renewable Energies, http://www.wire1002.ch.

models reported in the literature, accompanied with computational and handwork illustrations, are discussed in the book.

The eleven chapters are structured along logical lines of progressive thought. Chapter 1 deals with the concept of energy mix, including a more detailed book outline in the last section. Chapter 2 introduces terrestrial and satellite-based solar radiation measurements and surveys the largest solar radiation databases. Chapters 3–8 relate solar regime with weather parameters, describing and assessing various approaches for nowcasting solar irradiance and forecasting solar irradiation. Chapter 3 deals with the state of the sky assessment while Chap. 4 is focused on different ways to characterize the solar radiative regime and its stability in a given period of time. Chapter 5 surveys the algorithms for estimating solar radiation at the ground level, targeting the idea of their usage in nowcasting by inputting predicted values of weather parameters. Chapters 6 and 7 are devoted to statistical extrapolation of measurements, being focused on ARIMA (Chap. 6) and fuzzy logic (Chap. 7) forecasting of clearness index on short-time horizon. Chapter 8 proposes a simple way of predicting solar yield by using forecasted values of daily air temperature extremes in temperature-based models for solar irradiation. In the next two chapters, the issues of modeling the output of photovoltaic systems operating in specific weather conditions are addressed. Several models which translate the modules parameters from standard test conditions to real operating outdoor conditions are reviewed and illustrated in Chap. 9. In Chap. 10, a comparative assessment of the results reported in the literature regarding the forecasting of PV systems output are performed. Conclusions and perspectives are summarized in Chap. 11.

The authors hope this book gathers information that may be useful to both researchers in the field of solar radiation forecasting and engineers engaged in power grid control. Also, parts of the book may be used for teaching undergraduate and postgraduate students in related courses.

June 2012 The authors

Acknowledgments

The authors thank Dr. Alain Heimo (Meteotest, Chair COST Action ES1002 Weather Intelligence for Renewable Energies—WIRE) for support and encouragement.

Some results reported in this book were obtained by the authors when working to a grant of the Romanian National Authority for Scientific Research, CNCS—UEFISCDI, project number PN-II-ID-PCE-2011-3-0089 and to the European Cooperation in Science and Technology project COST ES1002.

Some models and testing procedures reported in this book were worked using data measured on the Solar Platform of the West University of Timisoara, developed with financial support from the Romanian Ministry of Research and Education under the frame of the National Research Program PN II, project PASOR 21039/2007.

The authors affiliated to the West University of Timisoara express special thanks to Professor Ion I. Cotaescu for his support.

Contents

Notations

Geographical coordinates and temporal reference	
ϕ	Latitude
L	Longitude
z	Altitude
j	Day within the year (Julian day)
ω	Hour angle
t	Solar time
t_l	Local time

Solar geometry	
δ	Declination angle
ε	Sun–Earth distance correction factor
h	Sun elevation angle
θ_z	Zenithal angle
θ	Incidence angle
β	Surface tilt angle
μ	Surface azimuth angle
μ_S	Sun azimuth angle

Atmospheric transmittance	
τ	Atmospheric transmittance
m	Air mass
β_A	Angstrom turbidity coefficient
l_w	Water vapour column content
l_{o_3}	Ozone column content
p	Atmospheric pressure
p_0	Normal atmospheric pressure
T	Air temperature
u	Relative humidity

Solar radiation	
G_{SC}	Solar constant
G_{ext}	Extraterrestrial solar irradiance
G	Global solar irradiance
G_d	Diffuse solar irradiance
G_b	Beam solar irradiance
H	Global solar irradiation
H_d	Diffuse solar irradiation
H_b	Beam solar irradiation
H_t	Total solar irradiation on a tilted surface
k_t	Clearness index (defined in respect to G_{ext})

Measures for the state of the sky	
σ	Relative sunshine
C	Total cloud amount
κ	Cloud shade
ξ	Sunshine number
ζ	Sunshine stability number

Photovoltaics	
I	Current
I_{SC}	Short-circuit current
V	Voltage
V_{OC}	Open-circuit voltage
P	Power
MPP	Maximum power point
P_m	Power in MPP
η	Efficiency
R_S	Serial resistance
R_p	Parallel resistance
F_f	Fill factor
A	Surface area
k_B	Boltzmann constant
X_{STC}	X measured in standard test conditions

Statistics	
RMSE	Root mean square error
MAE	Mean absolute error
MBE	Mean bias error
$\bar{X}$	Mean of X
Var	Variance
Skew	Skewness
Kurt	Kurtosis

Chapter 1
The Future of the Energy Mix Paradigm

1.1 Current Status in Photovoltaics

The photovoltaic (PV) effect was discovered in 1839 by the French physicist Edmond Becquerel (1820–1891). The first working solar cell was built by the American inventor Charles Fritts (1850–1903), who coated a selenium wafer with a thin layer of gold to form a junction. The device had an efficiency of about 1 %. The first modern solar cell, based on a diffused monocrystalline silicon p–n junction, was created in 1954 at the Bell Laboratories, USA, by Chapin et al. (1954). The efficiency of this cell was about 6 % and its cost was very high, about 250 $/W_p$. Today's commercial cells have roughly three times better efficiency at a hundred times smaller price. During the 1950 and 1960s, silicon solar cells have been widely developed for applications in space. In the 1970s, the energy crisis led to a sudden growth of interest and support for research in the PV sector and for starting the development of terrestrial applications. Various strategies were explored for producing more efficient PV devices, at the same time employing less expensive materials and technologies. During the 1990s, PV standalone and grid-connected systems expanded. The integration of PV generators into buildings turn into a most exciting application, since the cost of the PV system is in part offset by the savings in land costs and building materials that are functionally replaced by the PV panels. During the late 1990s, the PV industry was growing at a rate of 15–20 % per year (Shah et al. 1999), resulting in a massive reduction of the systems installation cost. The expansion of the PV market after 2000 is determined by the demands of PV power generation plants. As of 2011, most recent data prove the PV market growing at a very high annual rate of 30–40 % (Razykov et al. 2011), similar to that of the telecommunication and computer sectors.

The constant growth of PV market definitely forces down the price of a PV system. But this is not enough. In the future, the cost reduction of PV systems

M. Paulescu et al., *Weather Modeling and Forecasting of PV Systems Operation*, Green Energy and Technology, DOI: 10.1007/978-1-4471-4649-0_1,

should be accompanied by an increase of solar cells efficiency in order for the solar electricity price to become competitive on the market. Both issues are briefly addressed in the following.

1.1.1 Solar Cells Efficiency

While the theoretical thermodynamic limit of PV conversion efficiency is of ~93 %, the efficiency of a conventional p–n solar cell is theoretically limited to ~34 % (Shockley and Queisser 1961). This relatively low efficiency is determined by the loss of most of the incident flux of solar energy at the first step of energy conversion. Four mechanisms are involved: (1) reflection, (2) transmission, (3) incomplete absorption, and (4) thermalization of above bandgap energy excess.

(1) Using modern technology, reflection losses were reduced after the year 2000 to almost zero. The techniques include geometrical texturing schemes of semiconductor surface combined with appropriate thickness and refractive index of antireflection coatings. An evidence for this is the Passivated Emitter and Rear Locally Diffused (PERL) cell structure (Zhao et al. 1995), which include a double layer antireflection coating. The Reactive Ion Etching (RIE) procedure has been proven very useful to yield low-reflectance surface on multicrystalline silicon wafers. Using RIE, Ruby et al. (1999) reported a surface reflectance of less than 2 % for most of the usable portion of the solar spectrum. Therefore, the improvements of such techniques for reducing the cell reflectance are not expected to further generate a significant increase of solar cells efficiency.
(2) It is known that the semiconductor must be thick enough to absorb all incoming photons. This condition can be easily satisfied in semiconductors with direct bandgap but it is difficult to fulfill in semiconductors with indirect bandgap, like crystalline silicon, because of their low absorption coefficient. To enhance absorption for the crystalline silicon thin film, light-trapping schemes have been employed (Green 2002).
(3) Only photons with energy greater than the bandgap will be absorbed in the semiconductor material of a solar cell. Consequently, a smaller energy bandgap will absorb a wider band of the solar spectrum.
(4) On the other hand, only the energy equal to the bandgap of the semiconductor material is needed to generate the electron–hole pair. Since most of the absorbed photons have more energy, the excess energy will be lost by thermalization.

The issues (3) and (4) represent absolute physical limitations beyond which technical improvements of single bandgap solar cell are not possible. The first approach to minimize these limitations consists in choosing a semiconductor material with an optimal bandgap. A further increase of the solar cell efficiency is possible by using a multijunction structure (Cotal et al. 2010), e.g., cells stacked on

Table 1.1 Record efficiencies of terrestrial solar cells measured in standard test conditions (1000 W/m^2, AM1.5G spectrum (NREL 2012), 25 °C)

Cell type	Efficiency [%]	Test center	Date
Si crystalline	25.0 ± 0.5	Sandia [http://www.sandia.gov/]	03/1999
Si multicrystalline	20.4 ± 0.5	NREL [http://www.nrel.gov/]	05/2004
Si amorphous	10.1 ± 0.3	NREL	07/2009
GaAs (thin film)	28.3 ± 0.8	NREL	08/2011
$CuInGaSe_2$	17.4 ± 0.5	NREL	04/2009
CdTe	16.7 ± 0.5	NREL	09/2001
Photochemical DSSC	11.0 ± 0.3	AIST [http://www.aist.go.jp]	09/2011
Organic (thin film)	10.0 ± 0.3	AIST	10/2011
Multijunction GaInP/GaInAs/Ge	34.1 ± 0.2	FhG-ISE [http://www.ise.fraunhofer.de/]	09/2009

Source of data Green et al. (2012)

top of each other. By stacking cells in the order of their bandgaps, with the cell with the largest bandgap on the top, photons are filtered as they pass through the stack, ensuring that each photon is absorbed in the cell that can convert it most efficiently. A multijunction cell with a large number of cells, theoretically can reach an efficiency of 68.5 % (Tobias and Luque 2002).

In December 2011, the conversion efficiency of laboratory solar cells obtained by various technologies reaches relatively high values (Table 1.1), e.g., 25 % for crystalline silicone-based cells (Zhao et al. 1998) and 43.5 % for multijunction concentrated cells (source Green et al. 2012). The module efficiency is usually 1–3 % lower than the solar cell efficiency due to glass reflection, frame shadowing, non-unitary packaging factor (i.e., the loss of some cell surface due to the package and wiring). The best results for modules are slightly lower: 22.9 % is the best efficiency reached by a monocrystalline module and 18.2 % is the best efficiency of a multicrystalline module (Green et al. 2012). These records are very important since more than 90 % of today's solar cells production is based on crystalline silicon (Mason 2008). But, these laboratory solar cells and modules originate from sophisticated design and cannot be mass produced due to prohibitive costs. Commercial crystalline PV modules efficiency typically ranges from 12 to 16 %. An outstanding review of the actual PV technologies can be read in Razykov et al. (2011).

Thus, there is enough motivation to look toward new approaches in improving solar converter efficiency. In Green's vision (Green 2003), a *third generation of photovoltaics* will root from nanotechnology. It follows the crystalline (first-generation) and thin film (second-generation) technologies. In order to be competitive on the market, the third-generation solar cells should combine the low-cost of the second-generation with the higher efficiency of the first-generation or better.

Techniques based on various processes such as photon recycling (Badescu and Landsberg 1993) and band-to-band impact ionization (Landsberg et al. 1993;

Landsberg and Badescu 2002) have been proposed in the last 20 years to increase the efficiency of solar cells.

Many new types of solar cells are candidates for the basis of future technologies. Two of them are reminded here. (1) The multiple quantum well (MQW) solar cell, pioneered by Keith Barnham and colleagues from the Imperial College of London (Barnham et al. 2000). A critical review of MQW solar cell efficiency can be read in Anderson (2001). Two-scale models, which combine quantum and classic physics, estimate a conversion efficiency of about 40 % (for instance Paulescu et al. 2010). (2) The intermediate band solar cell concept, introduced by Luque and Marti (1997) with a theoretical demonstration that the insertion of an intermediate band between the valence band and the conduction band of a solar cell semiconductor material can increase the efficiency up to ~63 %.

The simplest way to implement a third-generation approach may consist in using existing solar cells coupled with up and down converters (Conibeer 2010), which are devices attached to the solar cells in order to increase their efficiency. A down converter (Trupke et al. 2002a) absorbs a single high-energy photon and emits two or more low-energy photons. Modeling of solar cells with down conversion of high energy photons, antireflection coatings and light trapping is discussed for instance in De Vos et al. (2009). An up converter (Trupke et al. 2002b) absorbs two or more sub-bandgap photons and emits a single high-energy photon. Realistic models of up conversion in solar cells (Badescu 2008; Badescu and Badescu 2009) demonstrate that their conversion efficiency may exceed 40 %.

With better optical and electrical characteristics of nanomaterials and the fast advance of nanotechnology, the near future can promote the nanostructured solar cells as a real competitor on the market.

From the supply point of view, in 2010 China and Taiwan cumulated 59 % of the solar cells worldwide production. Total cell production from the China/Taiwan region increased from 5.6 GW in 2009 to 14.1 GW in 2010, representing a year-over-year increase of 152 %. Europe is a net importer of PV devices and this trend will probably continue.

1.1.2 PV Market

In the last decade, the PV industry experienced a robust and constant growth and it is expected to continue in the years ahead. Figure 1.1 illustrates the contribution of the main actors to the global cumulative installed capacity.

At the end of 2009, the world's cumulative installed PV capacity was close to 23 GW while in 2010, almost 40 GW are installed to produce some 50 TWh of electricity every year. The EU is the actual leader with almost 30 GW in 2010. This represents about 75 % of the world's total cumulative PV capacity. Japan (3.6 GW) and the USA (2.5 GW) are next in the top. China (0.89 GW) is expected to become a major player in the coming years.

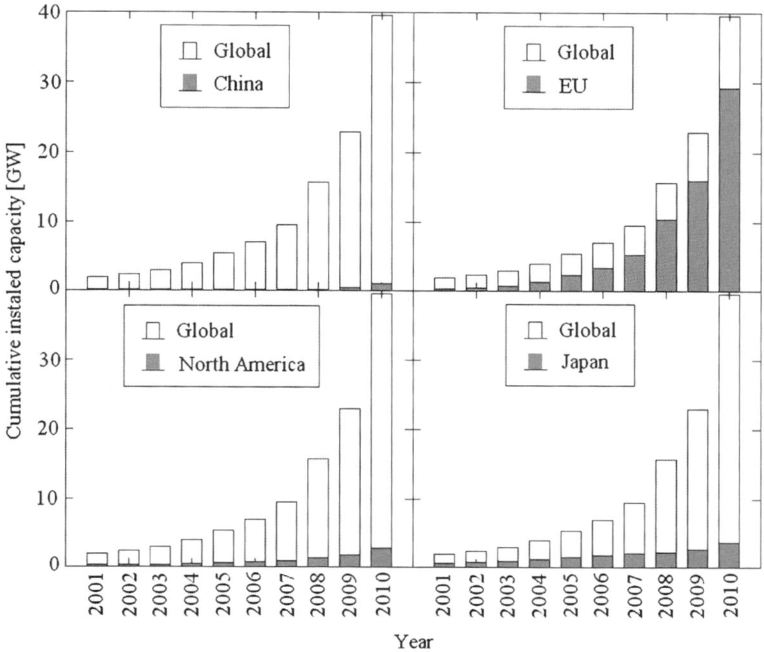

Fig. 1.1 Evolution of cumulative installed PV capacity through 2001–2010. *Source* of data EPIA (2011a)

As Fig. 1.1 shows, the total installed PV capacity in the world has multiplied by a factor of 22, from 1.79 GW in 2001 to 39.5 GW in 2010 with a yearly growth rate of 37.7 %. The PV sector is expected to stay one of the fastest growing of the economy. In terms of market the EU has developed from an annual market of less than 1 GW in 2003 to over 13 GW in 2010 (Fig. 1.2). Inside the EU the development is heterogeneous with Germany the leader (7.4 GW in 2010), followed by Italy (2.3 GW) and the Czech Republic (1.4 GW). The EU took this first position when Germany's market started to grow under the influence of an encouraging feed-in tariff on long-term contract with guaranteed grid access (0.18–0.24 euro/kWh in 2012, down from 0.45–0.57 in 2004) enforced by the German Renewable Energy Act. Under this law the energy market has started to turn away from fossil and atomic fuels, from centralized electricity structures toward renewable energy sources and a decentralized approach of energy production. One can also note from the above tariffs that, while the producers of solar electricity are offered viable prices, they also have to keep pace with the downward tendency in the cost of the PV-generated kWh by employing newest technology.

Indeed, over the last 20 years the price of PV electricity exhibited a downward trend and is expected to decline further in the years to come. PV system prices have declined accordingly and are expected to decrease in the coming years by 30–50 % depending on the segment. In Europe, the cost of PV electricity

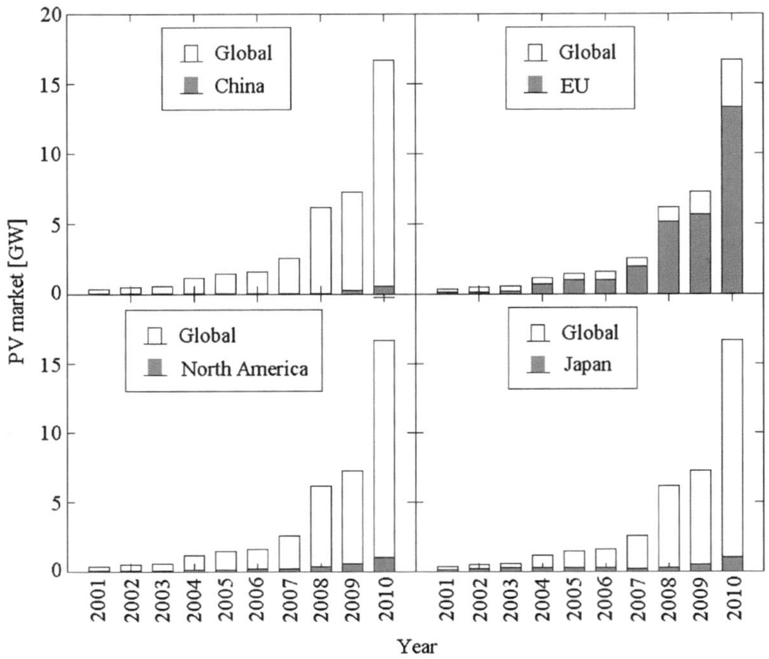

Fig. 1.2 Evolution of the annual PV market through 2001–2010. *Source* of data EPIA (2011a)

generation is expected to decrease from a range of 0.16–0.35 euro/kWh in 2010 to 0.08–0.18 in 2020 depending on system size and the solar resource at the site (EPIA 2011b). It is notable that, although solar electricity is still not cost-competitive with traditional power generation, the price gap to conventional electric energy tariffs is narrowing and is expected to close in around 2015. This, of course, is good news for the consumer who pays for the growth of the renewable energy sector with the electricity bill. In order to reach the ambitious environmental targets set by policy, it is expected that the PV and wind electricity generation growth to be continued in the next years.

1.2 The Energy Mix

The term energy mix refers to the distribution of various sources (fossil fuels, nuclear, biomass, wind, and solar energy) contributing to produce the electrical power delivered in the grid.

In 2010 total global power generating capacity was estimated at 4950 GW. Renewable capacities comprises about a quarter of total power generating capacity and supplies close to 20 % of global electricity. Figure 1.3 shows the share of energy supplies by different primary sources. Excluding hydropower, in 2010

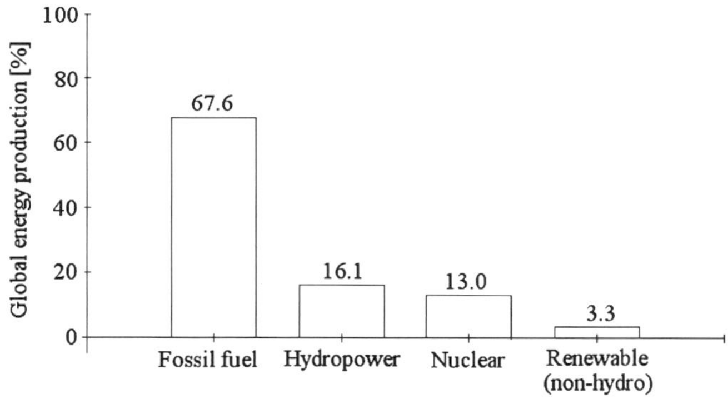

Fig. 1.3 Global energy production by different primary sources in 2010. *Source* of data REN21 (2011)

renewable energies capacity was of 312 GW (a 25 % increasing over 2009) and supplies 3.3 % from the total (REN21 2011). Wind and solar sources contribute to the global electricity production with less than 0.5 % but this sector is growing fast. Solar PV increased fastest of all renewable technologies during 2005–2010 (49, 72 % in 2010) followed by biodiesel (38 %, only 7 % in 2010) and wind (27, 25 % in 2010). PV electricity is estimated to have a contribution of 2 % of global electricity consumption by 2020.

The power generation capacities installed and cancelled in Europe in 2010 is presented in Fig. 1.4. PV was the leading renewable energy technology with an added 13.3 GW compared to 9.3 GW for wind. According to the source considered, the total installations for gas vary between 18 and 22 GW, representing a major increase compared to 2009.

Since the electric grid does not store any energy by itself, the energy network production and consumption must match perfectly. Any imbalance could cause grid instability or failures. Loads and generator availability both have a degree of variability and uncertainty. Standards and procedures have evolved over the past century to manage variability and uncertainty to maintain reliable operation of the electric grids. There are many different ways to manage variability and uncertainty. In general, grid operators use mechanisms including forecasting, scheduling, and economic dispatch to ensure performance that satisfies reliability standards in a least cost manner.

Hydroelectric, fossil fuelled, biomass, and nuclear power plants provide a stable output of electricity because they use a controlled primary source of energy. There are important differences between them leading to the following classification:

- Base-load generators (coal, nuclear) have the lowest costs per unit of electricity because they are designed for maximum efficiency and are operated continuously at high output (more than 80 %).
- Peaking generators (diesel, gas turbines) have short start-up times and are prepared to support the grid during peak hours. They have the highest costs per kWh (but lower construction costs).

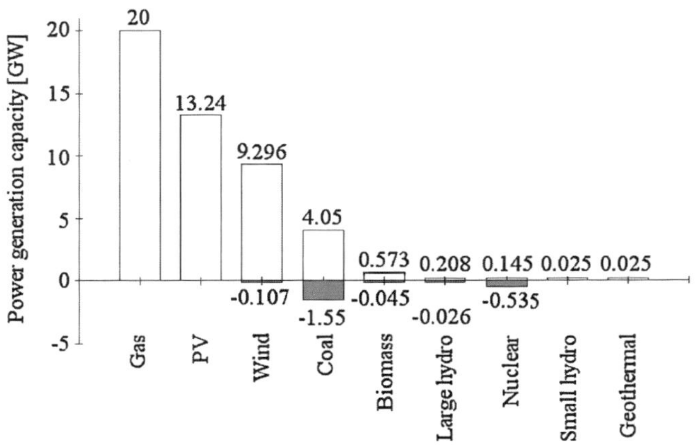

Fig. 1.4 Power generation capacities installed and canceled during 2010 in EU. *Source* of data EPIA (2011a)

- Intermediate generators (hydro, steam turbine plants running on natural gas or heavy fuel oil) provide inertial energy reserve, are capable of quick up- and down ramping to balance load variations (especially hydro), making them an important asset in a grid.

Every reliable energy network must have a mix of the above categories. Then, the additional challenge is to incorporate into the grid wind and solar energy generators, whose primary resource cannot be controlled. Because of the intermittent output they produce, these power sources constitute a threat for the stability of the electric supply. A grid that relies on large percent of electricity generated by such intermittent and irregular plants must be prepared to dispatch sudden changes in energy supply. Basically, other power plants (mainly envisaged are hydro and natural gas plants) have to react quickly to the variations of PV and wind energy sources.

The increase in gas installations (Fig. 1.4) has in fact a logical link with the increase of variable electricity sources such as PV and wind, while the number of coal power plants cancelled in 2010 resulted from the increase in investments in renewable energy, reducing the need for any additional capacities that are not flexible enough to integrate in the future power generation mix.

To conclude, it is necessary that the development of wind and solar capacities to be made along with the construction of new predictable power plants. These are required to absorb the fluctuating load and balance the intermittent supply. Secure electricity supplies depend on the operation of electric grid, which connect consumers to power plants. The fundamental requirement of network operation is to maintain electricity generation continuously equal to electricity demand despite the variation of demand and the variability of supplies from intermittent sources. This calls for an appropriate mix of generation sources. On the other hand, in order to integrate large amounts of fluctuant power plants into the electricity grids,

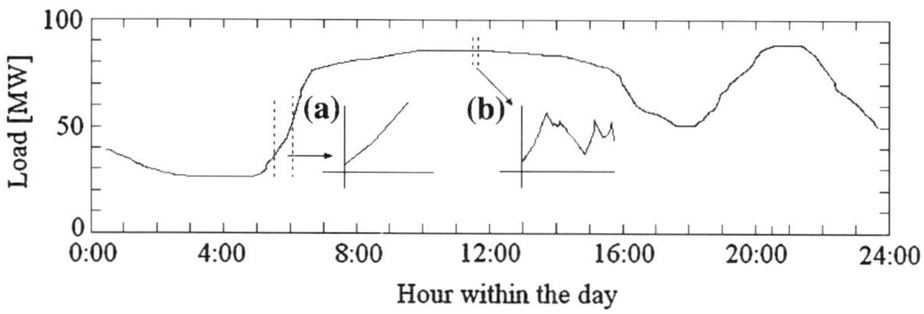

Fig. 1.5 Diagram (generic) of the load variation relevant to the operation of power systems. Inset is magnified the load variation on a time scale of **a** hour, **b** minutes

system operators need both to understand the variability of these systems and to be able to forecast this variability at different spatial and temporal scales.

1.3 Understating PV Systems Variability

The flexibility of a power plant is characterized in terms of parameters such as start-up time, shutdown time, or ramp rate. Power plants based on coal or gas fueled boilers have the longest start-up time, 8–48 h. Gas turbines have a start-up time of order 20 min while hydrogeneration can start almost instantly, in about 1 min.

Figure 1.5 illustrates the relevant time scales for the operation of power plants. Most system operators frequently use a day-ahead commitment process to assign generators to meet the next day's forecasted load. In the time of 10 minutes to hours time scale, operators will change the output of committed generators in order to track changes in load through the day. In the hour fraction time scale appropriate regulation reserves are scheduled in order to balance minute-by-minute the grid.

The response time of a PV plant is almost instantaneous; its output power follows the abrupt change in solar irradiance level due to passing clouds. The performance of PV plants significantly depends on the fact that direct solar radiation is incident or not on the PV arrays. Fast variation of solar radiation may generate the so-called "solar ramp" problem, which is one of the greatest obstacles in operating the power grid (Mills et al. 2011). The term refers to grid management when solar irradiance changes rapidly causing a massive shift in power. When the sun is uncovered by clouds, the direct solar radiation is suddenly incident on the whole PV modules array and the power generated increases rapidly causing an excess of power in the system. The grid operator must ramp downgeneration from other source in order to avoid the grid collapse. When the sun is covered by clouds, a sudden need for electricity occurs and the operator has to turn on other power sources. Solar thermal systems react to solar irradiance changes in minutes while the PV systems react in seconds. Since there are situations when the fluctuation on

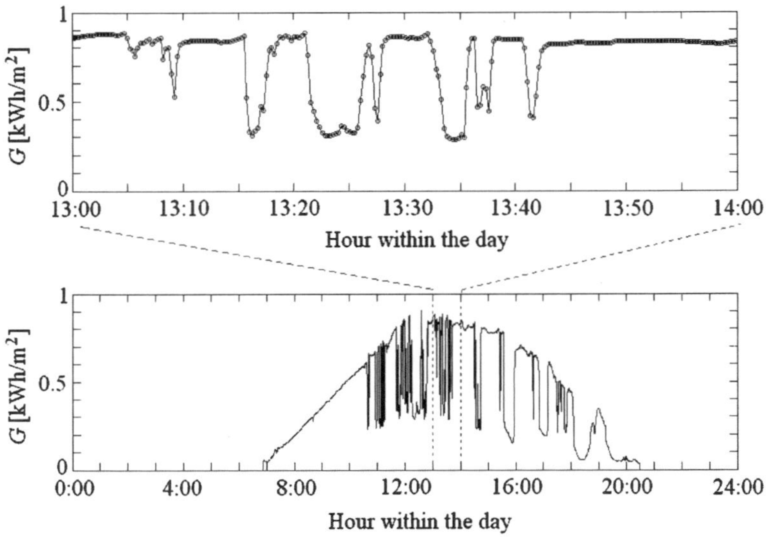

Fig. 1.6 Change in global solar irradiance G of 15 s lag. In the up side the selected area between 13:00 and 14:00 is magnified. Data recorded at Timisoara (45°46′N, 21°23′E, 85 m altitude), Romania in 20 Jul 2010, are displayed

solar radiative regime is on a time scale of minute or less (Tomson 2010; Mills et al. 2011), nowcasting of direct solar irradiance on very short time periods becomes an opportune research area.

Figure 1.6 shows the variation of solar irradiance during a day in the town of Timisoara, Romania (for localization see the map in Fig. 3.1). Large fluctuations of output power may occur in a PV plant located there, with time scales of seconds to minutes. This has to be managed by the grid operator in real time.

Changes in global solar irradiance at a point due to a passing cloud can exceed 60 % of the peak of solar irradiance in seconds. The time it takes for a passing cloud to shade an entire PV system depends on various factors, namely the PV system size and cloud speed. In Ref. Mills et al. (2011) it is showed that a 75 % ramp in 10 s measured by a pyranometer was associated with 20 % in the same 10-s ramp in a 13.2 MW PV plant in Nevada. A severe event that changed the output of a pyranometer by 80 % in 60 s led to a 50 % change in the same time of the power output.

On the other hand, PV systems monitoring at less than 1 min sampling (e.g., 10 s (Burger and Ruther 2005) and 15 s (Ransome and Funtan 2005; Ransome and Wohlgemuth 2005) show that hourly averaging of solar irradiance and PV modules temperature underestimates the delivered PV power in high irradiance conditions. Since the output of PV modules reacts rapidly to changes of global solar irradiance and their temperature changes slowly, PV modules will give higher power than calculated from hourly averages. These show that nowcasting the occurrence of direct irradiance on periods shorter than 1 min is very important for proper grid management.

The geographic area of interest for forecasting can vary from small regions where grid congestion must be managed to a large area over which electricity supply and demand must be balanced. Experience with managing wind energy indicates that gathering diverse wind farms to the same grid leads to a much smoother wind profile than would be expected from scaling the output of a single wind turbine (Holttinen et al. 2009). The same conclusion is also valid for aggregating the output of solar plants located in different sites (Mills et al. 2011). Managing variability is easier when several diverse fluctuating sources are aggregated to the transmission lines. This is in fact the same as at the consumer's end, the daily load shape that system operators use to plan for the real-time operation of the grid is radically smoother than the daily profile of an individual customer.

1.4 Book Outline

It already belongs to common sense that solar energy will play a major role in enhancing energy security while reducing energy-related CO_2 emissions, only the pace of this evolution being disputed. The facts presented above indicate that in the near future the percentage of solar electricity in the energy mix will continuously increase. Day after day, small or large solar systems are connected to the grid. Sometimes, aided by favorable policies, reality exceeds the most optimistic predictions. A good example is the amazing growth of the PV installed capacity in Czech Republic during 2010, from less than 1 GWp to more than 2 GWp. Another example could be Romania, where at 1 January 2012 the installed PV capacity was less than 2 MWp with the Governmental PV Systems Strategy targeting 260 MWp by 2020 (Iacobescu and Badescu 2012). Surprisingly, the year 2012 has already begun with 51 grid-connected PV projects summing up to 240 MWp in various stage of implementation, with a quarter of them planned to be operational before the end of 2012 (Nistorescu 2012).

In order to expand the insertion of solar power on the electric grid, solar resource assessment and forecasting the electric energy generated by solar plants are critical issues. The lesson learned with wind energy shows that accurate wind speed forecasts can substantially reduce grid integration costs (Saintcross et al. 2005). A review of current methods and recent advances in wind forecasting is reported in (Foley et al. 2012). Accurate solar irradiance and irradiation forecasting can be used for proper power grid operation and for scheduling conventional power plants. This should end reducing the solar systems integration cost (IEA 2007).

Many research projects probing ways to provide weather information for accurate forecasting the output power of PV plants are in progress. For example, the European COST Action ES1002 "Weather Intelligence for Renewable Energies" has two main lines of activity (WIRE 2011): (1) to develop dedicated post-processing algorithms coupled with weather prediction models and data measurement especially by remote sensing observations; (2) to investigate the difficult relationship between the highly intermittent weather-dependent power production and the energy distribution toward end users. The second goal will require from energy producers and distributors definitions of the requested forecast data and new technologies dedicated to the management of power plants and electricity grids.

The way toward accurate forecasting of solar plant output, from minute to days ahead, raises many challenges. This book covers the following two subjects: forecasting solar resource for the next minute up to 24 h ahead and modeling the output power of PV systems. In addition to this introductory chapter, the book comprises other nine chapters, as follows:

Chapter 2 is devoted to ground- and satellite-based broadband measurements of solar radiation. Radiometric quantities and instruments are summarized. The main surface solar radiation monitoring networks are reviewed and a survey of available databases is presented.

Chapter 3 deals with the state of the sky assessment. A number of existing relationships between clearness index and sunshine duration are tested. Best-fit correlations are also derived. The sunshine number, a Boolean random parameter stating whether the sun is covered or not by clouds, is defined. Statistical measures for the sunshine number are introduced. The dependence of the four statistical indicators on the cloud shade value has been evaluated by theory and by using measurements, respectively. The results are useful for those applications where the fluctuating nature of solar radiation has to be taken into account.

Chapter 4 is focused on different ways of characterizing both the radiative regime of a day and the stability of this regime and shows how the sunshine number can be used for day classifications. A new parameter, the sunshine stability number, is defined to quantify the stability of the radiative regime. Other measures based on disorder and complexity concepts, are introduced to properly quantify the daily fluctuations of global solar irradiance. The procedure to obtain a proper ARIMA model is described in detail. The solution for forecasting time series of sunshine number is based on ARIMA(0,*d*,0) models.

Chapter 5 surveys the algorithms for estimating the amount of solar energy collectable at the ground level on horizontal and inclined surfaces as well as on sun tracking surfaces. Two arguments motivate the insertion of this chapter in the book. First, some models estimate solar irradiance using meteorological parameters as entries. Employing forecasted parameters, these models may constitute functional tools in forecasting solar irradiance. Second, this chapter gives details concerning many physical quantities and equations applied through most chapters of the book.

Chapter 6 is focused on the practice of instantaneous clearness index nowcasting on very short time intervals and daily clearness index forecasting by using ARIMA modeling. Models constructions and their prediction accuracy are discussed.

Chapter 7 deals with forecasting clearness index via artificial intelligence (AI) techniques, very different approaches than classical statistics. First, several advances developed inside artificial intelligence are recapitulated. Second, artificial neural networks (ANN), probably the most used AI technique in PV power output forecasting, are reviewed. Then, fuzzy logic, a method with great potential in forecasting solar irradiance, is introduced. The chapter core consists of two fuzzy models, one for nowcasting solar irradiance and another for forecasting solar irradiation at daily lag, which are presented in detail.

Chapter 8 starts from two facts: air temperature is certainly the most measured surface meteorological parameter and accurate forecasting of air temperature is usually performed. Thus, a predicted value of air temperature may be used as entry in air temperature-based models for solar radiation aiming to forecast collectable solar energy. In this consideration, air temperature-based models for estimating global solar irradiance and irradiation are reviewed, assessing their accuracy. Numerous models consist of Ångström-type equations, in which the daily extremes of air temperature are used to give a measure of the state of the sky. Out of the ordinary, fuzzy logic is considered to relate the global solar irradiation to the daily amplitude of air temperature.

Chapter 9 switches on a different topic: conversion of solar radiation into electricity. Forecasting the output power of a PV plant involves the estimation of the conversion efficiency along to the prediction of solar irradiance. The main point here is the modeling of PV modules output in specific conditions of operation. This chapter summarizes four models of the voltage-current characteristic of a PV module and the way to solve the equations for calculating the power output. Several computational examples illustrate the methods.

Chapter 10 surveys recently reported results in forecasting the output power of PV plants.

Chapter 11 summarizes the main ideas presented in this book. Conclusions are drawn and perspectives are outlined.

References

Anderson NG (2001) On quantum well solar cell efficiencies. Physica E 14:126–131

Barnham KWJ, Ballard I, Connolly JG, Ekins-Daukes N, Kluftinger BG, Nelson J, Rohr C, Mazzer M (2000) Recent results on quantum well solar cells. J Mater Sci-Mater El 11(7): 531–536

Badescu V, Landsberg PT (1993) Theory of some effects of photon recycling in semiconductors. Semicond Sci Tech 8:1267–1276

Badescu V (2008) An extended model for up-conversion in solar cells. J Appl Phys 104:113120

Badescu V, Badescu AM (2009) Improved model for solar cells with up-conversion of low-energy photons. Renewable Energy 34:1538–1544

Burger B, Ruther R (2005) Site-dependent system performance and optimal inverter sizing of grid-connected PV systems. In Proceedings of 31th IEEE PVSC, Orlando, Florida, pp 1675–1678

Chapin DM, Fueller CS, Pearson GL (1954) A new silicon p-n junction photocell for converting solar radiation into electrical power. J Appl Phys 25:676–677

Conibeer G (2010) Up and down-conversion for photovoltaics. In: Badescu V, Paulescu M (eds) Physics of nanostructured solar cell. Nova Science Publishers, New York, pp 251–270

Cotal HL, Law DC, Nasser HK, Bedair SM (2010) Recent development in high efficiency multijunction solar cells. In: Badescu V, Paulescu M (eds) Physics of nanostructured solar cell. Nova Science Publishers, New York, pp 251–270

De Vos A, Szymanska A, Badescu V (2009) Modelling of solar cells with down-conversion of high energy photons, anti-reflection coatings and light trapping. Energy Convers Manage 50:328–336

EPIA (2011a) European Photovoltaic Industry Association. Global market outlook for photovoltaics until 2015. http://www.epia.org/publications/photovoltaic-publications-global-market-outlook.html

EPIA (2011b) Solar Photovoltaics—Competing in the Energy Sector. http://www.epia.org/publications/photovoltaic-publications-global-market-outlook.html

Foley AM, Leahy PG, Marvuglia A, McKeogh EJ (2012) Current methods and advances in forecasting of wind power generation. Renewable Energy 37:1–8

Green MA (2002) Lambertian light trapping in textured solar cells and light-emitting diodes: analytical solutions. Prog Photovoltaics 10(4):235–241

Green MA (2003) Third generation photovoltaics. Springer, Berlin

Green MA, Emery K, Hishikawa Y, Warta W, Dunlop ED (2012) Solar cell efficiency tables (version 39). Prog Photovoltaics 20:12–20

Holttinen H, Meibom P, Orths A, van Hulle A, Lange B, O'Malley M et al. (2009) Design and operation of power systems with large amounts of wind power. Final Report, Phase one 2006–2008. IEA WIND Task 25. http://www.vtt.fi/inf/pdf/tiedotteet/2009/T2493.pdf

Iacobescu F, Badescu V (2012) The potential of the local administration as driving force for the implementation of the national PV systems strategy in Romania. Renewable Energy 30:117–125

IEA (2007) IEA—energy technologies at the cutting edge. International Energy Agency. OECD Publication Service, Paris

Landsberg PT, Nussbaumer H, Willeke G (1993) Band–band impact ionisation and solar cell efficiency. J Appl Phys 74:1451–1452

Landsberg PT, Badescu V (2002) Solar cell thermodynamics including multiple impact ionization and concentration of radiation. J Phys D Appl Phys 35:1236–1240

Luque A, Marti A (1997) Increasing the efficiency of ideal solar cells by photon induced transitions at intermediate levels. Phys Rev Lett 78:5014–5017

Mason N (2008) Manufacturing technology: fabrication innovation. Nat Photonics 2:281–283

Mills A, Ahlstrom M, Brower M, Ellis A, George R, Hoff T, Kroposki B, Lenox C, Miller N, Milligan M, Stein J, Wan Y-h (2011) Understanding variability and uncertainty of photovoltaics for integration with the electric power system. IEEE Power Energ Mag 9(3):33–41

Nistorescu L (2012) Renasterea Banateana, 6805:1 (In Romanian)

NREL (2012) National Renewable Energy Laboratory. Extraterrestrial solar spectrum. Available online (accessed Jan 2012): http://rredc.nrel.gov/solar/spectra/am1.5/

Paulescu M, Tulcan-Paulescu E, Gravila P (2010) A hybrid model for quantum well solar cell. Int J Mod Phys B 24(14):2121–2133

Ransome S, Funtan P (2005) Why hourly averaged measurement data is insufficient to model PV system performance accurately. 20th European PVSEC, Barcelona, Spain. Available online: http://www.steveransome.com/PUBS/2005Barcelona_6DV_4_32.pdf

Ransome S, Wohlgemuth JH (2005) A summary of 6 years performance modeling from 100+ sites worldwide. In Proceedings of 31th IEEE PVSC, Orlando, Florida, pp 1611–1614

REN21 (2011) Renewable Energy Policy Network for the 21st Century. Renewables 2011 global status report, Paris. http://www.ren21.net/Portals/97/documents/GSR/REN21_GSR2011.pdf

Razykov TM, Ferekides CS, Morel D, Stefanakos E, Ullal HS, Upadhyaya HM (2011) Solar photovoltaics electricity: current status and future prospects. Sol Energy 85:1580–1608

Ruby DS, Zaidi SH, Roy M, Narayanan S (1999) Plasma texturing of silicon solar cells. In: Proceedings of 9th Workshop on Crystalline—Silicon Solar Cell Materials and Processes. Breckenridge

Saintcross J, Piwko R, Bai X, Clara K, Jordan G, Miller N, Zimberlin J (2005) The effects of integrating wind power on transmission system planning, reliability, and operations. Report of the New York state energy research and development authority, Albany, New York. http://www.nyserda.org/publications/wind_integration_report.pdf

Shah A, Torres P, Tscharner R, Wyrsch N, Keppner H (1999) Photovoltaic technology: the case for thin-film silicon cells. Science 285:692–698

Shockley W, Queisser HJ (1961) Detailed balance limit of efficiency of p-n junction solar cells. J Appl Phys 32:510–519

Tobias I, Luque A (2002) Ideal efficiency of monolithic, series-connected multijunction solar cells. Prog Photovoltaics Res Appl 10(5):323–329

Tomson T (2010) Fast dynamic processes of solar radiation. Sol Energy 84:318–323

Trupke T, Green MA, Würfel P (2002a) Improving solar cell efficiencies by down-conversion of high-energy photons. J Appl Phys 92:1668–1674

Trupke T, Green MA, Wurfel P (2002b) Improving solar cell efficiencies by up-conversion of sub-band-gap light. J Appl Phys 92:4117–4122

WIRE (2011) COST Action ES1002: weather intelligence for renewable energies. http://www.wire1002.ch/

Zhao J, Wang A, Altermatt P, Green MA (1995) Twenty-four percent efficient silicon solar cells with double layer antireflection coatings and reduced resistance loss. Appl Phys Lett 66: 3636–3638

Zhao J, Wang A, Green MA, Ferrazza F (1998) Novel 19.8 % efficient "honeycomb" textured multicrystalline and 24.4 % monocrystalline silicon solar cells. Appl Phys Lett 73:1991–1993

Chapter 2
Solar Radiation Measurements

2.1 Solar Radiation Components at the Ground Level

In a point at the top of Earth's atmosphere, the beam of nearly parallel incident sunrays is referred to as extraterrestrial radiation (ETR). ETR fluctuates about 6.9 % during a year (from 1412.0 Wm^{-2} in January to 1321.0 Wm^{-2} in July) due to the Earth's varying distance from the Sun. Figure 2.1 shows the spectral distribution of ETR at the mean Sun–Earth distance. The graph is plotted at low resolution with data from Gueymard and Myers (2008), which is detailed enough for many engineering applications, like forecasting of PV plants output. The extraterrestrial spectrum is also available online (NREL 2012), along with other reference radiation spectra. The integration of the extraterrestrial spectrum over all wavelengths defines the solar constant G_{SC}. Thus, G_{SC} represents the flux density of incoming solar radiation on a unitary surface perpendicular to the rays at the mean Sun–Earth distance. Since the Sun radiance varies to some extent over short and long periods (Fröhlich 1991), the solar constant does not remain steady over time. There is a variation of about $\pm$ 1 Wm^{-2} around the mean solar constant during a typical Sun cycle of 11 years (Gueymard and Myers 2008). Based on data collected over 25 years from terrestrial to space observations, the actual best estimate of the average solar constant is $G_{SC} = 1366.1$ Wm^{-2} (Gueymard 2004).

When the solar radiation flux passes through the Earth's atmosphere, its spectral distribution is modified by absorption and scattering processes. The complex effect experienced by solar radiation flux spectral distribution when transiting the Earth's atmosphere is also illustrated in Fig. 2.1, which displays the AM1.5G global spectrum defined by the Commision Internationale de l'Eclairage (CIE) and the American Society for Testing and Materials (ASTM) for testing the terrestrial solar cells. The standard assumes the following: the receiving surface is tilted 37° toward the equator, the solar zenith angle is 48°19', the total ozone column content is 0.34 cm atm, the Ångstrom turbidity coefficient at wavelength 0.5 μm is 0.084 and the water vapor column content is 1.42 g cm^{-2}. The Air Mass (AM)

M. Paulescu et al., *Weather Modeling and Forecasting of PV Systems Operation*, Green Energy and Technology, DOI: 10.1007/978-1-4471-4649-0_2,

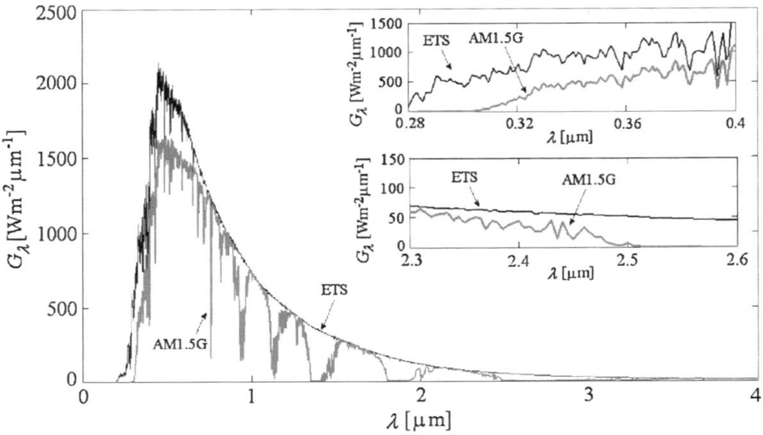

Fig. 2.1 Extraterrestrial solar spectrum (*ETS*) and terrestrial standard solar spectrum AM1.5G. Details of ultraviolet and infrared spectral domains are presented inset. G_λ is the extraterrestrial spectral flux density and λ is the photon wavelength

(see Chap. 5 for details), if approximated by the inverse of the cosine of the zenith angle, is $1/\cos(48°19') = 1.5$.

As a result of its passage through the atmosphere, the ETR is separated into different components. The beam component of solar radiation is that part of ETR which directly reaches Earth's surface. Scattering of the ETR in the atmosphere generates the diffuse component. A part of the solar radiation that is reflected by the ground may also be present in the total solar radiation. More precisely, the following quantities associated to solar radiation are commonly measured:

Direct beam irradiance (G_n) is the energy flux density (units: W/m^2) of the solar radiation incoming from the solid angle subtended by the Sun's disk on a unitary surface perpendicular to the rays.

Direct horizontal irradiance (G_b) differs from the direct beam irradiance in that it is measured on a flat horizontal plane. Lambert's cosine law states that the energy flux density on a plane surface is directly proportional to the cosine of the incidence angle. Since the incidence angle of the solar beam striking the horizontal ground is equal to sun the zenith angle θ_z (Fig. 2.2), then:

$$G_b = G_n \cos\theta_z \tag{2.1}$$

Diffuse irradiance (G_d) represents the energy flux density of the solar radiation incoming from the entire sky dome on a horizontal surface, excluding the direct beam coming from the Sun's disk.

Global irradiance (G) is the sum of the direct horizontal and diffuse components, given as:

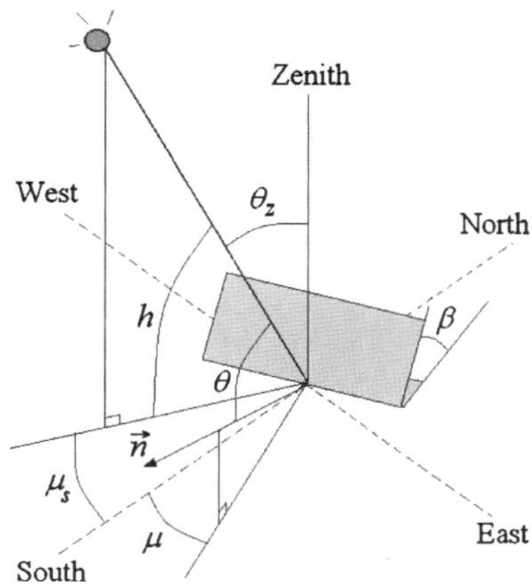

Fig. 2.2 Angles describing the position of the sun: θ_z—zenith angle; h—elevation angle, μ_s—azimuth angle. Angles describing the position of the surface: β—slope angle, μ—surface azimuth angle. The incidence angle θ is the angle between the sun direction and the surface's normal $\vec{n}$

$$G = G_b + G_d = G_n \cos\theta_z + G_d \tag{2.2}$$

The term "global" is associated to the fact that the solar radiation is received from the entire 2π solid angles of the sky vault.

The total irradiance (G_t) received by a surface tilted with an angle β in respect to the horizontal plane (Fig. 2.2) is the sum of beam flux density, diffuse flux density, and the additional flux density G_r of the solar radiation reflected from the ground, respectively. Usage of Eq. (2.2) yields:

$$G_t = G_n \cos\theta + R_d G_d + G_r \tag{2.3}$$

where θ is the incidence angle (i.e., the angle between the sun direction and the normal to the surface (Fig. 2.2), R_d is the conversion coefficient taking into account the sky view factor and G_r is the energy flux density of radiation reflected by the ground that is intercepted by the tilted surface. Models for estimating global solar irradiance on tilted surfaces differ generally in their treatment of R_d which is considered the main potential source of errors (see Chap. 5).

By summing up over a finite time period $\Delta t = t_2 - t_1$ one obtains the solar irradiation components:

$$H = \int_{t_1}^{t_2} G(t)dt \tag{2.4}$$

usually measured in J/m^2 or Wh/m^2. In Eq. (2.4), $G(t)$ stands for any of the above solar irradiance components, and consequently H refers to the corresponding solar irradiation component.

For proper characterization of the radiative regime the state of the sky should also be assessed. Two quantities are commonly used to describe the state of the sky. The most usual indicator is the total cloud cover amount C which represents the fraction of the celestial vault covered by clouds (estimated in tenths or oktas). The second quantity describing indirectly the state of the sky is the relative sunshine σ (also called sunshine fraction). It is defined as $\sigma \equiv s/S$, where S is the length of a given time interval and s is the bright sunshine duration during that interval. State of the sky assessment is treated at large in Chap. 3.

2.2 Ground Measurements of Solar Radiation

Radiometry is the science of electromagnetic radiation measurement. The generic device is named radiometer.

Each of the quantities defined in Sect. 2.1 are measured with a specific device; for instance, the pyrheliometer that measures the direct beam irradiance and the pyranometer that measures the horizontal beam and diffuse irradiances. Details on both instruments will be presented in the following.

2.2.1 Solar Radiometers

Detection of the optical electromagnetic radiation is primarily performed by conversion of the beam's energy in electric signals that subsequently can be measured by conventional techniques. Due to their nearly constant spectral sensitivity for the whole solar spectral range, radiometers equipped with thermal sensors are widely used to measure broadband solar irradiance. Temperature fluctuations (the instruments are placed outdoor and their temperature may vary between −20 and 70 °C), wind, rain, and snow are factors that affect the measurements. The minimization of these perturbations is a difficult task in the engineering of solar radiometers.

Pyrheliometer

The pyrheliometer is a broadband instrument that measures the direct beam component G_n of solar radiation. Consequently, the instrument should be permanently pointed toward the Sun. A two-axis Sun tracking mechanism is most often used for this purpose. The detector is a multi-junction thermopile placed at the bottom of a collimating tube (Fig. 2.3a) provided with a quartz window to protect the instrument. The detector is coated with optical black paint (acting as a full absorber for solar energy in the wavelengths range 0.280–3 μm). Its temperature is compensated to minimize sensitivity of ambient temperature fluctuations. The pyrheliometer aperture angle is 5°. Consequently, radiation is received from the Sun and a limited circumsolar region, but all diffuse radiation from the rest of the sky is excluded. A readout device is used to give the instant value of the direct beam irradiance. Its scale is adapted to the sensitivity of the particular instrument in order to display the value in SI units, Wm^{-2}.

For illustration, a picture of a Hukseflux DR01 First Class pyrheliometer (Hukseflux 2012) is presented in Fig. 2.3b.

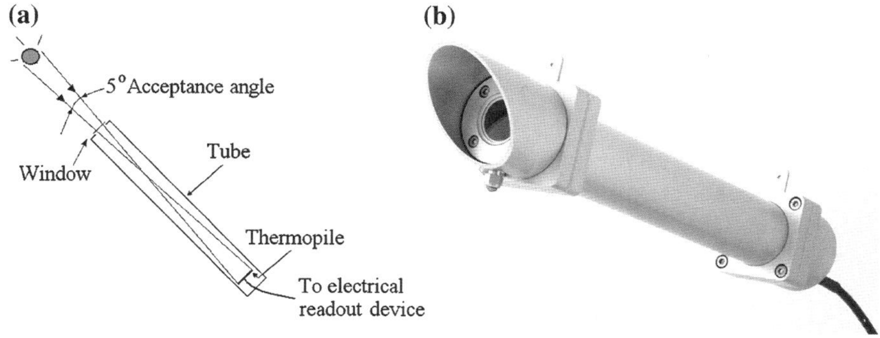

Fig. 2.3 **a** Schematic of a pyrheliometer. **b** Photo of Hukseflux DR01 first class pyrheliometer (Hukseflux 2012). (*Public license on Wikimedia commons*)

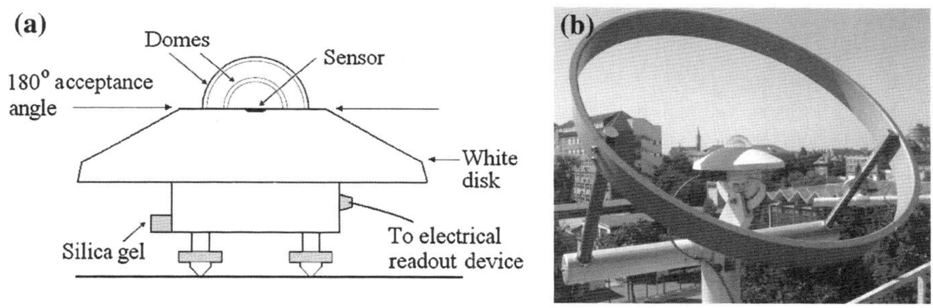

Fig. 2.4 **a** Schematic of a pyranometer. **b** First class pyranometer LPPYRA 12 (DeltaOHM 2012) equipped with shadow ring, mounted on the solar platform of the West University of Timisoara, Romania (SRMS 2012)

Pyranometer

Pyranometers are broadband instruments that measure global solar irradiance incoming from a 2π solid angle on a planar surface. A typical pyranometer is schematically represented in Fig. 2.4a. It consists of a white disk for limiting the acceptance angle to 180° and two concentric hemispherical transparent covers made of glass. The two domes shield the sensor from thermal convection, protect it against weather threat (rain, wind, and dust) and limit the spectral sensitivity of the instrument in the wavelength range 0.29–2.8 μm. A cartridge of silica gel inside the dome absorbs water vapor.

A pyranometer can be also used to measure the diffuse solar irradiance G_d, provided that the contribution of the direct beam component is eliminated. For this, a small shading disk can be mounted on an automated solar tracker to ensure that the pyranometer is continuously shaded. Alternatively, a shadow ring may prevent the direct component G_b from reaching the sensor whole day long (see Fig. 2.4b). Because the daily maximum Sun elevation angle changes day by day, it is necessary to change periodically (days lag) the height of the shadow ring.

Table 2.1 Characteristics of pyranometers, ISO 9060/1990 standard

ISO specification	Secondary standard	First class
WMO characteristics	High quality	Good quality
Response time (to reach 95 % of the final value)	<15 s	<30 s
Zero off-set response:		
—Response to 200 W/m^2 net radiation	7 Wm^{-2}	15 Wm^{-2}
—Response to 5 °C/h change in ambient temperature	±2 Wm^{-2}	±4 Wm^{-2}
Resolution	±1 Wm^{-2}	±5 Wm^{-2}
Stability (change in sensitivity per year)	±0.8 %	±1.5 %
Linearity (deviation from sensitivity at 500 Wm^{-2} over 100–1,000 Wm^{-2} irradiance range)	±0.5 %	±1 %
Directional response for beam radiation (error due when assuming that the normal incidence response at 1000 Wm^{-2} is valid for all directions)	±10 Wm^{-2}	±20 Wm^{-2}
Spectral selectivity (deviation of the product of spectral absorptance and transmittance, respectively, from the mean)		
ISO (0.35–1.5 μm)	±3 %	±5 %
WMO (0.3–3 μm)	±2 %	±5 %
Temperature response (maximum relative error due to any change of ambient temperature within a 50 °C interval)	±2 %	±4 %
Tilt response (percentage deviation from horizontal response when the tilt is changed from horizontal to vertical at 1,000 Wm^{-2})	±0.5 %	±2 %
Achievable uncertainty, 95 % confidence level		
WMO hourly totals	3 %	8 %
WMO daily totals	2 %	5 %

On the other hand, because the shadow ring also intercepts a part of the diffuse radiation, it is necessary to correct the measured values. The percentage of diffuse radiation intercepted by the shadow ring varies during the year with its position and atmospheric conditions (Siren 1987).

Self-calibrating absolute radiometers (Reda 1996) are used as primary standard, the other radiometers being calibrated against an absolute instrument. The uncertainty of the measured value depends on factors such as: resolution (the smallest change in the radiation quantity which can be detected by the instrument), nonlinearity of response (the change in sensitivity associated with incident irradiance level), deviation of the directional response (cosine response and azimuth response), time constant of the instrument (time to reach 95 % of the final value), changes in sensitivity due to changes of weather variables (such as temperature, humidity, pressure, and wind), long-term drifts of sensitivity (defined as the ratio of electrical output signal to the irradiance applied). All the above uncertainties should be known for a well-characterized instrument. Certain instruments perform better for particular climates, irradiances, and solar positions; therefore, the instruments should be selected according to their end use.

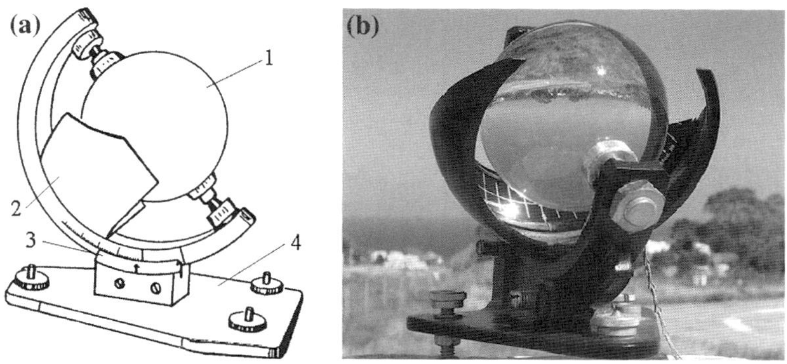

Fig. 2.5 **a** Schematic of Campbell-Stokes sunshine recorder. **b** Photo of a typical Campbell-Stokes sunshine recorder. (*Public license on Wikimedia commons*)

The accepted classification of pyranometers in respect to their quality is defined by the International Standard ISO 9060/1990 that is also adopted by the World Meteorological Organization (WMO 2008). ISO 9060 standard distinguishes three classes: the best is called (somewhat improperly) *secondary standard*, the second best is called *first class* and the third one—*second class*. Table 2.1 summarizes the characteristics of pyranometers of the first two levels of performance.

Sunshine duration measurement

According to (WMO 2008), sunshine duration in a given period is defined as the sum of the time intervals for which the direct solar irradiance exceeds the threshold of 120 Wm^{-2}. In practice, two methods are widely used for measuring sunshine duration—burning card method and pyranometric method—which will be briefly presented next.

Burning card method is based on the Campbell–Stokes sunshine recorder, which basic setup consists of a glass sphere mounted concentrically in a segment of a spherical bowl (Fig. 2.5). The support is adjustable so that the axis of the sphere may be inclined to the angle of the local latitude. The spherical bowl segment holds the recording card. The glass sphere focuses the direct beam solar radiation on to the card, burning a trace whenever the Sun is shining. The position and length of the trace indicate the starting time and duration of the sunshine interval. The errors of this recorder are mainly due to the dependence of burning initiation on card's temperature and humidity as well as to the overburning effect, especially in case of broken clouds (Kerr and Tabony 2004).

The pyranometric method implies measurement of global G and diffuse G_d solar irradiance; by subtraction, the direct beam solar irradiance is next derived, to be compared with the WMO threshold. Using the fundamental Eq. (2.2), the WMO sunshine criterion can be expressed as:

$$\xi(t) = \begin{cases} 1 & if \quad (G - G_d)/\cos\theta_z > 120\,\mathrm{W}/m^2 \\ 0 & otherwise \end{cases} \tag{2.5}$$

In Eq. (2.5) θ_z is the Sun zenith angle and ξ stands for the sunshine number (Badescu 2002), a Boolean variable stating whether the Sun is covered or not by clouds. Statistical properties of the sunshine number are investigated in Chap. 3 while methods to quantify the fluctuations of solar radiative regime by using the sunshine stability number are reported in Chap. 4.

The sunshine duration during a time interval Δt is obtained by multiplying Δt with the mean sunshine number $\bar{\xi}$ during Δt.

The errors in the pyranometric method stem from the errors of measuring global and diffuse solar irradiance, which are amplified at higher zenith angles (see Eq. 2.5). Choosing a high quality pyranometer is of primary importance to reduce the results uncertainty level. The usage of shading rings has as a consequence the undervaluation of the incident diffuse solar energy. Corrections are required to diminish this negative effect. Last but not least, the sampling frequency is important. At a higher measurement rate the sunshine duration can be evaluated more precise. At least one sample per minute is required to properly capture the fast changes of the solar radiative regime.

2.2.2 Surface Measurements

A comprehensive knowledge of the solar energy available in a location not only means its characterization by the total value, but also its temporal repartition, spectral distribution, and nature (direct or diffuse). Most countries have set up networks for measuring solar radiation but investments and maintenance costs for each site are not negligible. Thus, a national network often consists of relative small number of stations even in industrialized countries. Of course, the number of stations comprised in each network and measurement qualities varies from a network to another. Generally, global solar irradiation and sunshine duration are available on daily sums or monthly mean. Diffuse and direct components of solar irradiation, total cloud cover amount as well as data recorded on hourly basis are more rarely recorded. Only few stations around the world measure routinely the spectral distribution of solar radiation at the ground or the solar irradiance on tilted surfaces.

There is a sum of shortcomings associated with nowadays databases. The number of solar radiation stations is too small to achieve accurate global coverage. Consequently, interpolation and extrapolation of available data are used for estimation solar radiation in each point of the map. The errors in estimating solar irradiance in a given point increase with the distance from the closest station. The exact number of solar radiation stations measuring solar irradiance through the world is difficult to count, various sources giving different information. Even so, there are surely more than one thousand continental stations around the world that measure solar radiation. It was shown that daily solar irradiation measured at a station may be considered valid in an area of 30 km around (Cros and Wald 2003). Assuming that the 1,000 stations are equally distributed on land (146.94 millions of km^2), the probability of being in optimal vicinity of a station is less than 2 %.

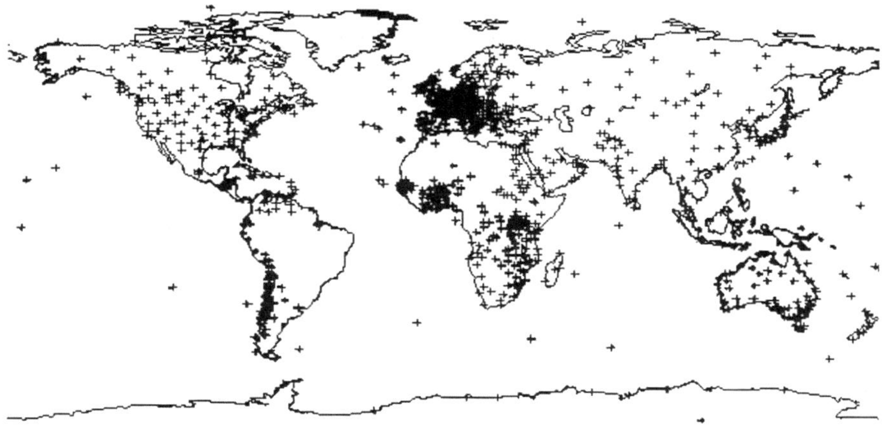

Fig. 2.6 Map of the stations contributing to the WRDC database (*Source* WRDC 2012b)

In other words, in 98 % of cases, the stations are too far to deliver accurate information to users. On the other hand, time characteristics of the data are often unsatisfactory. For example, beam solar irradiance, of vital interest in forecasting solar-thermal systems power output, is rarely available. Moreover, the databases store data in various formats and units using various time idioms. Thus, sometimes even data access and correct interpretation is a difficult task. It can be concluded that the solar radiation data from present databases are not matching many of the application requirements. There is enough room for efforts in integrating information systems to diminish the discrepancy between data availability and end user needs.

In the following, two large surface networks, World Radiation Data Center (WRDC) and Baseline Surface Radiation Network (BSRN) are summarized, while some national networks are only enumerated.

World Radiation Data Center (WRDC 2012a) collects data from the largest network for monitoring solar radiation, developed inside the WMO. WRDC is located at the Main Geophysical Observatory in St. Petersburg, Russia and serves as a central depository for solar radiation data collected at over one thousand measurement sites throughout the world (Fig. 2.6). The map presented in Fig. 2.6 shows that the spatial distribution of the stations is strongly heterogeneous through the world. The network is very dense in Western and Central Europe and Japan but there are large parts of the continents uncovered.

The system of collecting and archiving data is specified in the recommendations of the Meeting of Experts on the Future Activities of the WRDC (WMO 1983) and Resolution 6 (EC-XXXVI) of the World Meteorological Organization (WMO 1984). WRDC collects from WMO stations the results of measuring daily and hourly sums of radiation parameters with quality control check. WRDC performs a supplementary quality control and requests stations to confirm the data for total quality assurance. In addition to this basic information, WRDC receives

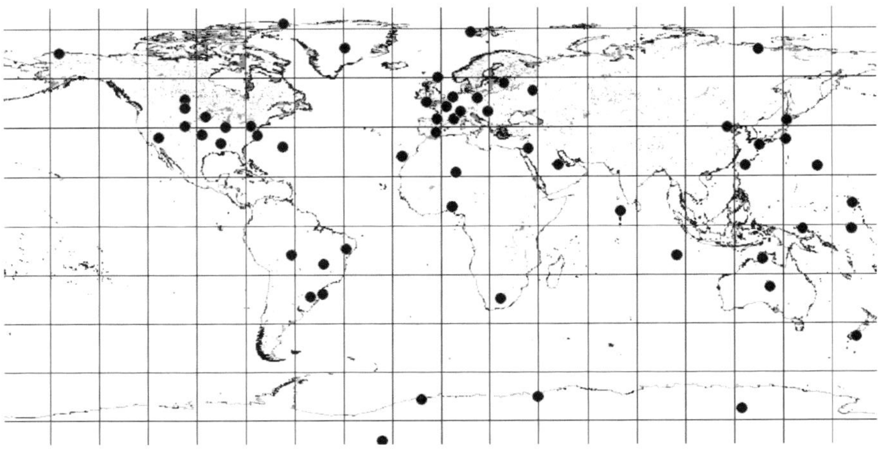

Fig. 2.7 Running and planning BSRN stations (BSRN 2012)

information on instruments and calibrations (supplementary information on this topic can be read in Ref. WRDC (2012b).

WRDC centrally collects, archives, and publishes radiometric data from the world to ensure the availability of these data for research by the international scientific community. WRDC archive contains mainly measurements of global solar irradiation, diffuse solar irradiation and sunshine duration in format of daily sums and monthly mean. Data collected from 1964 to 1993 are accessible online on the site of the US Department of Energy's, National Renewable Energy Laboratory (WRDC 2012b) and data collected from 1994 to present are accessible on the site of Main Geophysical Observatory, St. Petersburg, Russia (WRDC 2012a).

Baseline Surface Radiation Network (BSRN 2012). BSRN is a project of the Radiation Panel from the Global Energy and Water Cycle Experiment (GEWEX 2012) as part of the World Climate Research Program (WCRP 2012). The project is designated for detecting important changes in the Earth's radiation field at the Earth's surface which may be related to climate changes. Currently, about 40 stations located in contrasting climatic zones, covering a latitude range from 80°N to 90°S (see Fig. 2.7), are measuring solar and atmospheric radiation with instruments of the highest available accuracy and with high time resolution (1–3 min). The BSRN stations also contribute to the Global Atmospheric Watch (GAW 2012).

The solar radiation data are stored together with surface and upper-air meteorological observations in an integrated database. The BSRN database has been developed for storing high quality radiation data. It adds its own quality control process to that from the member organizations collecting data. The BSRN measurements are used to validate the radiation schemes in global climate models and satellite algorithms.

In addition to global networks there are also national networks which are repository for solar radiation data.

An important national solar radiation monitoring network has been developed by the US National Oceanic and Atmospheric Administration, namely NOAA's Surface Radiation (SURFRAD 2012). Independent measures of upwelling and downwelling solar and infrared radiation are the primary measurements; auxiliary observations include direct and diffuse solar irradiance, photosynthetically active radiation, UVB, and meteorological parameters. Quality controlled data are packed into daily files that are distributed almost in real time by anonymous FTP and HTTP protocols (SURFRAD 2012). Quality assurance built into the design and operation of the network and a good data quality control ensures that a continuous, high accuracy product is released. Another program is the US Department of Energy Atmospheric Radiation Measurement (ARM 2012) with the aim to increase the knowledge about the interaction between clouds and atmospheric radiative fluxes. From the observational perspective, the focus is on measuring the solar and thermal infrared radiative fluxes at Earth's surface and of all the atmospheric quantities that affect those fluxes. The Cooperative Network for Renewable Resource Measurements (CONFRRM 2012) is another important source of solar radiation data. CONFRRM is a joint effort between the US National Renewable Energy Laboratory and other agencies to conduct long-term solar radiation and wind measurements at selected locations in the United States. CONFRRM provides high quality data for determining site-specific resources, as well as data for the validation and testing of models to predict available resources based on meteorological or satellite data. Quality control and quality assurance occur before and during data collection and include procedures such as the proper selection and installation of instruments and data acquisition equipment, as well as regular maintenance and calibration. Collected data are also post processed, to check in a final quality assessment whether a data value is reasonable, too small, too large, or missing.

All European countries have established and maintain national networks for solar radiation measurements and contributing to WRDC database. The spatial density of stations in the national networks and their quality varies from a country to country. For example, in Spain, which is an important actor in the photovoltaic research and market, there are 52 stations currently contributing to WRDC. Taking into account that the surface of Spain is 504,030 km^2, the stations spatial density is roughly 1–10,000 km^2. In Eastern Europe, in Romania, the national meteorology network comprises 150 meteorological stations but less than 10 maintain a long-term global solar irradiation database and contribute to WRDC. The stations spatial density is 1–24,000 km^2, 2.5 times lower than in Spain.

In Europe, there are several integrated information systems, where databases are supplemented by post processing products like maps and software, available online or on CD-ROMs. Two examples follow.

Photovoltaic Geographical Information System (PVGIS 2012) is a research and demonstration instrument for geographical assessment of solar resource and solar systems in the context of distributed energy generation (Suri et al. 2005). The server, operated by the Joint Research Center of the European Commission, offers

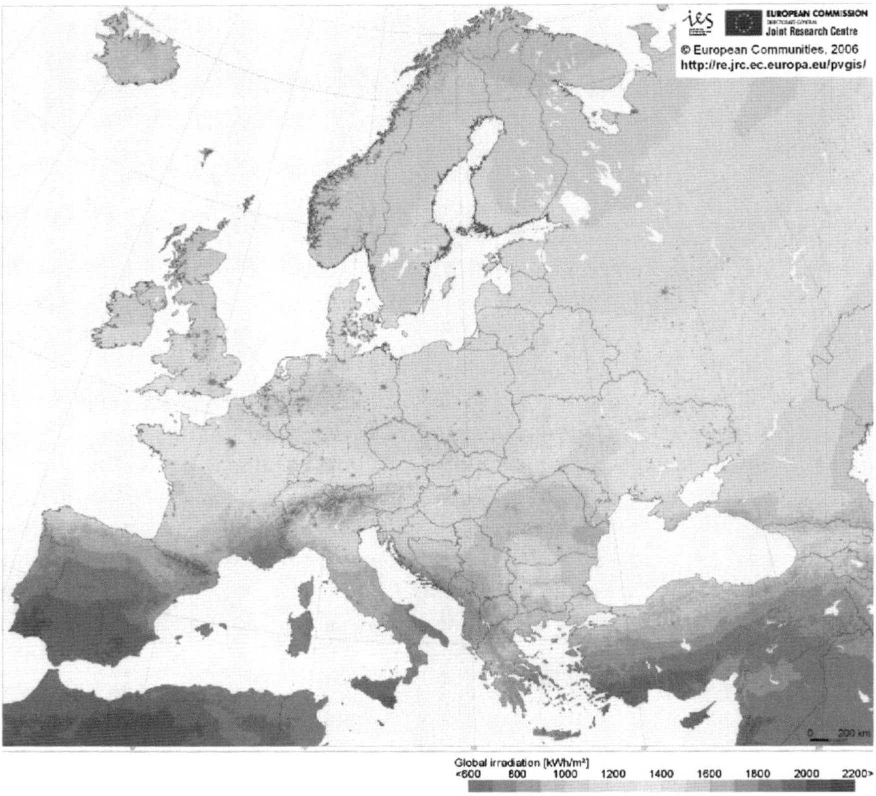

Fig. 2.8 Yearly global solar irradiation over Europe incident on optimally inclined (*equal to local latitude*) south-oriented surface (Suri et al. 2007)

map-based query of basic statistics of solar radiation, temperature, and the other data for two regions: Europe and Northern Africa. For Europe the database is based on an interpolation of ground station measurements (1 km grid, period 1981–1990). Figure 2.8 shows an example of European solar radiation map provided by the PVGIS. For the Mediterranean Basin and Africa the maps are developed by processing the HelioClim-1 database (2 km grid resolution, period 1985–2004).

METEONORM (METEONORM 2012) is a comprehensive climatologic database for solar energy applications combined with a synthetic weather generator. It contains a large database of ground station measurements collected from various sources (more than 8,300 are listed for the version 7). Two time periods of the measurements are available: (1) 1961–1990 and 1996–2005 for temperature, humidity, precipitation, and wind speed, and (2) 1961–1990 and 1981–2000 for radiation parameters. Enhanced satellite data are used to improve the estimations for areas with low density of weather stations. The METEONORM outputs are climatologic averages and derived products for any point on earth, estimated by

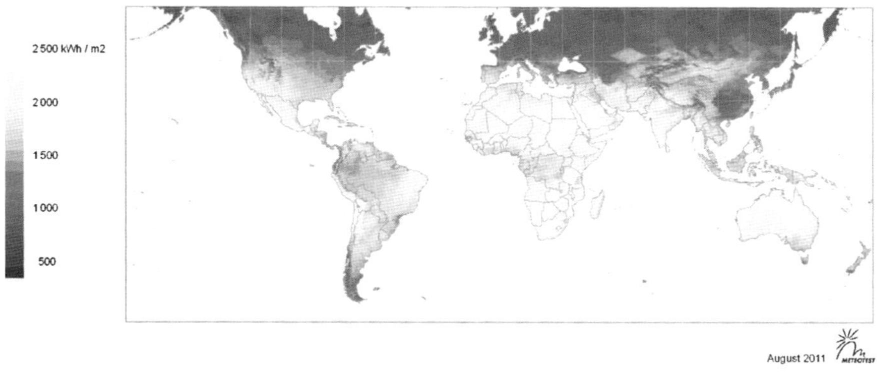

Fig. 2.9 Yearly global solar irradiation over the world, period 1996–2005, grid size 0.33° uncertainty 7 %. *Source* METEONORM (http://www.mwteonorm.com)

interpolation at very high resolution (0.1–1 km). A world map of yearly sum of daily global solar irradiation generated by the METEONORM software is presented in Fig. 2.9.

2.3 Solar Radiation Derived from Satellite Observation

In the previous section, we have seen that even if the availability of ground solar radiation data is on the rise, the spatial density of stations is still far too low. In order to fill the gaps in surface measured data, approaches based on satellite observation turn into a suitable alternative. In the last decades, a number of methods for estimating solar radiation from satellite data have been developed (e.g. Vonder Haar et al. 1973; Moser and Rachke 1984; Cano et al. 1986; Nunez 1993; Zelenka et al. 1999; Perez et al. 2002; Schillings et al. 2004; Rigollier et al. 2004; Janjai et al. 2005; Martins et al. 2007; Janjai 2010).

Most of the methods for deriving solar radiation from satellite observations employ meteorological geostationary satellite images. The geostationary satellites, orbiting at about 36,000 km, can offer a temporal resolution of up to 15 min and a spatial resolution of up to 1 km. The meteorological satellites collect images over a large area and with high time resolution allowing identification and forecasting the clouds evolution. This information is further processed, leading to the prediction of spatial variability of solar radiation at the ground level. Compared to ground measurements, satellite-derived hourly irradiation has been shown to be the most accurate option for locations that are further away with more than 25 km from a ground station (Zelenka et al. 1999). Thus, processing data collected by satellites can be a viable solution for forecasting solar radiation at the ground, aiming to proper operate the power grid.

A brief review of the satellite-based approaches for deriving solar radiation followed by a survey of available databases is presented next.

2.3.1 Satellite Based Models for Deriving Solar Radiation

The equation that governs the satellite-based models is derived from the interaction of ETR with the Earth-atmosphere system. This way, a part of ETR (G_{ext}) is reflected (G_r), other part is absorbed (G_a), and the remaining (G_g) is absorbed by the ground. Enclosed in brackets is the notation for the corresponding energy flux density. The energy conservation gives:

$$G_{ext} = G_r + G_a + G_g \tag{2.6}$$

Expressing the energy absorbed by the ground as a function of global solar irradiance $G_g = (1-\rho)G$, where ρ is the surface albedo, Eq. (2.6) becomes:

$$G = \frac{1}{1-\rho}(G_{ext} - G_r - G_a) \tag{2.7}$$

Equation (2.7) represents the basis of all models developed for retrieving solar irradiance from satellite images. In Eq. (2.7), G_{ext} is well defined by astronomical equations (see Chap. 5) and G_r is measured by the satellite radiometer. The methods for estimating ground data differ by the way in which G_a and ρ are estimated.

Three models are summarized below: Heliosat model (HelioClim 2012), Operational Model (Perez et al. 2002), and Janjai model (Janjai et al. 2005).

Heliosat method (HelioClim 2012) converts images acquired by meteorological geostationary satellites, such as Meteosat (Europe), GOES (USA), or GMS (Japan), into data and maps of solar radiation received at ground level.

The development of Heliosat method is an ongoing effort of the Center for Energy and Processes, Ecole des Mines de Paris/Armines, France (CEP 2012).

The original model proposed by Cano et al. (1986) was improved through different versions. The basic idea is that the cloud cover amount over a given area statistically determines the global solar irradiance for that area. Thus, the processing takes two steps. A cloud cover index is derived for each pixel of the original satellite image and subsequently used in a second step for estimation of the global solar irradiance.

The cloud cover index is a basic concept in retrieving solar irradiance from satellite images. This parameter has been defined in Cano et al. (1986) as:

$$n = \frac{\rho - \rho_g}{\rho_c - \rho_g} \tag{2.8}$$

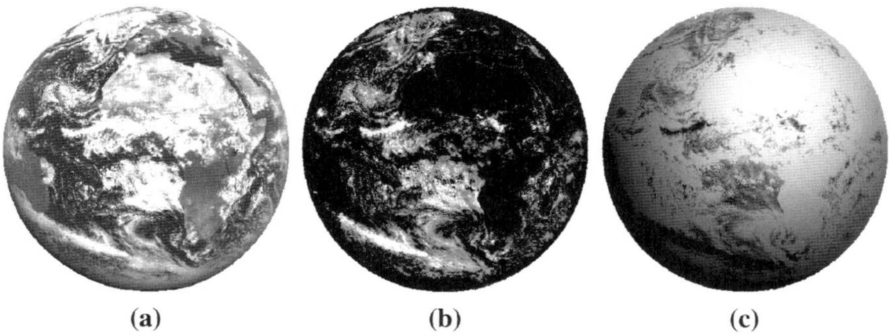

Fig. 2.10 Illustration of the processed images by using the Helisat-2 model (http://www.helioclim.org/heliosat/): **a** Raw meteosat data (August 1, 1992, 11h30). **b** Derived cloud cover index. **c** Derived hourly solar global irradiation. *Image courtesy* Lucien Wald

where ρ is the albedo measured by the satellite, ρ_c is the cloud albedo and ρ_g is the ground albedo. The cloud cover index ranges from 0 to 1 and may be interpreted as the percentage of the cloud cover amount per pixel.

Heliosat-1 model consists basically in a linear correlation of cloud cover index n and instantaneous clearness index k_t (Diabate et al. 1988):

$$k_t = an + b \tag{2.9}$$

were a and b are empirical parameters.

Heliosat-2 version (Rigollier et al. 2004) is dealing with atmospheric extinction and cloud extinction separately. The clear sky solar irradiance is calculated with the ESRA model (Rigollier et al. 2000), briefly summarized in Chap. 5 of this book. It uses at input only the Linke turbidity factor, a parameter that signifies the number of clear and dry atmospheres required to yield the observed extinction. Inputs to the method Heliosat-2 are not numerical counts of the satellite image, like in Heliosat-1. These counts are calibrated and thus converted into radiances. The Heliosat-1 construction of cloud cover index has been in principle preserved in the updated version Heliosat 2. Figure 2.10 presents an example of the processed Earth images by using the Helioclim-2 method.

Heliodat-3 version uses a new type of correlation scheme based on a radiative transfer code (Mueller et al. 2004), which makes use at input of atmospheric parameter information (clouds, ozone, and water vapor) retrieved from the Second Generation of Meteosat satellites databases (spatial resolution 1/3 km; temporal resolution 15 min; spectral channels 12).

Operational Model (Perez et al. 2002) is also a development of the Cano model. It has been elaborated at the Atmospheric Sciences Research Center (ASRC 2012) of the University of Albany, USA and applied to GOES satellite images. Cloud cover index is used to modulate the global solar irradiance estimated under clear sky using atmospheric turbidity as entry. The estimation of clear sky global solar irradiance G_0 is based on the Kasten (1980) model with some modifications to exploit an atmospheric turbidity coefficient formulation, which removes the dependence on solar geometry (Ineichen and Perez 2002). The model equation is:

$$G_0 = c_1(z)G_{\text{ext}} \exp[-mc_2(z)(c_3(z) + c_4(z)(T_L - 1))] \exp\left[0.01m^{1.8}\right] \cos\theta_z \quad (2.10)$$

where $c_1(z) = 0.0509z + 0.868$, $c_2(z) = 0.0392z + 0.0387$, $c_3(z) = \exp(-z/8)$ and $c_4(z) = \exp(-z/1.25)$. In Eq. (2.10), z is the site altitude (in km), θ_z is the zenith angle, m is the altitude-corrected AM and T_L is the Linke turbidity factor. Details on the computations of these quantities are given in Chap. 5.

For each area (e.g. pixel), the global hourly irradiance is estimated by adapting G_0 to the actual cloud cover amount by means of cloud index:

$$G = G_0(0.0001\, G_0 f(n) + 0.9)f(n) \quad (2.11)$$

where $f(n) = 2.36n^5 - 6.2n^4 + 6.22n^3 - 2.36n^2 - 0.58n + 1$.

The model can also exploit operationally available snow cover resource data, while deriving local ground specular reflectance characteristics from the stream of incoming satellite data.

The operation of the model on a geographic scale, either for the preparation of maps or site/time-specific time series requires some degree of logistics and information processing. This includes several layers of gridded information: (1) Raw satellite pixels obtained via direct processing of primary GOES satellite images. The achievable resolution of visible channel GOES image could approach 0.01°. (2) Terrain elevation. (3) Climatological Linke turbidity. (4) Snow cover (5) Specular correction factor.

Janjai model (Janjai et al. 2005) has been developed for mapping solar irradiation in a tropical environment. An improved variant has been reported in Janjai (2008) and is briefly summarized below. Apart from the previous models, the Janjai model does not focus on the calculation of solar irradiance for each hourly satellite image. It is motivated by the fact that the state of the sky may experience high fluctuation in this time interval (especially in the tropics). This implies serious limitations on the ability of satellites to map irradiance by using only several scan images. Instead, the regional cloud structure emerges after daily averaging. Thus, the proposed method is aimed to the calculation of long-term averages of daily radiation using long-term satellite data. All parameters involved in the satellite model are calculated on monthly average basis. The outcome of the calculation belongs to solar radiation climatology which is usually required for generating a map for solar energy applications.

In this model, the incident solar radiation which enters the earth's atmosphere is scattered back to the outer space by air molecules and clouds with the cloud atmospheric albedo ρ_A and by atmospheric aerosols having an albedo ρ_{aer}. The rest of the radiation continues to travel downward and is absorbed by ozone, mixed gases, water vapor, and aerosols with absorption coefficients α_O, α_g, α_w and α_{aer}, respectively. Upon reaching the surface, the radiation flux is reflected back by the ground with the albedo ρ_g. As it passes upward through the atmosphere, it is further depleted by aerosol scattering (ρ_{aer}) and aerosols absorption (α_{aer}) and

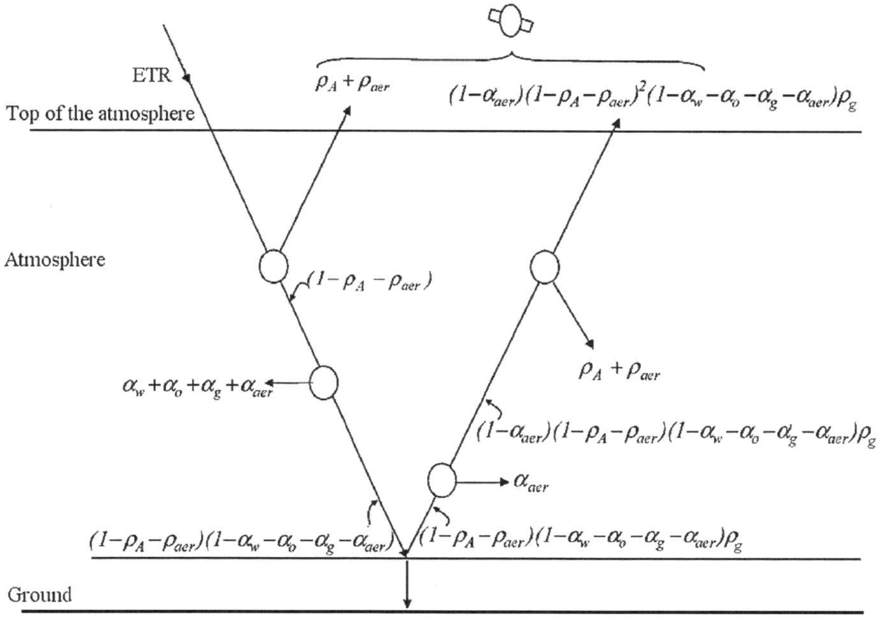

Fig. 2.11 Schematic diagram showing the radiation budget in the atmosphere as it is showing by the satellite in a spectral band. (ρ_A is the cloud—atmospheric albedo, ρ_{aer}. is the atmospheric aerosols albedo and ρ_g is the ground albedo. α_O, $\alpha_{g\alpha}$, α_w and α_{aer} stand for absorption coefficients of ozone, mixed gases, water vapor and aerosols, respectively. After Janjai (2008), with permission

cloud-atmospheric scattering $\langle\rho_A\rangle$. No further absorption due to water vapor, ozone, and gases is considered as it is assumed that the spectral irradiance in the absorption band of these atmospheric constituents has been absorbed completely during the downward travel of the solar flux. These processes are schematically shown in Fig. 2.11.

The albedo of the earth-atmosphere system detected by the satellite in a spectral band (e.g., 0.55–0.72 μm for GOES 9) can be written as:

$$\rho_{EA} = \rho_A + \rho_{aer} + (1 - \rho_A - \rho_{aer})^2 \big(1 - \alpha_o - \alpha_w - \alpha_g - \alpha_{aer}\big)(1 - \alpha_{aer})\rho_g \tag{2.12}$$

This model employs satellite data to estimate cloud-atmospheric albedo in the satellite band, which is converted into a broadband cloud-atmospheric albedo using global radiation measured at ground stations. This broadband albedo and surface recorded supplementary data are used to map daily global radiation. The preparation of satellite data, the determination of the model coefficients, the conversion procedures, and a model accuracy assessment are detailed in Janjai (2008).

2.3.2 Online Available Databases

Solar radiation information from satellite data provided by web-based systems and databases are mainly solar radiation maps. Some websites also act as data servers. Four frequently accessed websites involving solar radiation derived from satellite images are reviewed in the following.

Satel-Light—European Database of Daylight and Solar Radiation (Satel-Light 2012) was probably one of the first web sites to provide solar radiation data. The Satel-Light project was funded by the European Union from 1996 to 1998. The methodology is based on the Heliosat model. Images produced by the Meteosat satellite every half hour are the only one source of information. Satel-Light service covers Europe and a small region of the North Africa. The information provided concerns solar irradiance and daylight statistical data in terms of monthly means of hourly and daily values from the period 1996 to 2000.

SoDa—The project Solar Data (SoDa 2012) is an effort to consolidate different databases through a web server containing solar radiation parameters and other relevant information: long-term time series of daily irradiation, climatic data and derived quantities, simulation of radiation under clear skies, and simulation of different solar power systems. The Quality Assessment benchmark of the SoDa database consists of weighing the estimated values against the corresponding ground station measurements. Data are disseminated by the SoDa website through the SoDa Service. A service or a resource can be a database (e.g., solar radiation database, temperature database), an algorithm that performs on data to create new information or an application that provides information that can be directly used in practice. The SoDa database is processed by MINES ParisTech—ARMINES (HelioClim 2012).

NASA Surface Meteorology and Solar Energy (SMSE 2012) is a large archive of over 200 satellite-derived meteorological and solar radiation parameters. The data are available on a 1° longitude by 1° latitude equal-angle grid covering the entire globe (64,800 regions). The data are generated using the GEOS 4 dataset, with a resolution of 1.25° longitude and 1° latitude. Interpolation is used to produce $1 \times 1°$ regions. The solar radiation data are generated using the Pinker and Laszlo (1992) algorithm. Cloud information is taken from the International Satellite Cloud Climatology Project DX dataset on an equal area grid with an effective 30×30 km pixel size. The output data are generated on a nested grid with a resolution of one degree latitude globally and longitudinal resolution ranging from 1° in the tropics to 120° at the poles. This is regridded to a 1° equal-angle grid (360 longitudes by 180 latitudes). The regridding method is done by replication wherein any grid region that is larger than $1 \times 1°$ is subdivided into $1 \times 1°$ regions, each taking the same value as the original.

Data quality control is carried out by comparison with ground measured data. Regression analysis of SMSE versus BSRN monthly averaged values of global solar irradiation for the time period July 1983 to June 2006 shows a relative root mean square error of 10.25 % and the relative mean bias error of −0.01 % (SMSE 2012).

Data can be retrieved from the SMSE server (SMSE 2012), where they are organized in groups fitting various solar applications, e.g., sizing and pointing solar modules, solar cooking, tilted solar modules, cloud information, and so on.

The *Australian Bureau of Meteorology* (ABM 2012) maintains an online service (Climate Data Online), which provides historical data of daily global solar irradiation in a variety of formats (table, graphs). The data are derived from satellite images taken by the Geostationary Meteorological Satellite GMS-5, Geostationary Operational Environmental Satellite (GOES-9), and MTSAT-1R and MTSAT-2 satellites. The process of displaying the data uses the latitude and longitude of the Bureau's ground observation stations to retrieve the solar radiation values for a point within Australia. A basic site summary and topographic map of the area around the Bureau station is available online (ABM 2012).

2.4 Data Assessment Related to PV Power Forecasting

A PV plant project goes through two stages: (1) development and (2) PV plant exploitation. Different kinds of information on solar resource are needed in each stage.

In the development stage reliable solar radiation statistics are required for plant location, system design, and for feasibility study. In most cases, monthly averages values of solar irradiation or TMY—typical meteorological year (e.g. Janjai and Deeyai 2009) are enough. The overview made during this chapter on the existing databases of solar resource data shows that they contain adequate information for the development stage of a power plant. In most situations the data, either raw or completed by estimations, may convene the needs of users. Questions may arise on quality issues and uncertainty related to the methods of measurement and data processing, which varies from station to station and from a database to another. Even if the years after 2000 brought a real progress in data quality assurance, improvement is still needed. The fact that uncertainties in solar data can make the difference between profit and loss stresses the importance of the data quality issue.

In the exploitation stage of a power plant, forecasting the output power is required for proper operating the power grid. Moreover, real-time data flow is critical for optimizing supply/demand patterns in the power grid with multiple fluctuating sources, PV plants and/or wind farms. Thus, accurate forecasting of solar irradiance became a mandatory task. Depending on the time horizon, different forecasting methods are considered. In the shortest time domain ("Nowcasting", 0–3 h), numerical weather models perform badly: the forecast has to be based on extrapolations of real-time measurements. In the second time domain ("Short-Term Forecasting", 3–6 h), numerical weather models are coupled with post-processing modules in combination with real-time measurements. In the third time domain ("Forecasting", 6–72 h or more), only the numerical weather model in combination with post-processing modules and satellite information are recommended (Lara-Fanego et al. 2012). For the reason that this book is mostly dedicated to nowcasting solar radiation and the topics are focused on statistical and artificial intelligence approaches, numerical weather prediction (NWP) models are acknowledged briefly in Sect. 2.5.

Existing surface databases are of little use to forecast solar irradiance since the information contained in them is far from real time. Consequently, it is optimal for each PV plant to have its own solar irradiance and meteorological parameters measuring station to provide the basis for time series forecasting. On the other hand, solar irradiance data retrieved from satellite images are of major importance in evaluating temporal and spatial changes of the solar resource, most important in regions with multiple and large PV capacities. Flow maps of the direct beam radiation occurring at the ground can be performed at a time resolution of 15 min.

2.5 Numerical Weather Prediction Models

NWP uses the current weather conditions as input into mathematical models describing the processes occurring in the atmosphere to predict the weather for a certain future period. Historically, NWP starts with the work of V. Bjerknes and L. F. Richardson in the beginning of the last century (Lynch 2008). The first operational forecast has been produced in 1954 by Carl-Gustav Rossby's group at the Swedish Meteorological and Hydrological Institute (Harper et al. 2007). The first successful climate model has been developed in 1956 by N. A. Phillips (Philips 1956; Cox 2002) while the first general circulation climate model has been developed about 1970 at NOAA Geophysical Fluid Dynamics Laboratory (NOAA 2008). Several limited area (regional) models have been proposed and implemented in the 1970 and 1980s (Shuman 1989). Starting in the 1990s, model ensemble forecasts have been used to extend NWP forecasting period (Molteni et al. 1996; Toth and Kalnay 1997).

A comprehensive overview of techniques used in numerical weather forecasting can be found in Coiffier and Sutcliffe (2012). This book presents a short history of NWP and their evolution followed by a step-by-step description of the various model equations and how to solve them numerically. This book outlines the main elements of a meteorological suite and the theory is illustrated throughout with practical examples.

2.5.1 NWP Categories

The NWP models may be classified in General Circulation Models (GCMs), which describe processes at global level, and regional or local models, whose application surface area is restricted. Local NWP models use atmospheric reanalyzes as initial and boundary conditions for the model run, which then realistically downscale (using physical equations) to a more accurate physical resolution. The NWP model that downscales reanalyzes data is termed a mesoscale model. Because the mesoscale models run over a smaller area than global scale models, the physics can include additional details. Therefore, given enough computing power, these models can be used to forecast various meteorological parameters over a wide area with high temporal and spatial resolution.

Among the most popular GCMs are Global Forecast System (GFS), ECMWF (a model developed by the European Center for Medium-Range Weather Forecasts), UKMO (developed by the United Kingdom Met Office) and GME (the model of Deutsche Wetterdienst) while among the regional/local models one may quote HRM (High Resolution Model), High-Resolution Limited Area Model (HIRLAM), WRF-Nonhydrostatic Mesoscale Model (WRF-NMM), WRF-ARW (Advanced Research WRF), The Unified Model and MM5 (Mesoscale Model) (Santos-Munoz et al. 2009; Isvoranu 2011). Generally, the local models are using boundary and initial conditions obtained by running GCMs. However, the local models may also be run by using boundary and initial conditions provided by the user.

WRF model has a wide range of physical parameterizations, which allow setting the model to better describe the physical processes based on model domain, resolution, location, and application (Ruiz-Arias et al. 2008).

Selection of NWP models is based on the following criteria: (1) performance; (2) cost; (3) popularity; (4) easy access to ground and satellite meteorological data to be used as boundary and initial conditions. European researchers prefer to use models such as HRM and HIRLAM while researchers in USA and Canada prefer WRF and MM5 (Grell et al. 1998). Most models developed in Europe are based on a semi-Lagrangean treatment of the primitive equations. North-American models are usually based on the Eulerian approach. Presently, all models allow both deterministic and probabilistic simulations (Isvoranu 2011).

The performance of diverse NWP models has been analyzed in several papers. For instance, results provided by WRF-NMM, UKMO, and GME have been compared in Santos-Munoz et al. (2009). Also, the models WRF and ECMWF have been analyzed in Matsangouras et al. (2011). These studies show a better performance for WRF model. There is, indeed, an increased popularity of this model in Europe (Isvoranu 2011).

Since individual weather prediction models have their specific strengths and weaknesses in certain weather situations, the results of several different models are usually compared with practice. Another approach is to use the same model, run it with different boundary conditions and compare the results (ensemble forecast). An important parameter is the forecast uncertainty which is derived by using statistical methods or ensemble forecasts and still represents a challenging and complex task due to the chaotic nature of the weather. The output of the weather prediction models is then post-processed and improved with statistical combination of past and/or online measurement data. To achieve this goal, a number of statistical methods can be used to reduce the systematic forecast errors (e.g., model output statistics, Kalman filter, fuzzy logic, and neural network).

2.5.2 NWP for Renewable Energy Forecasting

Wind power forecasts have a high level of quality driven by the strong development of wind energy in countries like Germany, Denmark, or Spain. There exists already a number of providers of operational forecasts, but the forecast systems are mainly focusing on large wind park areas in flat terrain and mid-latitudes, though potential for wind energy production is also very high in mountainous/hilly areas and arctic conditions (e.g., in northern Scandinavia, the Alps, and in South-Eastern Europe). Unlike the wind power, solar yield forecast is still on an early state. The basic principles of solar power forecasts are more or less similar to wind power forecasts but other parameters need to be considered. There are mainly two different ways for solar electricity production: by solar-thermal power plants and by PV plants. The solar-thermal systems use the direct-normal solar irradiance to convert solar energy through focusing receivers into heat, which is then used to drive a thermodynamic cycle and thereby produce electricity. PV systems enable direct conversion of global horizontal irradiance into electricity through semiconductor devices.

Different approaches to forecast irradiance can be taken depending on the target forecasting time. For very short time forecasts (up to 6 h, nowcasting), approaches based on extrapolating the solar radiation field from cloud motion have been proposed (Heinemann et al. 2006). These forecasts are meant for solar field control in solar-thermal and PV plants. In addition, statistical techniques have been proposed for forecasting solar irradiance with up to 24 h (Mellit and Pavan 2010). NWP models are the basis of solar yield forecasts with up 48 h time horizon, the time range useful for grid integration and decision making in the energy market.

Evaluation studies on the MM5 mesoscale model reliability for estimating global solar irradiance were carried out by Zamora et al. (2005) in some locations in USA. Heinemann et al. (2006) evaluated the MM5 model global irradiance forecasts in Germany for lead time up to 48 h. Lorenz et al. (2009a) evaluated several NWP-based GHI forecasts in Europe. Overall, results showed relative RMSE values of about 40 % for Central Europe and about 30 % for Spain. Lorenz et al. (2009b) evaluated hourly global irradiance forecasts, based on ECMWF model, for power prediction of PV systems in Germany. They reported relative RMSE values of about 35 % for single stations for a 24 h horizon forecasts. Remund et al. (2008) evaluated different NWP-based global irradiance forecasts in the USA, reporting relative RMSE values ranging from 20 to 40 % for a 24 h forecast horizon. Similar results were reported by Perez et al. (2009), evaluating NWP-based irradiance forecasts in several places in the USA.

Regarding the forecasts of direct-normal solar irradiance, an additional problem comes into the scene, since direct solar irradiance is not provided by NWP models. Consequently, an additional processing procedure is needed, such as those proposed by Breitkreuz et al. (2009) and Ruiz-Arias et al. (2010). Breitkreuz et al. (2009) proposed a model (called AFSOL) based on the combined use of information provided by a NWP model, an air-quality model and remote sensing retrievals. Ruiz-Arias et al. (2010) propose a new regressive model for the estimation of the hourly diffuse solar irradiation under all sky conditions. The model is based on a sigmoid function and uses the clearness index and the optical air mass as predictors. Lara-Fanego et al. (2012) presented a comprehensive evaluation study of the reliability of direct normal and global forecasts based on the WRF mesoscale model.

Wittmann et al. (2008) analyzed the Spanish premium feed-in tariff model in a case study for a solar-thermal power plant in Andalusia, based on direct-normal solar irradiance forecasts provided by the AFSOL model. This system provides two options for the producers: either transfer the yield to the power distribution company with the electricity sale price stated as a single regulated tariff (tariff model), or sell on the free market at the going market price plus a premium (premium model), which, for solar electricity, is set at 250 % of its average electricity tariff. Operators of installations are obliged to provide the distributor with a forecast of the electricity they intend to feed into the grid the next day by at least 11:00 h (local time) of the previous day. Penalties are established for deviations: the cost of deviation is 10 % of the spot market prices applied to the forecast deviations when the permitted tolerance (20 % for solar and wind power) is exceeded. Interestingly, results proved that the economical benefits that can be achieved based on these forecasts are strongly depended on the time of the day at which forecast deviations take place.

Currently, operational weather forecast models are used with spatial resolutions of a few kilometers. Major National Weather Services intend to implement models with even higher spatial resolution during the next years. The consequences of such downscaling remain to be investigated. At present time, one major short coming is the accurate prediction of the 3D cloud field which is crucial for a good solar production forecast. Furthermore, the implementation of ad hoc measurement data, especially yielded by remote sensing techniques such as satellites, ceilometers, or LIDAR (Light Detection and Ranging) is expected to provide a large potential to improve the forecasts.

References

ABM (2012) Australian bureau of meteorology. http://www.bom.gov.au/

ARM (2012) Department of energy atmospheric radiation measurement. http://www.arm.gov/

ASRC (2012) Atmospheric sciences research center of the university of albany, USA. http://www.asrc.albany.edu/

Badescu V (2002) A new kind of cloudy sky model to compute instantaneous values of diffuse and global solar irradiance. Theor Appl Climatol 72(127–136):2002

Breitkreuz H, Schroedter-Homscheidt M, Holzer-Popp T, Dech S (2009) Short-range direct and diffuse irradiance forecasts for solar energy applications based on aerosol chemical transport and numerical weather modeling. J Appl Meteorol Climatol 48:1766–1779

BSRN (2012) Baseline surface radiation network. http://www.bsrn.awi.de/

Cano D, Monget JM, Guillard H, Albuisson M, Regas N, Wald L (1986) A method for the determination of the global solar radiation from meteorological satellite data. Sol Energy 37:31–39

CEP (2012) Center for Energy and Processes, Ecole des Mines de Paris/Armines, France. http://www.scep.ensmp.fr/

Coiffier J, Sutcliffe C (2012) Funcamentals of numerical weather prediction. Cambridge University Press, Cambridge

CONFRRM (2012) Cooperative networks for renewable resource measurements. http://rredc.nrel.gov/solar/new_data/confrrm/

Cox JD (2002) Storm watchers, John Wiley & Sons, p. 210

Cros S, Wald L (2003). Survey of the main databases providing solar radiation data at ground level. In: Goossens R.i (ed.) Proceedings of the 23rd EARSeL Annual Symposium Remote Sensing in Transition, Ghent, Belgium, 2–4 June 2003, pp. 491–497

DeltaOHM (2012) Pyranometers, albedometers, net irradiance meter. Manual. http://www.Deltaohm.Com/ Ver2008/Uk/Pyra02.Htm

Diabate L, Demarco H, Michaud-Regas N, Wald L (1988) Estimating incident solar radiation at the surface from images of the eath transmitted by geostationary satellites: the heliosat project. Int J Sol Energy 5:261–278

Fröhlich C (1991) History of solar radiometry and the world radiation reference. Metrologia 28:111–115

GAW (2012) Global atmospheric watch. http://www.wmo.int/pages/prog/arep/gaw/gaw_home_en.html

GEWEX (2012) Global energy and water cycle experiment. http://www.gewex.org/

Grell G, Dudhia J, Stauffer D (1998) A Description of the fifth-generation penn state/NCAR mesoscale model (MM5). NCAR tech. note, NCAR/TN-398 + STR, USA

Gueymard CA (2004) The sun's total and spectral irradiance for solar energy application and solar radiation models. Sol Energy 76:423–453

Gueymard CA, Myers DR (2008) Solar radiation measurement: progress in radiometry for improved modeling. In: Badescu V (ed) Modeling solar radiation at the earth surface. Springer, Berlin

Harper K, Uccellini LW, Kalnay E, Carey K, Morone L (2007) 2007: 50th Anniversary of operational numerical weather prediction. Bull Am Meteorol Soc 88(5):639–650

Heinemann D, Lorenz E, Girodo M (2006) Forecasting of solar radiation. In: Wald L, Suri M, Dunlop ED (eds) Solar energy resource management for electricity generation from local level to global scale. Nova Science Publishers, New York

HelioClim (2012) HelioClim server. Centre Energetique et Procedes of Ecole des Mines de Paris. http://www.helioclim.org/heliosat/index.html

Hukseflux (2012) Radiation measurement sensors. Available online http://www.hukseflux usa.com/radiation-measurement.phpAccessed January 2012

Ineichen P, Perez R (2002) A new airmass independent formulation for the linke turbidity coefficient. Sol Energy 73(3):151–157

Isvoranu D (2011) Internal report for project PN-II-ID-PCE-2011-3-0089: Nowcasting and forecasting of PV power plant operation. Polytechnic University of Bucharest, Romania

Janjai S, Laksanaboonsong J, Nunez M, Thongsathitya A (2005) Development of a method for generating operational solar radiation maps from satellite data for a tropical environment. Sol Energy 78:739–751

Janjai S (2008) Generation of solar radiation maps from long-term satellite data. In: Badescu V (ed) Modeling solar radiation at the earth surface. Springer, Berlin

Janjai S, Deeyai P (2009) Comparison of methods for generating typical meteorological year using meteorological data from a tropical environment. Appl Energy 85:528–537

Janjai S (2010) A method for estimating direct normal solar irradiation from satellite data for a tropical environment. Sol Energy 84(9):1685–1695

Kasten F (1980) A simple parameterization of two pyrheliometric formulae for determining the linke turbidity factor. Meteorol Rundsch 33:124–127

Kerr A, Tabony R (2004) Comparison of sunshine recorded by Campbell–Stokes and automatic sensors. Weather 59:90–95

Lara-Fanego V, Ruiz-Arias JA, Pozo-Vazquez D, Santos-Alamillos FJ, Tovar-Pescador J (2012) Evaluation of the WRF model solar irradiance forecasts in Andalusia (southern Spain). Sol Energy. doi:10.1016/j.solener.2011.02.014

Lorenz E, Remund J, Muller SC, Traunmuller W, Steinmaurer G, Pozo D, Ruiz-Arias JA, Fanego VL, Ramirez L, Romeo MG, Kurz C, Pomares LM, Guerrero CG (2009a) Benchmarking of different approaches to forecast solar irradiance. In: 24th European Photovoltaic Solar Energy Conference, Hamburg, Germany, pp. 21–25

Lorenz E, Hurka J, Heinemann D, Beyer HG (2009b) Irradiance forecasting for the power prediction of grid-connected photovoltaic systems. IEEE J Sel Top Appl Earth Obs Remote Sens 2(1):2–10

Lynch P (2008) The origins of computer weather prediction and climate modeling. J Comput Phys 227(7):3431–3444

Martins FR, Pereira EB, Abreu SL (2007) Satellite-derived solar resource maps for Brazil under SWERA project. Sol Energy 81:517–528

Matsangouras IT, Nastos PT, Pytharoulis I (2011) Synoptic-mesoscale analysis and numerical modeling of a tornado event on 12 February 2010 in northern Greece. Adv Sci Res 6:187–194

Mellit A, Pavan AM (2010) A 24 h forecast of solar irradiance using artificial neural network: application for performance prediction of a grid-connected PV plant at trieste. Italy Sol Energy 84(5):807–821

METEONORM (2012) METEONORM software. http://www.meteonorm.com

Molteni F, Buizza R, Palmer TN, Petroliagis T (1996) The ECMWF ensemble prediction system: methodology and validation. Q J R Meteorol Soc 122(529):73–119

Moser W, Rachke E (1984) Incident solar radiation over Europe from METEOSAT data. J Climate Appl Meteorol 23:166–170

Mueller RW, Dagestad KF, Ineichen P, Schroedter-Homscheidt M, Cros S, Dumortier D, Kuhlemann R, Olseth JA, Piernavieja G, Reise C (2004) Rethinking satellite-based solar irradiance modeling: The SOLIS clear-sky module. Remote Sens Environ 91:160–174

NOAA (2008) National oceanic and atmospheric administration. The first climate model http://celebrating200years.noaa.gov/breakthroughs/climate_model/welcome.html

NREL (2012) National renewable energy laboratory. Extraterrestrial solar spectrum. http://rredc.nrel.gov/solar/spectra/am0

Nunez M (1993) The development of a satellite-based insolation model for the tropical western pacific ocean. Int J Climatol 13:607–627

Perez R, Ineichen P, Moore K, Chain C, Kmiecik M, George R, Vignola F (2002) A new operational for satellite-derived irradiances: description and validation. Sol Energy 73(5):307–317

Perez Y, Ramos-Real FJ (2009) The public promotion of wind energy in Spain from the transaction costs perspective 1986–2007. Renew Sust Energ Rev 13(5):1058–1066

Phillips NA (1956) The general circulation of the atmosphere: a numerical experiment. Q J Roy Meteorol Soc 82(352):123–154

Pinker RT, Laszlo I (1992) Modeling surface solar irradiance for satellite applications on the global scale. J Appl Meteorol 24:389–401

PVGIS (2012) Photovoltaic geographical information system. http://re.jrc.ec.europa.eu/pvgis

Reda I (1996) Calibration of a solar absolute cavity radiometer with traceability to the world radiometric reference. Tech. Rep. NREL TP-463-20619, National Renewable Energy Laboratory, Golden, Colorado

Remund Y, Perez R, Lorenz E (2008) Comparison of solar radiation forecasts for the USA. In: 23rd European Photovoltaic Solar Energy Conference, Valencia, Spain, pp. 3141–3143

Rigollier C, Bauer O, Wald L (2000) On the clear-sky models of the ESRA-European radiation atlas, with respect to the heliosat method. Sol Energy 68:33–48

Rigollier C, Lef'evre M, Wald L (2004) The method Heliosat-2 for deriving shortwave solar radiation from satellite images. Sol Energy 77:159–169
Ruiz-Arias JA, Pozo-Vazquez D, Sanchez–Sanchez N, Montavez JP, Hayas-Barru A, Tovar-Pescador J (2008) Evaluation of two MM5-PBL parameterizations for solar radiation and temperature estimation in the south-eastern area of the Iberian peninsula. Il Nuovo Cimento 31(5–6):825–842
Ruiz-Arias JA, Alsamamra H, Tovar-Pescador J, Pozo-Vazquez D (2010) Proposal of a regressive model for the hourly diffuse solar radiation under all sky conditions. Energy Convers Manage 51(5):881–893
Santos-Munoz D, Wolff J, Santos C, García-Moya JA, Nance L (2009) Implementation and validation of WRF model as ensemble member of a probabilistic prediction system over Europe. 10th Annual WRF Users' Workshop, Boulder, USA, 23–26 June 2009
Satel-Light (2012) European database of daylight and solar radiation. http://www.satel-light.com
Schillings C, Mannstein H, Meyer R (2004) Operational method for deriving high resolution direct normal irradiance from satellite data. Sol Energy 76:475–484
Siren KE (1987) The shadow band correction for diffuse irradiation based on a two-component sky radiance model. Sol Energy 39:433–438
SMSE (2012) NASA surface meteorology and solar energy. http://eosweb.larc.nasa.gov/sse/
SRMS (2012) Solar Platform of the West University of Timisoara, Timisoara, Romania. http://solar.physics.uvt.ro/srms
SODA (2012) Solar radiation data. http://www.soda-is.com/eng/index.html
Shuman FG (1989) History of numerical weather prediction at the national meteorological center. Weather Forecast 4(3):286–296
SURFRAD (2012) Surface radiation network, U.S. national oceanic and atmospheric administration. http://www.srrb.noaa.gov/surfrad/index.html
Suri M, Huld T, Dunlop ED (2005) PV-GIS: a web-based solar radiation database for the calculation of PV potential in Europe. Int J Sustain Energ 24:55–67
Suri M, Huld TA, Dunlop ED, Ossenbrink HA (2007) Potential of solar electricity generation in the European Union member states and candidate countries. Solar Energy 81:1295–1305. http://re.jrc.ec.europa.eu/pvgis/
Toth Z, Kalnay E (1997) Ensemble forecasting at NCEP and the breeding method. Mon Weather Rev 125(12):3297–3319
Vonder Haar T, Raschke E, Bandeen W, Pasternak M (1973) Measurements of solar energy reflected by the earth and atmosphere from meteorological satellites. Sol Energy 14:175–184
WCRP (2012) World climate research programme. http://wcrp.wmo.int/wcrp-index.html
Wittmann M, Breitkreuz H, Schroedter-Homscheidt M, Eck M (2008) Case-studies on the use of solar irradiance forecast for optimized operation strategies of solar thermal power plants. IEEE J Sel Top Appl Earth Obs Remote Sens 1(1):18–27
WMO (1983). Report of the meeting of experts on the future activities of the world radiation centre, Leningrad 28 February—1 March 1983. WCP-48, WMO Geneva
WMO (1984). Resolution 6 (EC-XXXVI)—International collection and publication of radiation data. WMO-No.631, Geneva
WMO (2008) Guide to meteorological instruments and methods of observation. world meteorological organization—No. 8 (7th edn.) Chapters 7 and 11
WRDC (2012a). World Radiation Data Center, Main Geophysical Observatory, St. Petersburg, Russia. http://wrdc.mgo.rssi.ru/
WRDC (2012b) World Radiation Data Center, WRDC online archive, National Renewable Energy Laboratory, US Department of Energy. http://wrdc-mgo.nrel.gov/
Zamora RJ, Dutton EG, Trainer M, McKeen SA, Wilczak JM, Hou YT (2005) The accuracy of solar irradiance calculations used in mesoscale numerical weather prediction. Mon Weather Rev 133:783–792
Zelenka A, Perez R, Seals R, Renne D (1999) Effective accuracy of the satellite-derived hourly irradiance. Theor Appl Climatol 62:199–207

Chapter 3
State of the Sky Assessment

In this chapter, we shall describe various indicators to characterize the state of the sky. Since most specific information about these indicators used in this book refer to Romania, some information about this country is useful. Romania is located in Southeast Europe, between 43° 37'07" N and 48°15'16" N and 20° 15'44" and 29° 42'24" E. Its area is 237,500 km^2 of which 30 % is mountains (heights over 800 m asl), 37 % is hills and plateaus (heights between 200 and 800 m asl), and 33 % is fields. The territory of the country is halved by the Carpathian chain (Fig. 3.1). As a result of the atmospheric circulation and the modifications the Carpathian chain imposes on it, the Romanian territory mostly belongs to the temperate-continental climate (Zamfir et al. 1994).

Most data used in this chapter refer to the six locations shown in Table 3.1. These localities were classified by using the Ivanov's index of continentality I (%) given by (Badescu 1991; Badescu and Zamfir 1996):

$$I = \frac{\Delta T + \Delta T_y + 0.25(100 - u)}{0.36\,\phi + 14} \cdot 100 \tag{3.1}$$

where ΔT [°C] is the difference between the average air temperature from the warmest and coldest months of the year, ΔT_y [°C] is the difference between the maximum and minimum air temperature values during the yearly average day, u[%] is the yearly average value of the air relative humidity and ϕ [degrees] is the latitude of the location. The climate is temperate continental when I is greater than 120 % and is weak continental and weak maritime when I is between 100 and 120 % and less than 100 %, respectively.

Section 3.1 deals with traditional indicators for the state of the sky, such as the total cloud cover amount and the relative sunshine. Section 3.2 defines another indicator, namely the sunshine number and shows its statistical properties. Auto-correlation properties of the sunshine number are presented in Chap. 4.

M. Paulescu et al., *Weather Modeling and Forecasting of PV Systems Operation*,
Green Energy and Technology, DOI: 10.1007/978-1-4471-4649-0_3,

3.1 Traditional Indicators for the State of the Sky

The performance of solar energy conversion devices is strongly dependent on the incoming flux of solar radiation, which, in turn, is strongly dependent on the state of the sky. Larger amounts of radiation are received when the sky is free of clouds. Moreover, when clouds are present, the incident radiation depends on the cloud types. Three quantities are commonly used to describe the state of the sky. They are shortly described next.

A usual indicator is the total cloud cover amount C (sometimes called cloudiness degree or point cloudiness), which represents the fraction of the celestial vault covered by clouds. The total cloud cover amount is essentially an instantaneous quantity. A daily averaged total cloud cover amount may be computed. The days are sometimes classified according to this average value. This is justified by the observed persistence of the total cloud cover amount.

The second quantity describing (indirectly) the state of the sky during the daytime is the relative sunshine σ (also called bright sunshine fraction, or simply sunshine fraction). Its definition is $\sigma \equiv s/S$, where S is the length of a given time interval and s is the bright sunshine duration during that interval. Usually, S is the time duration between sunrise and sunset (in hours) while s is the number of daily bright sunshine hours. Shorter S intervals are also used. Of course, a low σ value is an (indirect) indication of a high cloud cover amount. The relative sunshine is measured in a fewer meteorological stations than the total cloud cover amount C.

The third common method to describe (indirectly) the state of the sky is by means of the so-called daily global clearness index k_t which is defined as the ratio between the daily global solar irradiation on the Earth's surface and outside the atmosphere, respectively. Sometimes, global clearness index is defined with respect to the solar irradiation incident on the Earth's surface assuming clear sky conditions. This is an indirect method since the cloudiness degree of the day does not appear explicitly. However, it is expected that the smallest and the larges values of k_t are associated with overcast and clear days, respectively. The k_t index may be used as a parameter in relationships describing the performance of solar energy conversion devices (see, e.g., Knight et al. 1991; Gansler et al. 1994; Nakada et al. 2010).

3.1.1 Cloud Amount

Of particular importance are the sequential properties of the total cloud cover amount. Indeed, accurate simulations of solar energy conversion devices operation require as input appropriate meteorological data recordings. Obtaining continuous series of data is a difficult task for the solar energy engineer or scientist. However, a very promising solution is the synthesis of meteorological data. The literature on this subject in connection with solar energy applications started with works by

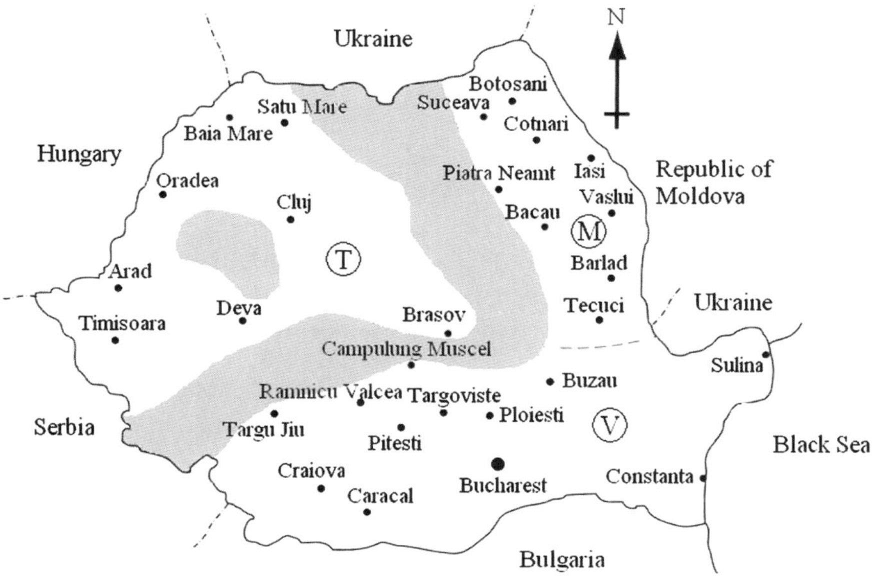

Fig. 3.1 Map of Romania showing 29 meteorological stations. The dashed area shows the Carpathian Mountains. The three historical provinces are: *T*—Transylvania, *M*—Moldavia; *V*—Valahia

Table 3.1 Six Romanian localities provided with radiometric stations

Localities	Latitude [°N]	Longitude [°E]	Altitude [m]	Climatic index *I* [%]
Bucharest	44.499	26.217	90	131.9
Constanta	44.166	28.617	52	112.2
Cluj-Napoca	46.783	23.567	410	121.6
Iasi	47.166	27.600	102	129.9
Timisoara	45.766	21.250	85	130.9
Sulina	45.149	29.667	3	97.3

Aguilar et al. (1988), Aguilar and Collares–Pereira (1992), Zabara and Yianoulis (1992), Boland (1995), and continues nowadays, e.g. Polo et al. (2011). In this section, we refer to simple autoregressive models to generate series of daily averaged total cloud cover amount values.

We are using measurements performed in two Romanian localities (i.e. Bucharest and Iasi, see Fig. 3.1). The climate of both localities is temperate-continental (Badescu 1991) (see Table 3.1). Data measured during around 1200 particular days from January and July are used in the analysis. In Bucharest, we use data collected in the years 1960–1969 while in Iasi the data are collected during 1964–1973 (RMHI 1974). The total cloud cover amount was evaluated at 6.00, 9.00, 12.00, 15.00, and 18.00 local standard time (LST) in July and at 9.00, 12.00, and 15.00 LST in January.

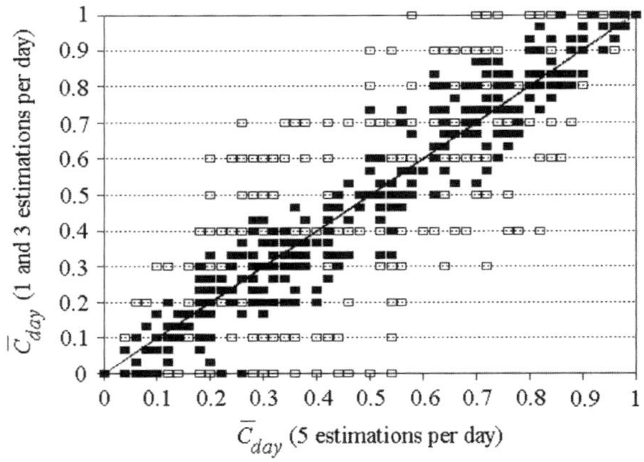

Fig. 3.2 Daily averages of total cloud cover amount $\bar{C}_{\text{day}}$ computed by using five observations per day versus daily averages computed by using three (*filled squares*) or one (*empty squares*) observations per day. Data recorded in Bucharest during July are used

The daily average of total cloud cover amount $\bar{C}_{\text{day}}$ is simply computed as the arithmetic mean of the three or five *instantaneous* estimations. Note that other ways of averaging the daily cloudiness were also identified (Badescu and Zamfir 1996). We are using here the arithmetic mean, however, because of its simplicity. Characterization of the days according to their $\bar{C}_{\text{day}}$ value is justified by the persistence of cloud cover amount, especially during the days with high cloud cover amount (Davies and McKay 1982; Young et al. 1995). This persistence is considered to be the main reason for the good performance of some solar radiation computing models which use $\bar{C}_{\text{day}}$ as input when cloud cover is estimated at intervals of 3 or even 6 h (Davies et al. 1988; Young et al. 1995).

It is known that some meteorological parameters (such as relative sunshine or short-period beam irradiation) have a frequency distribution function (FDF) which is unimodal or bimodal/skewed in respect with the period in the year (Festa and Ratto 1993, p. 23). This applies in the case of total cloud cover amount, too. Consequently, two months from different seasons are used here. They are July (representative for the months with unimodal FDFs) and January (for the months with bimodal FDFs).

Figure 3.2 shows that a relatively small difference exists between the values $\bar{C}_{\text{day}}$ computed in July at Bucharest by using, on one hand, all the five estimations during the day and, on the other hand, only three estimations (namely at 9.00, 12.00 and 15.00 LST). This confirms the already stated persistence of cloud cover amount. Computing $\bar{C}_{\text{day}}$ by using a single estimation per day (performed at 12.00 LST) increases considerably the spreading of data (Fig. 3.2). In the following, the values $\bar{C}_{\text{day}}$are computed by using the maximum available data, i.e., three estimations per day in January and five estimations per day in July.

Figures 3.3 and 3.4 show $\bar{C}_{\text{day}}$ time series obtained by using observation data in Bucharest and Iasi. The days of the January' s and July' s are numbered consecutively, so 1st of January in 1961 at Bucharest is the 32nd day in Fig 3.3a. In January, the sky is covered most part of the day in both Bucharest and Iasi (Figs. 3.3a and 3.4a, respectively). However, in Iasi there is smaller cloudiness variability. In July, many

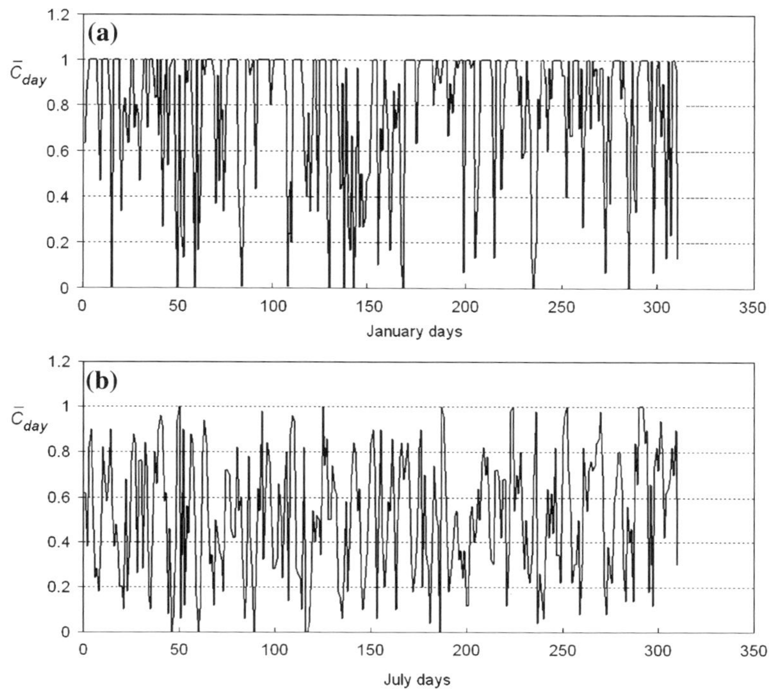

Fig. 3.3 Time series of daily average total cloud cover amount $\bar{C}_{\mathrm{day}}$ values computed by using observation data from ten years (1960–1969) in Bucharest. **a** January; **b** July. The days are numbered consecutively (e.g. 1st July 1961 is the 32nd day in the abscissa)

days have covered sky (Figs. 3.3b and 3.4b). The number of days with clear and overcast sky is practically the same. Again, the persistence of cloud cover is more obvious at Iasi than at Bucharest.

3.1.1.1 Modeling Based on Knowledge the Mean and Variance of Cloudiness Time Series

Measured time series of the same meteorological quantity are characterized by autocorrelation. For a given system, the autocorrelation between two successive observations will depend on their time lag.

The daily average of total cloud cover amount $\bar{C}_{\mathrm{day},j}$ may be regarded as a random variable with mean $<\bar{C}_{\mathrm{day}}>$ and variance $\sigma^2_{\bar{C}_{\mathrm{day}}}$ during the given time interval (i.e. ten January's or July's at Bucharest and nine at Iasi). One denotes by *Z*(*j*) the standardized (zero mean and unit variance) *excess* of the total cloud cover amount in the day *j*, defined as:

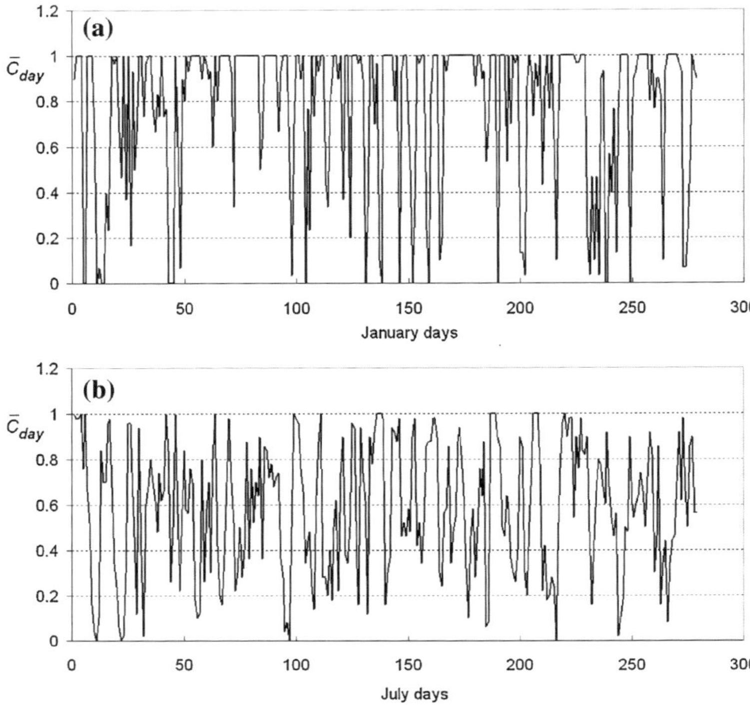

Fig. 3.4 The same as Fig. 3.3 at Iasi (years 1964–1972)

$$Z(j) = \frac{\bar{C}_{\text{day},j} - <\bar{C}_{\text{day}}>}{\sigma_{\bar{C}_{\text{day}}}} \tag{3.2}$$

The dth day-lag autocorrelation coefficient $\rho(d)$ is computed by using a general formula given in Gordon and Reddy (1988):

$$\rho(d) = \frac{\text{Cov}(Z(j), Z(j+d))}{\sqrt{\sigma^2(Z(j)) \cdot \sigma^2(Z(j+d))}} \tag{3.2a}$$

where Cov denotes covariance. Then, $\rho(d)$ is given by:

$$\rho(d) = \frac{W(d)}{W(0)} \tag{3.3}$$

where

$$W(d) = \frac{1}{J-d} \sum_{j=1}^{J-d} [Z(j)][Z(j+d)] \tag{3.4}$$

Here, the sum involves a number of J "effective" days from the time interval (a day j is "effective" if the day $j + d$ belongs to the same year).

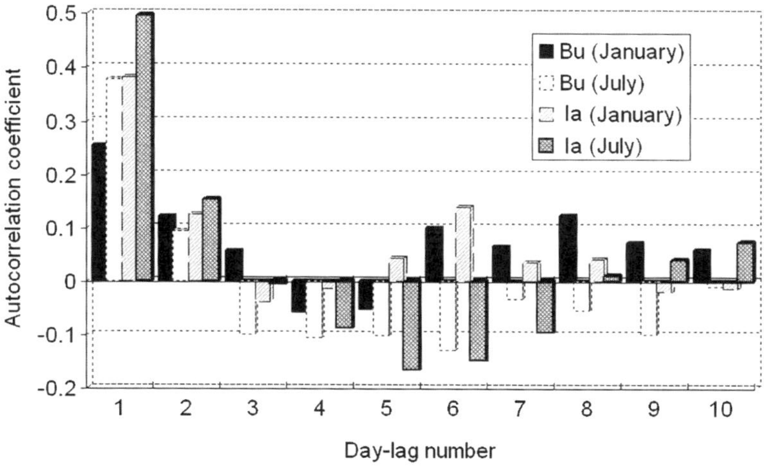

Fig. 3.5 The dth day-lag autocorrelation coefficient $\rho(d)$ of the observed $\bar{C}_{day}$ data ($d = 1$ to 10) for Bucharest (Bu) and Iasi (Ia) in January and July

Figure 3.5 shows the autocorrelation coefficient $\rho(d)$ ($d = 1$–10) for both Bucharest and Iasi in January and July. As expected $\rho(d)$ decreases by increasing the (small) day-lag number d. Generally, the current day total cloud amount is correlated with the cloud cover amount from the past two days only. Except is for January in Bucharest where there is a dependence on the last three days. The autocorrelation of total cloud cover amount is more important in July than in January and at Iasi than at Bucharest. These results agree with comments outlined when comparing Figs. 3.3 and 3.4, respectively.

The above results suggest the usage of second-order autoregressive (AR) models for synthetic generation of $\bar{C}_{day}$ time series. Note that first- and second-order AR processes were found suitable to generate time series of solar irradiation and irradiance (see e.g. Aquilar and Collares–Pereira 1992; Skartveit and Olseth 1992; Festa and Ratto 1993).

A second-order AR process is now used to generate time series of $\bar{C}_{day,j}$ for Bucharest and Iasi in both January and July. First, one reminds some basic facts. An AR process of order s for the variable Z, such as that defined in Eq. (3.2) is (Gordon and Reddy 1988):

$$Z(j) = \Phi_1 Z(j-1) + \Phi_2 Z(j-2) + \ldots + \Phi_s Z(j-s) + a(j) \tag{3.5}$$

where Φ_s are the partial autocorrelation coefficients and $a(j)$ is a random white noise. For a first-order process the partial autocorrelation coefficient Φ_1 is:

$$\Phi_1 = \rho(1) \tag{3.6}$$

where ρ is the autocorrelation coefficient, while for a second-order process one has:

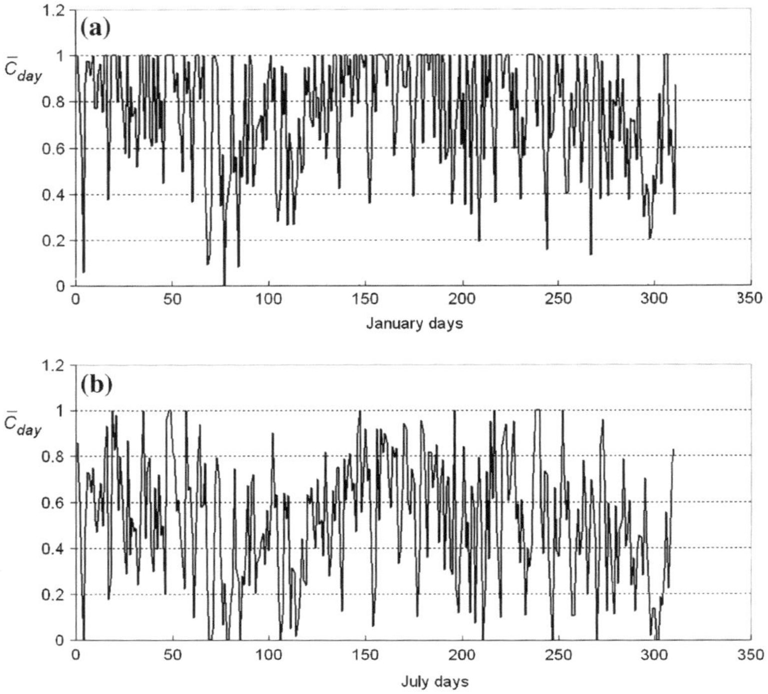

Fig. 3.6 Samples of time series of daily average of total cloud cover $\bar{C}_{\text{day}}$ values generated by using a second order autoregressive process at Bucharest (years 1960–1969): **a** January; **b** July

$$\Phi_1 = \rho(1)\frac{1-\rho(2)}{1-\rho^2(1)}, \quad \Phi_2 = \frac{\rho(2)-\rho^2(1)}{1-\rho^2(1)} \tag{3.7}$$

The white noise a should have zero mean, be normally distributed and have a variance σ^2 as follows. For a first-order model:

$$\sigma^2(a) = 1 - \rho^2(1) \tag{3.8}$$

while for a second-order model:

$$\sigma^2(a) = 1 - \Phi_1^2 - \Phi_2^2 - \frac{2\Phi_1^2\Phi_2}{1-\Phi_2} \tag{3.9}$$

In case of a first-order AR process the beginning of the series is:

$$Z(0) = a(0), \quad Z(1) = \Phi_1 Z(0) + a(1) \tag{3.10a, b}$$

while for a second-order process one uses:

$$Z(-1) = a(-1), \quad Z(0) = \Phi_1 Z(-1) + a(0) \tag{3.11a, b}$$

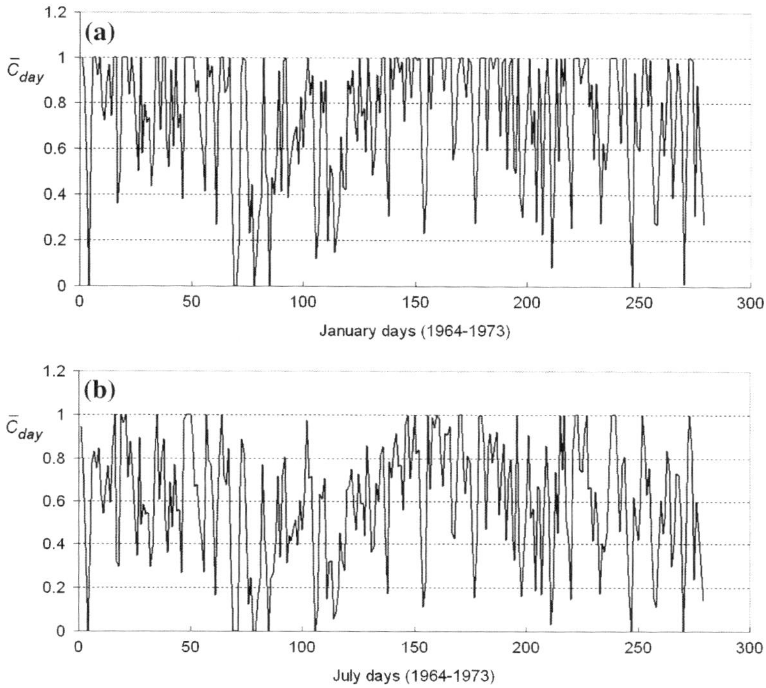

Fig. 3.7 The same as Fig. 3.6 at Iasi (years 1964–1972)

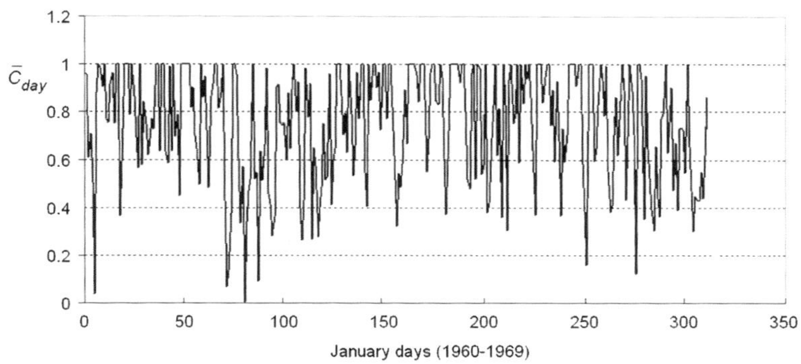

Fig. 3.8 Sample of time series of $\bar{C}_{\text{day}}$ values obtained by using a first order autoregressive model at Bucharest in January (years 1960–1969)

$$Z(1) = \Phi_1 Z(0) + \Phi_2 Z(-1) + a(1) \tag{3.11c}$$

Inspection of the observational data shows that the FDF of $\bar{C}_{\text{day}}$ is often nonGaussian. To overcome the non-normality problem a standard Gaussian mapping procedure is used (see, e.g., Festa and Ratto 1993, p. 175). The results are, however, poorer than those obtained without this mapping technique. Only results obtained without using a mapping technique are discussed in this section.

Samples are shown in Figs. 3.6 and 3.7. They refer to virtual days from the years 1960–1969 (Bucharest) and 1964–1972 (Iasi), respectively, since they were generated by using as input (estimated or measured) values of $<\bar{C}_{\mathrm{day}}>$ and $\sigma_{\bar{C}_{\mathrm{day}}}$ from these time periods. The generated series can be compared with the observation data given in Figs. 3.3 and 3.4. The synthesized values are close to the observed data in July in case of both Bucharest and Iasi (compare Figs. 3.3b and 3.4b, on one hand, and Figs. 3.6b and 3.7b on the other hand). In January, too, there is a good concordance between the synthetic and observed data when days with a high cloud cover amount are considered. However, the number of synthetic days with clear sky is smaller than that of days with observed clear days for both Bucharest and Iasi (compare Figs. 3.3a and 3.4a, on one hand, and Figs. 3.6a and 3.7a on the other hand).

A natural question could arise: is it an important decrease in accuracy when using first-order instead of second-order AR models?

Figure 3.8 shows a sample of synthetic data obtained by using a first-order AR model applied in January at Bucharest. These results can be compared with those obtained by using a second-order AR process in the same locality and month (Fig. 3.6a). The differences are small. Other computations, not shown here, confirm the fact that first- and second-order AR models have comparable performance, in the case we studied. Consequently, the above and below remarks about the performance of the second-order AR models apply to a large extent to the first-order processes, too.

The above statement is supported by the FDF of the synthetic $\bar{C}_{\mathrm{day}}$ data. One denotes by "class 0.1" those $\bar{C}_{\mathrm{day}}$ values which belong to the range 0.0–0.1. Similarly, the "class 0.2" comprises daily average total cloud amount data between 0.11 and 0.20 while the class 1.00 refers to data between 0.91 and 1.00.

Figure 3.9 shows the FDFs of $\bar{C}_{\mathrm{day}}$ values obtained at Bucharest by using observed and synthetic data. In the last case both first- and second-order AR models were considered. There is little difference between the performances of the two AR models. Also, there is a reasonable agreement between the observed and synthetic data (for statistical indicators see Fig. 3.10).

The FDFs in January have a marked skewness (i.e., asymmetry with respect to the mean). This feature, along with the bimodality, is known to be typical not only for cloud cover but also for relative sunshine data and short-period beam irradiation distributions (Festa and Ratto 1993, p. 23). U-shaped cloud cover frequency distributions were reported in early works by Olseth and Skartveit (1984, 1987). Later, these authors concluded that low standard deviations yield narrow unimodal distributions while increasing standard deviations yield bimodal (or skewed) distributions (Skartveit and Olseth 1992). Our results confirm their conclusion. Indeed, the unimodal distribution of July (Fig. 3.9b) is associated with a small standard deviation (Fig. 3.10b) while the skewed distribution of January (Fig. 3.9a) is associated with the slightly larger standard deviation of Fig. 3.10a.

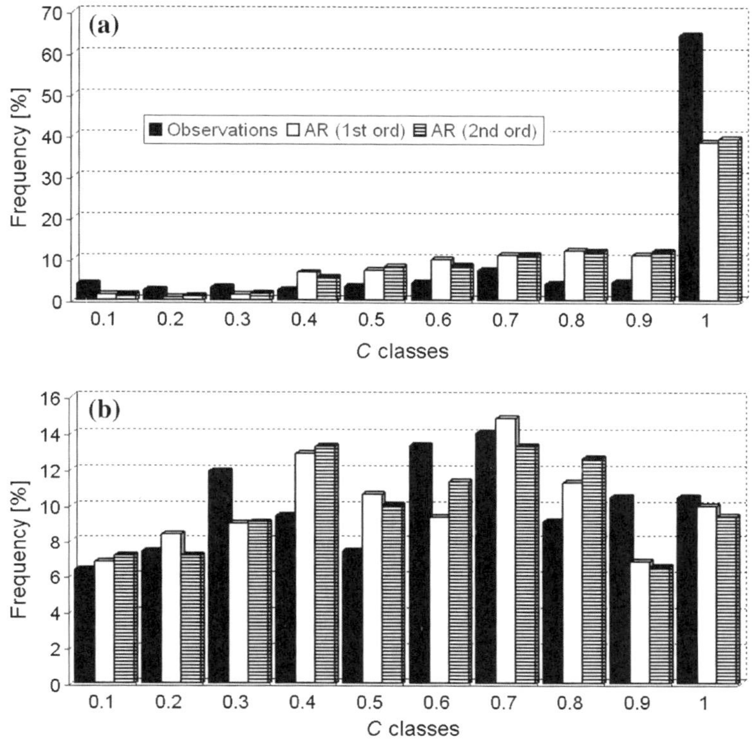

Fig. 3.9 Frequency (probability) distribution functions for $\bar{C}_{\text{day}}$ values obtained at Bucharest by using observed and synthetic data. Both first- and second-order AR processes were considered. **a** January; **b** July

The random character of the generated data makes the agreement between the synthetic and the observed data to depend on the generated sample. The performance of an AR process is more appropriately understood by studying a set of many samples. In order to have a more complete image about models performance one compares the first statistical moments of the observed and synthetic series, respectively. Four moments will be considered here for the series of n_{day} days from the time interval under consideration (310 days in case of both January's and July's). They are the mean $<\bar{C}_{\text{day}}>$, the standard deviation $\sigma_{\bar{C}_{\text{day}}}$, the skewness and the kurtosis (see Appendix A for definitions).

Figure 3.10 shows the first four moments of the observed series $\bar{C}_{\text{day}}$ (first sample) and a number of 25 synthetic series (samples 2–26) generated by using a second-order AR process at Bucharest. There is good agreement between the mean and the standard deviation of the observed and synthetic series, respectively. This satisfies the usual requirements concerning the data needed for solar energy applications. In January, the values of skewness and kurtosis for the synthetic time series are obviously different from those obtained for the observed time series.

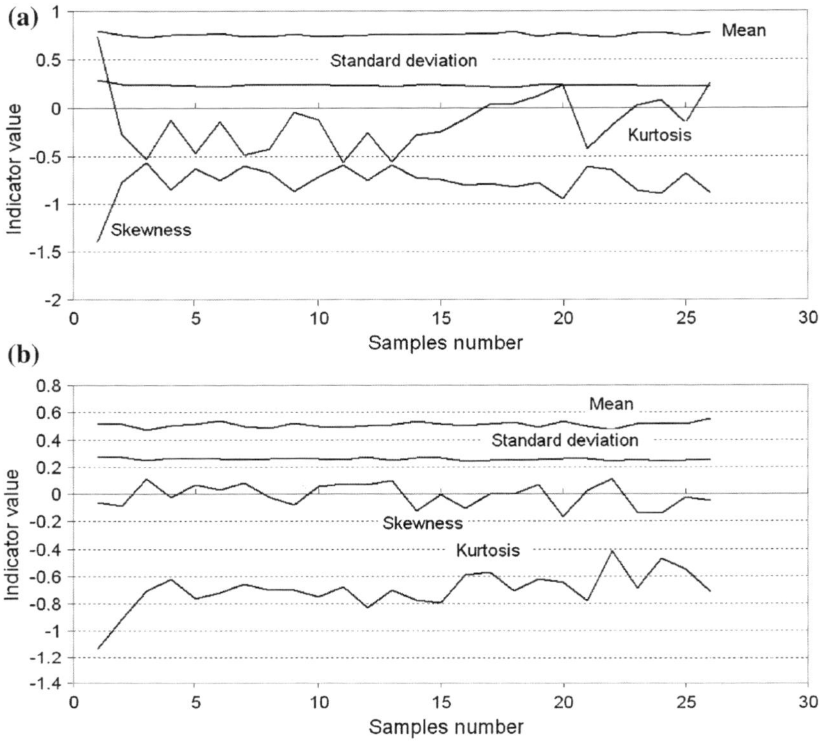

Fig. 3.10 First four statistical moments of the observed series $\bar{C}_{\text{day}}$ (sample #1) and a number of 25 synthetic series (samples #2 to #26) generated by using a second-order AR process at Bucharest: **a** January; **b** July

However, the skewness value of both observed and generated time series is always negative, indicating clustering of data to the right (i.e., toward large values of total cloud cover amount). This feature is more obvious in case of the observed time series. The kurtosis value of the generated series is usually negative, indicating that the FDFs are flatter than a normal distribution. The same indicator is positive when the observed time series are considered, showing these series are sharper than a Gauss curve.

In July there is a good concordance between the skewness values of the observed and generated data, respectively. The skewness value lies somewhere around zero showing in both cases the FDFs are nearly symmetrical around the mean. The kurtosis value of both observed and generated series is always negative indicating the FDFs are flatter than a normal distribution. However, the FDFs of the synthetic data are sharper than those of the observed data. For more details see Badescu (1997).

3.1.1.2 Modeling Based on Knowledge of Long-term Average Cloudiness

The above models need $<\bar{C}_{\text{day}}>$ and $\sigma_{\bar{C}_{\text{day}}}$ as input values. A simpler model to generate time series of daily averaged cloudiness data has been proposed in Badescu (2000). The model requires as input the long-term mean value of the cloudiness only. This value can be found easily on a monthly basis in the reports of many meteorological institutions. The models are based on a very simple auto-regressive relationship for the stochastic variable x (Festa and Ratto 1993, p. 159):

$$x_n = \Phi x_{n-1} + r_n \tag{3.12}$$

Here r_n is a random variable (usually normally distributed) while Φ is a parameter. By increasing Φ the dependence of x_n on the preceding value x_{n-1} increases and the autocorrelation of the stochastic variable x increases, too. For higher autocorrelation the weight of the random variable is expected to diminish (generally in a non-linear manner). These arguments prompted us to propose the following simple model to generate daily averaged total cloud cover amount data:

$$\bar{C}_{\text{day},n} = \Phi\, \bar{C}_{\text{day},n-1} + (1-\Phi)^m r_n \tag{3.13}$$

where m is a parameter. By using Eq. (3.13) one easily finds Φ as a function of the mean value $<\bar{C}_{\text{day}}>$ for a given location and a given time period:

$$\Phi \cong 1 - \left(\frac{\langle \bar{C}_{\text{day}} \rangle}{\bar{r}}\right)^{\frac{1}{m-1}} \tag{3.14}$$

where $\bar{r}$ is the mean value of the random variable r. A number of tests proved that the performance of the autoregressive process Eq. (3.13) is slightly improved by increasing the value of the parameter m. Here the value $m = 20$ is used.

After preliminary computations we inferred that the following conjecture have to be adopted in order the process Eq. (3.13) gives the best performance:

1. $r_n = \alpha_n$ in those months where the FDF is unimodal (July);
2. $r_n = 1 - \alpha_n^4$ in the months where the FDF is bimodal or skewed (January).

In both cases α_n is a random number uniformly distributed between 0 and 1. The simple model Eq. (3.13) is used to generate time series of daily averaged total cloud amount.

Samples are shown in Figs. 3.11 and 3.12. They refer to virtual days from the years 1960–1969 (Bucharest) and 1964–1972 (Iasi), respectively, since they are generated by using as input the mean of the observed values $<\bar{C}_{\text{day}}>$ from these time periods. The results can be compared to the observation data from Figs. 3.3 and 3.4.

A more relevant comparison can be made by computing the first-lag autocorrelation coefficient $\rho(1)$ for both the observation and synthetic time series during the ad hoc time interval comprising ten January's or July's. $\rho(1)$ is given by Eq.

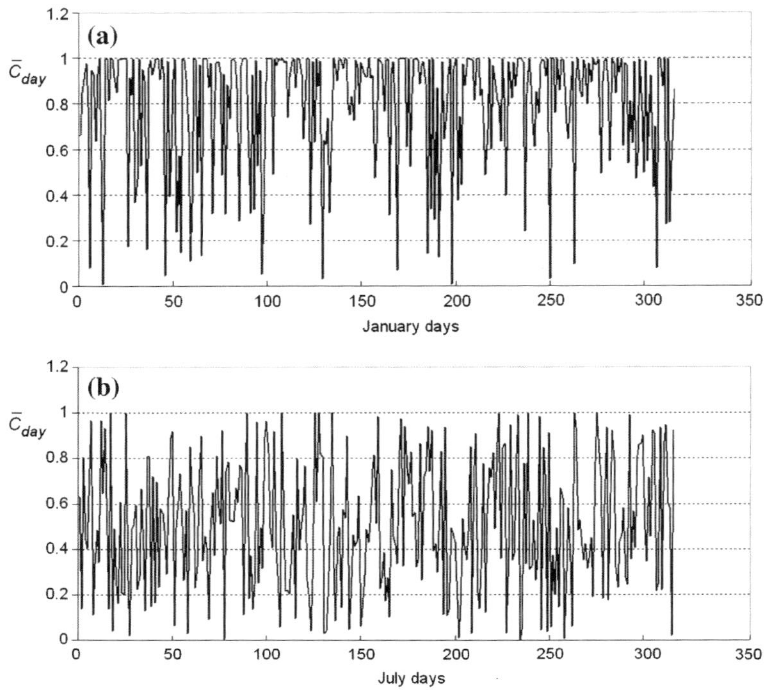

Fig. 3.11 Samples of time series of $\bar{C}_{\text{day}}$ values obtained by using the model Eq. (3.13) at Bucharest (years 1960–1969): **a** January; **b** July

(3.3) for $d = 1$. Table 3.2 shows the results. At Bucharest, the model performance is acceptable in January and it is poorer in July.

A short discussion follows concerning the FDF of $\bar{C}_{\text{day}}$. Figures 3.13 and 3.14 show the FDFs of the $\bar{C}_{\text{day}}$ values obtained at Bucharest and Iasi by using observed and synthetic data, respectively.

In January there is good agreement between the two sorts of FDFs. This applies for both localities (see Figs. 3.13a and 3.14a). When July data are considered the concordance between the FDFs based on synthetic and observed data, respectively, is poorer in case of Bucharest (Fig. 3.13b) but it is reasonably good in Iasi (Fig. 3.14b) (for statistical indicators see Figs. 3.15 and 3.16).

Figure 3.9 shows the FDFs of the synthetic $\bar{C}_{\text{day}}$ time series generated by two autoregressive models which use as input the mean $< \bar{C}_{\text{day}} >$ and the standard deviation $\sigma_{\bar{C}_{\text{day}}}$ (for given locality and time period). The present model leads to better agreement between the observed and synthetic FDFs (compare, on one hand, Figs. 3.13a and 3.9a and, on the other hand, Figs. 3.13b and 3.9b). This is obvious for January and less obvious for July.

Figures 3.15 and 3.16 show the first four moments of the observed series $\bar{C}_{\text{day}}$ (first sample) and a number of 25 synthetic series (samples #2 to #26), generated

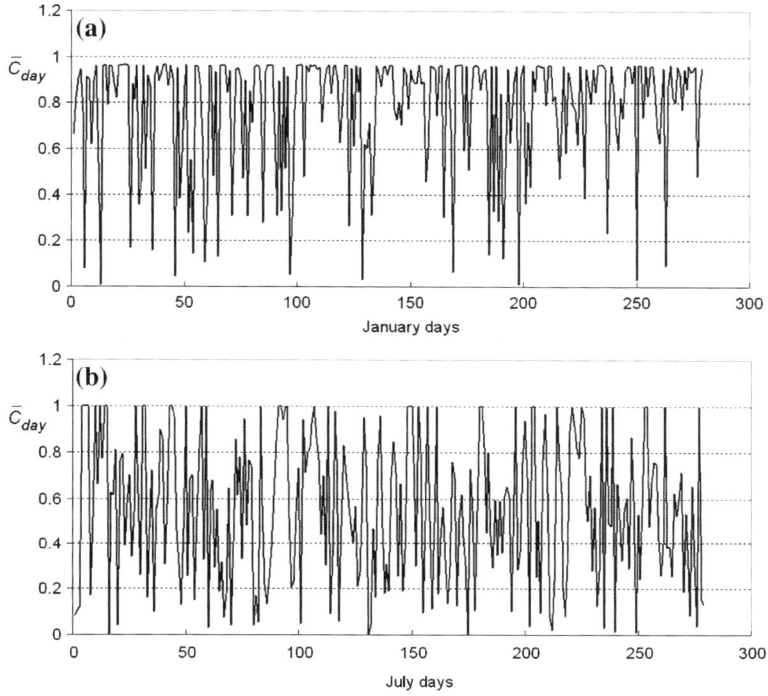

Fig. 3.12 The same as Fig. 3.11 for Iasi (years 1964–1972)

Table 3.2 First-lag day autocorrelation coefficient $\rho(1)$ computed by using observed and synthetic $\bar{C}_{\text{day}}$ data

	Bucharest		Iasi	
	January	July	January	July
Observation data	0.25	0.37	0.38	0.49
Synthetic data	0.24	0.17	0.20	0.12

by using the model Eq. (3.13) at Bucharest and Iasi. There is good agreement between the mean and the standard deviation of the observed and synthetic series, respectively. This satisfies the usual requirements concerning the quality of the data needed for the design of solar energy devices.

The skewness and kurtosis depend on the sample. The dependence is stronger in January than in July (compare Figs. 3.15a and 3.16a, on one hand, and Figs. 3.15b and 3.16b, on the other hand) and is relatively similar for both localities. In January, the skewness and kurtosis values for the synthetic series oscillate around the value of the observed series (see Figs. 3.15a and 3.16a).

In January, the skewness of both observed and generated series is negative, indicating clustering of data to the right (i.e., toward large values of total

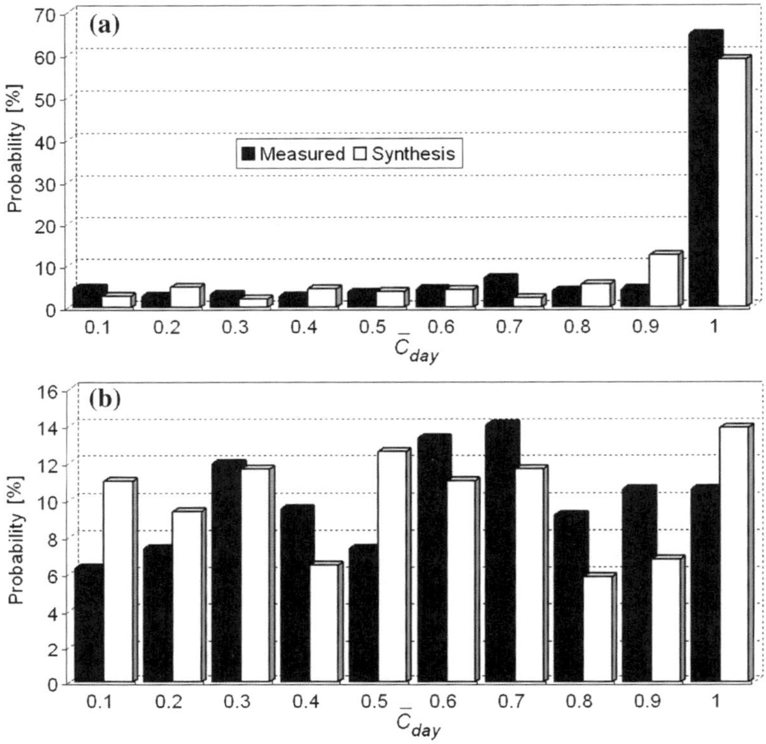

Fig. 3.13 Frequency (probability) distribution functions for $\bar{C}_{\mathrm{day}}$ values obtained at Bucharest (years 1960–1969) by using measured and synthetic data: **a** January; **b** July

cloud amount). The kurtosis of both types of series is usually positive, showing these FDFs are sharper than a Gauss curve. In July, there is good concordance between the skewness values of the observed and generated time series, respectively. The skewness value lies somewhere around zero showing in both cases the FDFs are nearly symmetrical around the mean. The kurtosis of both observed and generated series is negative indicating the FDFs are flatter than a normal distribution.

In the following, the performance of the present model is compared with that of an autoregressive model which uses as inputs the mean $<\bar{C}_{\mathrm{day}}>$ and the standard deviation $\sigma_{\bar{C}_{\mathrm{day}}}$. Both models have a comparable performance, when the mean and the standard deviation of the synthetic values are considered (compare Figs. 3.15 and 3.10). The present model is obviously better if one looks to the higher two statistical moments (skewness and kurtosis). The present model gives the best results in the cold season (January) (compare Figs. 3.15a and 3.10a).

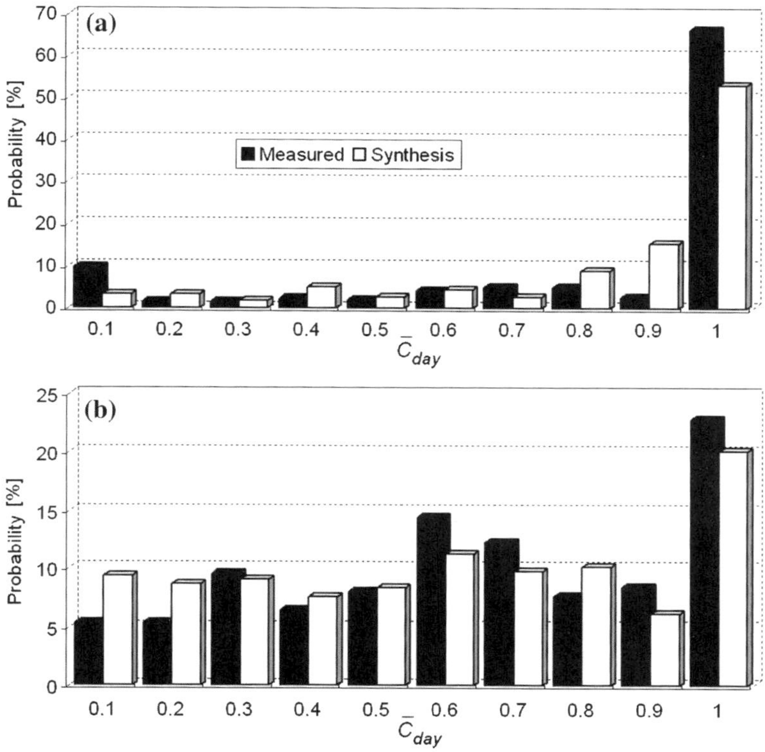

Fig. 3.14 The same as Fig. 3.13 for Iasi (years 1964–1972)

3.1.1.3 Summary and Discussion

Analysis of observed data for ten years in two Romanian localities shows that the daily averaged total cloud amount $\bar{C}_{\text{day}}$ in the current day depends, generally, on the cloud cover amount from the past two days, only. So, first- or second-order AR processes may be used for synthetic data generation. The results obtained by using a Gaussian mapping procedure were poorer than those obtained without this technique. The second-order models are only slightly better than the first-order models.

The unimodal frequency distribution of $< \bar{C}_{\text{day}} >$ in July is associated with a small standard deviation while the skewed distribution from January is associated with a large variance. The autoregressive models generate data whose mean and standard deviation are close to those of the observed data. This is good enough in case the synthetic data is needed for usual applications of solar energy. A good agreement between the skewness values of the observed and synthetic time series occurs during the warm season (July). No concordance was emphasized between the kurtosis values for the observed and generated series of daily averaged total cloud cover amount.

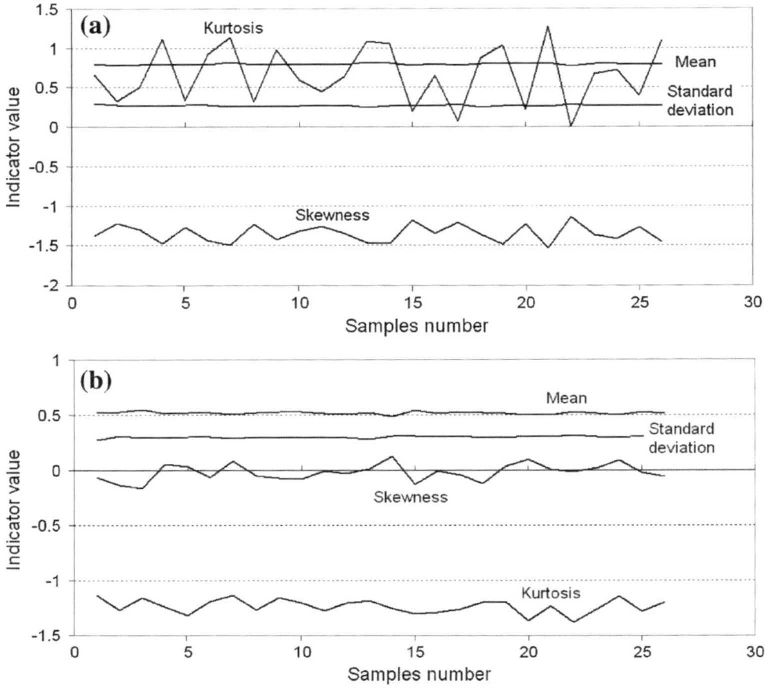

Fig. 3.15 First four statistical moments of the observed series $\bar{C}_{\rm day}$(sample #1) and a number of 25 synthetic series (sample #2 to #26) generated by using the model Eq. (3.13) at Bucharest (years 1960–1969): **a** January; **b** July

Series of $\bar{C}_{\rm day}$ values can be generated by using the autoregressive model Eq. (3.13). The model uses as input the long-term mean value of the daily averaged total cloud amount only. At Bucharest, the first-day lag autocorrelation coefficient of the synthetic series is in reasonable concordance with the observed time series in July. The concordance is poorer in January. During the cold season (January) there is a good agreement between the FDF of the observed and generated time series, respectively. This applies for both localities analyzed. When warm season (July) data is considered, the concordance between the FDFs based on the two types of time series is poorer in case of Bucharest and it is reasonably good in Iasi.

The autoregressive model Eq. (3.13) leads to better agreement between the observed and synthetic FDFs, as compared to similar PDFs obtained by using first- or second-order usual AR processes. Equation (3.13) generates time series whose mean and standard deviations are very close to those of the observed data. The higher statistical moments (i.e. skewness and kurtosis) show a dependence on the sample, as expected.

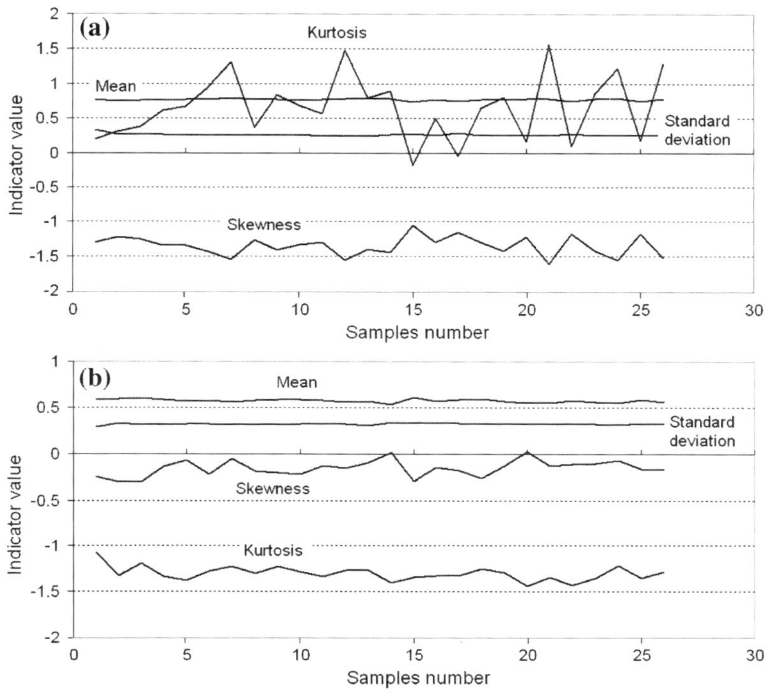

Fig. 3.16 The same as Fig. 3.15 for Iasi (years 1964–1972)

3.1.2 Relative Sunshine

Many authors reported on the empirical relationship between cloud cover and bright sunshine (Reddy 1974; Harrison and Coombes 1986; Paulescu and Schlett 2004). Their studies are of special interest for the solar radiation computing methods developed on the basis of long-term averages of bright sunshine (see e.g. IEA 1984) because sunshine records are not always kept at weather stations but long-term records of observed total cloud cover are available for most stations of the world. Empirical relationships between cloud cover C and bright sunshine σ for the Romania climate and latitudes have been reported in Badescu (1990). The complement of σ is often called cloud shade $\kappa = 1 - \sigma$. The three simple relationships we previously analyzed have the following forms:

$$\kappa = a_1 + b_1 C \qquad (3.15)$$

$$\kappa = a_2 C + b_2 C^2 \qquad (3.16)$$

$$\kappa = a_3 C + b_3 C^2 + c_3 C^3 \qquad (3.17)$$

where a_i, b_i (i = 1,2,3) are regression coefficients whose values were determined by a least squares fit of the Eqs. (3.15–3.17) to the observed values of cloud shade

Table 3.3 The accuracy of the Eq. (3.16) when applied in Valahia, Moldavia and Transylvania: $m_{k,\mathrm{obs}}$—the average cloud shade value, *MBE*—mean bias error, s_D—second centered moment of error distribution, *MAE*—mean absolute error, d_2—Willmott's index of agreement

	Valahia	Moldavia	Transylvania
$m_{k,obs}$	0.5479	0.5812	0.5798
MBE	−0.001	−0.001	−0.001
s_D	0.0706	0.0691	0.0671
MAE	0.056	0.053	0.053
d_2	0.9510	0.9433	0.9515

κ_{obs}. Once a_i, b_i were obtained, the three equations were used to compute new values of cloud shade κ_{comp}. In this section we refer exclusively to the simple relationship given by Eq. (3.16). Note that this nonlinear formula was also preferred by Harrison and Coombes (1986).

Meteorological data were collected from 29 Romanian weather stations to give a broader coverage of the country in both latitude and longitude (Fig. 3.1). In computations we used 1740 pairs of (C, κ) monthly average values from a five-year interval, from 1967 to 1972. A number of 348 multi-year monthly average values resulted.

The accuracy by which the Eq. (3.16) evaluates the $N = 1740$ monthly average values of κ was verified by using three usual statistical indicators of accuracy, namely the mean bias error (*MBE*), the second centered moment of the error distribution (s_{D}), the mean absolute error (*MAE*), and Willmott's index of agreement d_2 (see Appendix A for definitions).

We applied Eq. (3.16) to compute cloud shade by using only meteorological data from each of the three historical provinces of Romania. Table 3.3 shows the results. The empirical relationship we tested has the same performance on the whole Romanian territory. This can be verified by means of any of the four indicators. Note that *MBE*, s_D, and *MAE* agree with the fact that the simple relationship Eq. (3.16) has the worst performance in Valahia. However, the index of agreement d_2 yields another accuracy hierarchy.

First, we analyzed the accuracy of the regression formula that we determined by using the whole set of data, when applied in each of the 29 weather stations of Fig. 3.1 (case 1 in Table 3.4). Next, we tested the four statistical indicators to study the accuracy of Eq. (3.16) when applied in other areas than the one where the regression coefficients were determined. To be clearer, first we obtained a set of regression coefficients by fitting Eq. (3.16) to data associated to a station of Table 3.4. Then, Eq. (3.16) with that set of coefficients was tested for accuracy in other stations of Table 3.4. We compared with the accuracy of formulas obtained by using only data from the respective weather stations (case 2 in Table 3.4). All the four indicators recognize that, generally, the regression formulas obtained in case 2 are more accurate. However, some nonconcordances are observed in a few particular situations. So, for both cases 1 and 2, the indicators *MBE*, s_D, and *MAE* show Eq. (3.16) provides the best performance at Bacau. On the other hand, the

Table 3.4 The accuracy of the Eq. (3.16) when applied in different weather stations

Localities	*MBE* Case 1/Case 2	s_D Case 1/Case 2	*MAE* Case 1/Case 2	d_2 Case 1/Case 2
Botosani	−0.018/−0.001	0.048/0.048	0.041/0.038	0.973/0.976
Suceava	0.019/−0.000	0.057/0.056	0.047/0.042	0.941/0.949
Cotnari	−0.036/−0.000	0.048/0.048	0.048/0.037	0.959/0.973
Iasi	0.038/−0.000	0.069/0.069	0.062/0.053	0.925/0.945
Bacau	−0.008/−0.000	0.043/0.043	0.033/0.032	0.975/0.976
Vaslui	−0.002/−0.001	0.057/0.056	0.043/0.042	0.967/0.970
Birlad	0.092/−0.001	0.081/0.073	0.099/0.068	0.855/0.940
Pt Neamt	0.017/−0.000	0.062/0.062	0.051/0.050	0.928/0.931
Tecuci	−0.037/−0.001	0.042/0.043	0.047/0.034	0.970/0.982
Tg Jiu	−0.042/−0.000	0.079/0.079	0.067/0.065	0.915/0.933
Craiova	0.013/−0.001	0.072/0.070	0.059/0.057	0.957/0.963
Caracal	−0.032/−0.000	0.063/0.062	0.059/0.050	0.955/0.968
Rm Vilcea	−0.021/−0.000	0.060/0.060	0.049/0.048	0.951/0.956
Pitesti	−0.016/−0.000	0.056/0.055	0.046/0.043	0.960/0.967
Tirgoviste	0.037/−0.000	0.071/0.070	0.068/0.056	0.908/0.933
Ploiesti	−0.003/−0.001	0.057/0.057	0.045/0.045	0.966/0.966
Cl Muscel	−0.034/−0.000	0.046/0.043	0.047/0.035	0.940/0.960
Bucarest	0.034/−0.000	0.081/0.079	0.072/0.065	0.930/0.947
Buzau	0.020/−0.001	0.062/0.062	0.053/0.050	0.956/0.961
Constanta	−0.007/−0.000	0.048/0.048	0.039/0.038	0.983/0.984
Sulina	−0.055/−0.003	0.070/0.072	0.071/0.056	0.946/0.963
Satu Mare	0.025/−0.001	0.065/0.064	0.055/0.051	0.947/0.957
Baia Mare	0.015/−0.000	0.065/0.065	0.055/0.052	0.951/0.957
Oradea	0.001/−0.001	0.081/0.071	0.066/0.066	0.930/0.929
Cluj	−0.021/−0.001	0.054/0.053	0.047/0.044	0.964/0.971
Arad	−0.017/−0.000	0.066/0.065	0.056/0.052	0.949/0.959
Timisoara	0.003/−0.001	0.066/0.065	0.053/0.052	0.957/0.959
Deva	0.000/−0.000	0.072/0.072	0.057/0.057	0.941/0.940
Brasov	0.008/−0.001	0.055/0.055	0.042/0.043	0.955/0.955

Case 1—regression coefficients determined by using the whole set of data. Case 2—regression coefficients determined by using only data from the respective weather station. For values of *MBE*, s_D, MAE and d_2, at the level of the three Romanian provinces see Table 3.3

index of agreement d_2 estimated the best results at Constanta. The four indicators are in good concordance showing Birlad as the weather station with the worst results, when case 1 is considered. However, in case 2 the index of agreement indicates Piatra Neamt having the worst performance.

A last question regards the accuracy of Eq. (3.16) when applied in other time periods than the one when the regression coefficients were determined. To be clearer, first we obtained a set of regression coefficients by fitting Eq. (3.16) to data associated to the year of Table 3.5. Then, Eq. (3.16) with that set of coefficients was tested for accuracy in other year of Table 3.5. First, the empirical relationship obtained by using the whole set of data was applied during six different years (Table 3.5, case 1). The results were compared with those obtained with the

Table 3.5 The accuracy of the Eq. (3.16), when applied in different years

Years	MBE Case 1/Case 2	s_D Case 1/Case 2	MAE Case 1/Case 2	d_2 Case 1/Case 2
1967	−0.015/−0.000	0.063/0.063	0.051/0.050	0.952/0.954
1968	0.020/−0.000	0.075/0.073	0.060/0.056	0.945/0.955
1969	0.005/−0.001	0.060/0.060	0.047/0.046	0.973/0.974
1970	0.010/−0.001	0.071/0.070	0.056/0.055	0.947/0.949
1971	−0.006/−0.001	0.069/0.069	0.054/0.054	0.947/0.946
1972	−0.003/−0.000	0.072/0.072	0.058/0.058	0.910/0.911

Case 1—regression coefficients determined by using the whole set of data. Case 2—regression coefficients determined by using only data from the respective year. For values of *MBE*, s_D, MAE and d_2, at the level of the three Romanian provinces see Table 3.3

regression formulas determined by using only data from the respective years (case 2 of Table 3.5).

As expected, the best results were obtained in the case 2. In both cases 1 and 2 the indicators *MBE*, s_D, and *MAE* show Eq. (3.16) having the best and the worst performances in the years 1969 and 1968, respectively. The index of agreement is always in good concordance with the other indicators.

Second, we analyzed the accuracy of the formula we obtained by using the complete set of data, when applied in all the year's months (case 1 of Table 3.6). We compared with the accuracy of the formulas that we determined by using only data from the respective months (Table 3.6, case 2). Generally, the index of agreement is in good concordance with the other indicators showing the best results are obtained in case 2. However, d_2, s_D, and *MAE* disagree each other concerning the months with the best and the worst results, respectively. This is not surprising but a confirmation of the usual perception that no better statistical indicator exists.

3.1.3 Clearness Index

In this section, the daily clearness index k_t is defined as the ratio H/H_{ext}, where H and H_{ext} are the monthly average of the daily solar irradiation and the daily extraterrestrial solar irradiation on a horizontal surface, respectively.

Five meteorological databases are used now. The first two databases are RADGLOB (Danescu et al. 1980) and ROMINSO (Danescu et al. 1980). They contain multi-year monthly average values of solar global daily irradiation and monthly mean daily sunshine duration, respectively, for six Romanian localities listed in Table 3.1. Two additional databases are GLBLUIAS (Erhan 1979) and NEBLUIAS (Erhan 1979). They contain monthly values of solar global irradiation and cloud cover amount, respectively, at Iasi during the years 1964–1975. The last database (ROMETEO) contains average monthly values for various meteorological parameters (bright sunshine hours, cloud cover amount, wind speed, precipitation,

Table 3.6 The accuracy of the Eq. (3.16) when applied in different months

Months	*MBE* Case 1/Case 2	s_D Case 1/Case 2	*MAE* Case 1/Case 2	d_2 Case 1/Case 2
January	−0.065/0.000	0.052/0.040	0.070/0.032	0.848/0.944
February	−0.019/−0.000	0.051/0.046	0.043/0.037	0.884/0.892
March	0.013/−0.001	0.053/0.048	0.042/0.037	0.930/0.933
April	0.050/0.000	0.058/0.049	0.059/0.038	0.825/0.886
May	0.061/0.000	0.057/0.050	0.067/0.038	0.732/0.822
June	0.049/−0.000	0.060/0.054	0.062/0.043	0.775/0.825
July	0.018/−0.001	0.058/0.048	0.046/0.036	0.871/0.880
August	0.026/−0.000	0.061/0.045	0.050/0.035	0.871/0.903
September	−0.001/−0.000	0.054/0.048	0.043/0.039	0.941/0.942
October	−0.041/−0.001	0.049/0.046	0.053/0.036	0.913/0.945
November	−0.034/−0.000	0.049/0.046	0.048/0.036	0.927/0.948
December	−0.067/−0.000	0.053/0.054	0.071/0.043	0.836/0.939

Case 1—regression coefficients determined by using the whole set of data. Case 2—regression coefficients determined by using only data from the respective month. For values of *MBE*, s_D, MAE and d_2, at the level of the three Romanian provinces see Table 3.3

atmospheric temperature, and pressure and air relative humidity) in 29 Romanian localities presented in Fig. 3.1. Here, we are using data on bright sunshine hours at Iasi (years 1967, 1968, 1970–1971).

Figure 3.17 shows the isolines of the monthly mean daily values of k_t over the Romanian territory as they result from measurements (databases RADGLOB and ROMINSO) in January and July. During January, k_t ranges between 0.38 and 0.44 (Fig. 3.17a). It decreases with increasing latitude in Middle and Eastern Romania but in Western Romania one can see the influence of the Adriatic Sea located 400 km west of Romania. As expected, the clearness index is higher in summer when its dependence on latitude is obvious (Fig. 3.17b). There is a Black Sea influence in Eastern Romania where isolines direction becomes parallel to the seashore.

3.1.3.1 Clearness Index and Relative Sunshine

A large number of sunshine-based models are of Ångström-Prescott type (see Chap. 5), i.e., they express the monthly mean daily clearness index k_t as a function of the relative sunshine σ. For a survey showing more than 90 models see e.g. (Festa and Ratto 1993). Table 3.7 shows the relationships we selected and tested under the climate and latitudes of Romania. A few comments concerning their previously reported accuracy are given below.

When tested in Turkey, the model by Ogelman et al. has a mean bias error of 3.8 % in Adana and 6.9 % in Ankara (Festa and Ratto 1993) while when tested for 41 locations in Saudi Arabia the *MBE* varied between 2.34 and 26.1 % with a mean of 10.28 % (Rehman 1998). The model by Akinoglu and Ecevit (1990) was

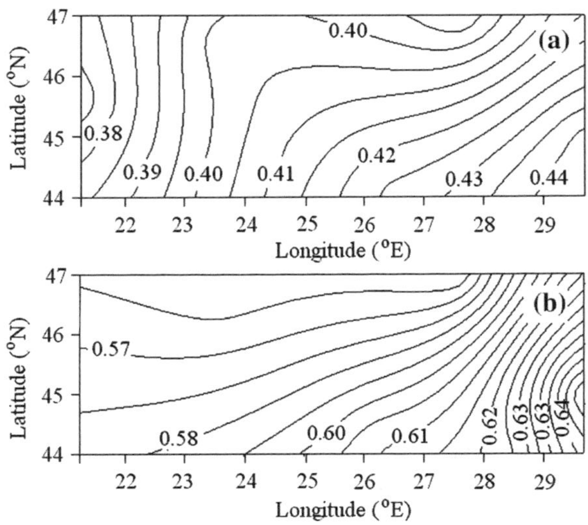

Fig. 3.17 k_t isolines based on multi-yearly average measurements in Romania: **a** January; **b** July

tested for 59 localities from Europe, Asia, Africa, and North America and an average *MBE* of 8.8 % was found (Festa and Ratto 1993). Further testing in Saudi Arabia shows the *MBE* varied between 2.17 and 25.8 % with a mean of 10.2 % (Rehman 1998). The models by Schuepp—ENEC (1986) and WMO (1981) were tested for seven localities in Ethiopia. The mean absolute error varied in case of Schuepp model (ENEC 1986) between 3.8 and 10.1 % while the WMO (1981) model performed slightly worse (*MBE* between 4.4 and 17.0 %) Drake and Mulugetta (1996). The model by Rietvelt (1978) was tested extensively. A *MBE* between 1.2 % at Maseru and 9.1 % at Albuquerque was reported in Fig 2.16 of (Festa and Ratto 1993). A Canadian study (Halouani et al. 1993) reported a mean bias error between −26.5 % in Inoucdjouac and 33.3 % in Coral Harbour and a root mean square error between 3.5 % in Toronto and 41.8 % at Cambridge Bay A. In Saudi Arabia the already quoted paper (Rehman 1998) found a MBE ranging from 2.65 to 30.8 % with a mean of 11.51 %. The model by Alnaser (1989) was analyzed in Qatar using data for 27 years (Alnaser 1993). The *MBE* varied between 5.0 % in September and 12.5 % in March. The model by Bahel et al. (1986) was tested in Saudi Arabia where its *MBE* varied between 3.04 and 19.4 % with a mean of 10.41 % (Rehman 1998). The Canadian study (Halouani et al. 1993) reported for the Gariepy (1980) model a *MBE* varying between -22.0 % in Sable Island and 10.7 % in Eureka and a *RMSE* varying between 4.2 % in Ottawa and 25.0 % in Sable Island.

Verification based on multi-year monthly average values.

The first seven relationships in Table 3.7 were tested by using the databases RADGLOB and ROMINSO. The following linear regression was obtained through a least-square method:

$$k_t = 0.2609 + 0.5138\sigma \tag{3.18}$$

Table 3.7 Existing relationships between monthly average clearness index k_t and relative sunshine σ

Number	Relationship	References	Country
1	$k_t = 0.195 + 0.676\sigma - 0.142\sigma^2$	Ogelman et al. (1984)	Turkey
2	$k_t = 0.145 + 0.845\sigma - 0.28\sigma^2$	Akinoglu and Ecevit (1990)	Turkey
3	$k_t = 0.25 + 0.5\sqrt{0.5\sigma(1+\sigma)}$	Schuepp (ENEC 1986); Drake and Mulugetta (1996)	Etiopia
4	$k_t = 0.29 + 0.49\sigma$	WMO (1981)	–
5	$k_t = 0.18 + 0.62\sigma$	Rietvelt (1978)	Yugoslavia, Sweden, Belgium, USA
6	$k_t = 0.2843 + 0.4509\sigma$	Alnaser (1989)	Bahrain
7	$k_t = 0.175 + 0.552\sigma$	Bahel et al. (1986)	Saudi Arabia
8	$k_t = a + b\sigma$ $a = 0.3791 - 0.0041T - 0.0176w$ $b = 0.481 + 0.0043T + 0.0097w$	Gariepy (1980)	Canada

T = monthly average ambient temperature [°C]; w = monthly average precipitation [cm]

This regression is derived from the databases RADGLOB and ROMINSO and the same databases were used to compare the models. Therefore, the regression Eq. (3.18) must obviously give the best results (Fig. 3.18). The second and third best models are the Alnaser and WMO relationships, respectively.

Figure 3.19 shows that the accuracy of Alnaser and WMO relationships, as well as that of Eq. (3.18), is smaller at small values of k_t (less than 0.5, roughly). The WMO model tends to overestimate k_t. This agrees with the mean bias error found in Fig. 3.18. The overestimation is higher at smaller k_t. The Alnaser model and the regression [Eq. (3.18)] have less important, slightly negative, bias error.

Figure 3.20 is based on multi-year monthly average values of the clearness index obtained by using measurements and Eq. (3.18), respectively, for all the six locations of Table 3.1 (databases RADGLOB and ROMINSO). The latitude and longitude of these localities cover almost whole surface of Romania. A negative relative error means that Eq. (3.18) underestimates the value of the clearness index. Generally, the relative error associated to Eq. (3.18) slightly depends on latitude (Fig. 3.20).

The dependence on longitude is even weaker. Its dependence on the month is, however, stronger (Fig. 3.21). Generally, the regression Eq. (3.18) underestimates k_t in the first half-year and overestimates it in the rest of the year. The dependence of the relative error on the month is obviously more important than its dependence on latitude.

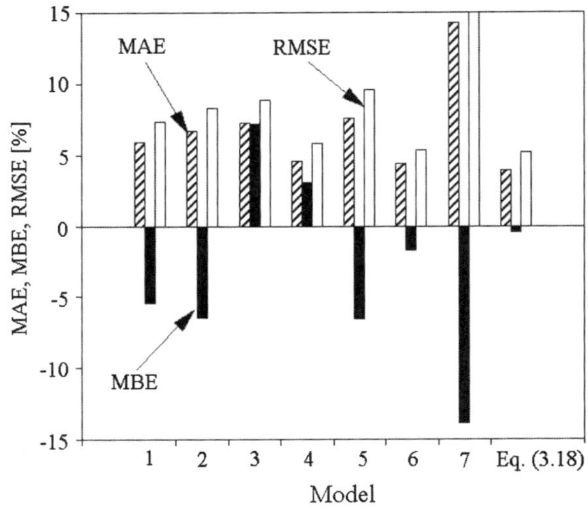

Fig. 3.18 Mean absolute error (*MAE*), mean bias error (*MBE*), and root mean square error (*RMSE*) of seven relationships between the monthly average clearness index k_t and relative sunshine (see Table 3.7). The accuracy indicators of the regression Eq. (3.18) are also shown. The databases RADGLOB and ROMINSO were used

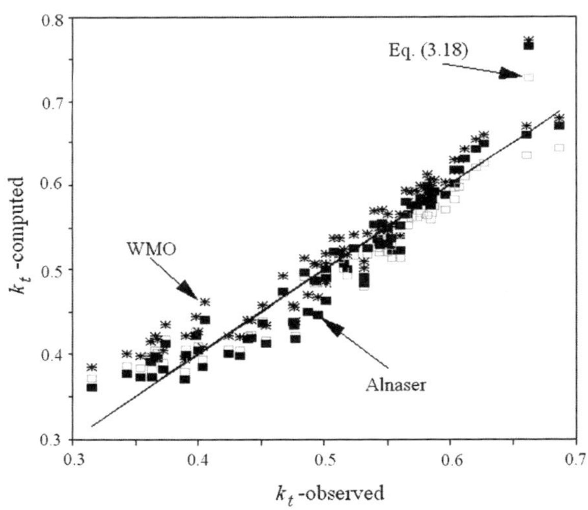

Fig. 3.19 Computed versus observation-based values of the monthly average clearness index k_t for the relationships ranked the first and the second best from Table 3.7. Values associated to the regression Eq. (3.18) are also shown. The databases RADGLOB and ROMINSO were used

Verification based on monthly average values

A second verification was made by using the databases GLBLUIAS and ROMETEO. All the eight models of Table 3.7 were tested. Their accuracy is comparable with that of previous results (compare Figs. 3.22 and 3.18).

The following linear regression was obtained through a least-square method:

$$k_t = 0.2228 + 0.5710\sigma \tag{3.19}$$

The regression Eq. (3.19) has the best performance (Fig. 3.22). The second and third ranked are the Alnaser and Ogelman et al. models, respectively. Generally, the accuracy is lower when using as input monthly averages from particular years instead of *multi-yearly* monthly averages (compare Figs. 3.22 and 3.18,

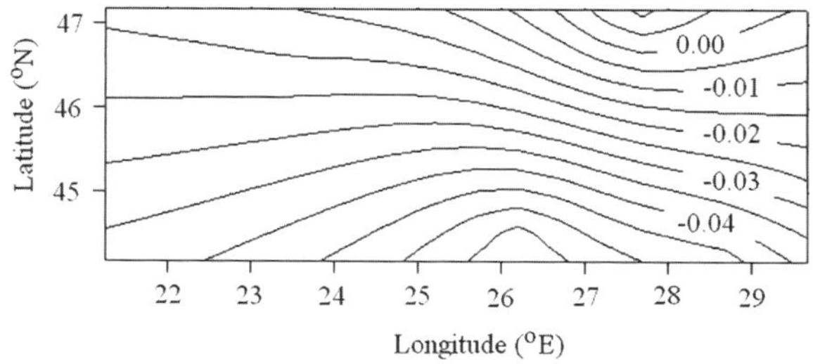

Fig. 3.20 Dependence on latitude and longitude of the relative error associated to the regression Eq. (3.18). The databases RADGLOB and ROMINSO were used

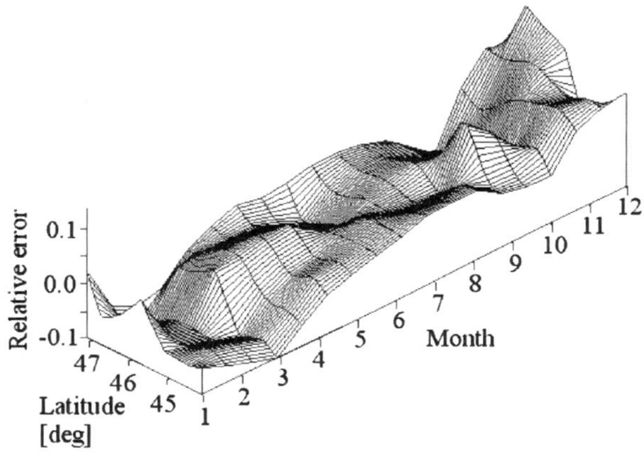

Fig. 3.21 Dependence on latitude and month of the relative error associated to the regression Eq. (3.18). The databases RADGLOB and ROMINSO were used

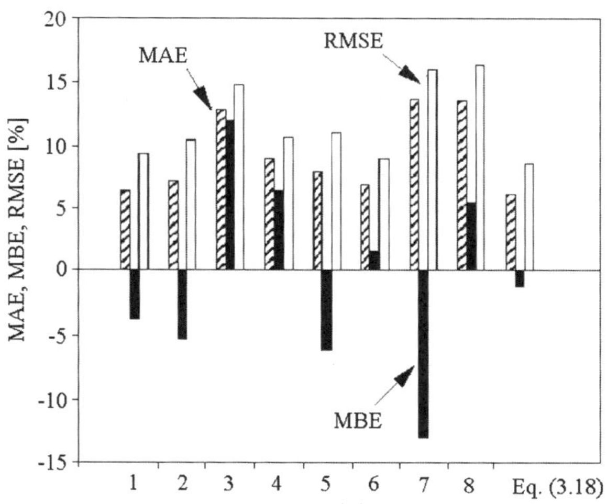

Fig. 3.22 Mean absolute error (*MAE*), mean bias error (*MBE*), and root mean square error (*RMSE*) of eight relationships between the monthly average clearness index k_t and relative sunshine (see Table 3.7). The accuracy indicators of the regression Eq. (3.19) are shown. The databases GLBLUIAS and ROMETEO (Iasi data) were used

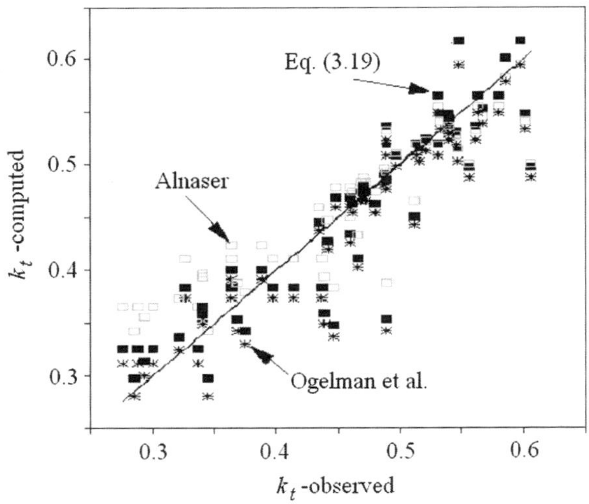

Fig. 3.23 Computed versus observation-based values of the monthly average clearness index k_t for the relationships ranked the first and the second best from Table 3.7. Values associated to the regression Eq. (3.19) are also shown. The databases GLBLUIAS and ROMETEO (Iasi data) were used

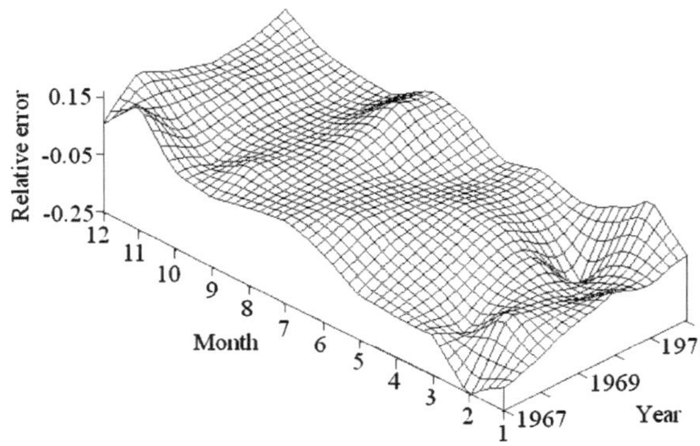

Fig. 3.24 Dependence on year and month of the relative error associated to the regression Eq. (3.19). The databases GLBLUIAS and ROMETEO (Iasi data) were used

respectively). The error dispersion is higher in the first case (compare Figs. 3.23 and 3.19, respectively). The accuracy is lower at small k_t values (Fig. 3.23).

The accuracy of Eq. (3.19) seems not to depend significantly on the year (Fig. 3.24). A certain overestimation of k_t in the second half of the year is again observed.

3.1.3.2 Clearness Index and Total Cloud Amount

The databases GLBLUIAS and NEBLUIAS allowed us to obtain the following linear, quadratic and cubic best-fit relationships, respectively:

$$k_t = 0.8683 - 0.6254C \quad (\text{RMSE} = 12.72\,\%) \tag{3.20}$$

$$k_t = 0.6379 + 0.07749C - 0.5180C^2 \quad (\text{RMSE} = 12.33\,\%) \tag{3.21}$$

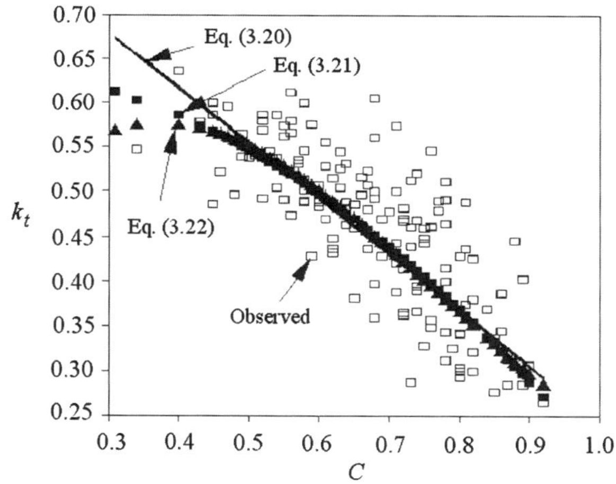

Fig. 3.25 First degree, second degree, and third degree fits of the monthly average clearness index k_t as function of total cloud amount. The databases GLBLUIAS and NEBLUIAS were used

$$k_t = 0.2382 + 2.0385C - 3.60377C^2 + 1.5722C^3 \quad (\text{RMSE} = 12.35\,\%) \tag{3.22}$$

The accuracy of Eqs. (3.20–3.22) is smaller than that using the number of bright sunshine hours as input (compare the RMSE of Eqs. (3.20–3.22), on one hand, and the RMSE of Eq. (3.19) from Fig. 3.22, on the other hand). The three best fits have comparable accuracy for k_t values higher than 0.5 (Fig. 3.25). The second and third degree fits have better performance at smaller k_t values.

3.1.3.3 Summary and Discussion

A number of existing relationships between clearness index and bright sunshine duration are tested. When multi-years average monthly mean values of bright sunshine hours are used as input the first ranked models are Alnaser (1989) and WMO (1981). In case monthly mean values of bright sunshine hours are used in calculation the best models are those from Ogelman et al. (1984) and Alnaser (1989). Best-fit correlations are also derived. They are given by Eqs. (3.18–3.22) and use bright sunshine hours or fractional total cloud amount as input.

The dependence of the sunshine-based regressions Eqs. (3.18–3.22) on month is stronger that their dependence on latitude. They generally underestimate the clearness index in the first half-year and overestimate it in the rest of the year. The accuracy of the regressions Eqs. (3.18–3.22) seems not to depend significantly on the year.

3.2 Sunshine Number

For an observer placed on the Earth's surface, the sunshine number $\xi(t)$ is defined as a time-dependent random Boolean variable, as follows:

$$\xi(t) = \begin{cases} 0 & \text{if the sun is covered by clouds at time t} \\ 1 & \text{otherwise} \end{cases} \tag{3.23}$$

Let us consider a time moment t during the day-time and a time interval Δt centered on t. We assume the distribution of the clouds over the sky, as well as the dynamics of this distribution, are not known. Then, $\xi(t)$ may be considered as a random variable during the time interval Δt. The probability for the sun being covered by clouds during Δt is denoted as $p(\xi = 0,\ \Delta t)$ and the probability that the sun will shine during the same time period is denoted as $p(\xi = 1,\ \Delta t)$. Because ξ is a Boolean variable, the two probabilities are related by the following normalization condition:

$$p(\xi = 0, \Delta t) + p(\xi = 1, \Delta t) = 1 \tag{3.24}$$

Measures for the probabilities $p(\xi = 0,\ \Delta t)$ and $p(\xi = 1,\ \Delta t)$ are now introduced. One denotes by $s(\Delta t)$ the number of time units with the sun shining during the time interval Δt. Then, the probability $p(\xi = 1,\ \Delta t)$ may be defined as usual by the ratio between $s(\Delta t)$ and Δt:

$$p(\xi = 1,\ \Delta t) = \frac{s(\Delta t)}{\Delta t} = \sigma(\Delta t). \tag{3.25}$$

Here, $\sigma(\Delta t)$ is the common relative sunshine for the time interval Δt. As already stated, the quantity

$$\kappa(\Delta t) = 1 - \sigma(\Delta t) \tag{3.26}$$

is called cloud shade (Harrison and Coombes 1986; Badescu 1991). Since $\kappa(\Delta t)$ is a function of the relative sunshine $\sigma(\Delta t)$, it depends on the relative position of the clouds and the sun, as seen by the observer on the sky. It does not depend on the clouds type, as far as the clouds are thick enough to stop the sun light.

A measure for the probability $p(\xi = 0,\ \Delta t)$ has been introduced by using results of integral geometry and geometrical probabilities (Badescu 1992, 2008):

$$p(\xi = 0, \Delta t) \approx C(\Delta t), \tag{3.27}$$

where $0 \le C(\Delta t) \le 1$ is the total cloud cover amount averaged on the time interval Δt. Equations (3.25) and (3.27) show that, for appropriate values of Δt, these probabilities may be computed by using measurements performed routinely by meteorological stations. Finally, use of Eqs. (3.24–3.27) gives:

$$C(\Delta t) \approx 1 - \sigma(\Delta t) = \kappa(\Delta t) \tag{3.28a, b}$$

Equation (3.28a,b) is a popular relationship used by many models of computing solar radiation on cloudy sky. Note that the usual attitude is to postulate Eq. (3.28a,b) but this relationship and the assumptions necessary to derive it were proved rigorously in Badescu (1992). Equation (3.28a,b) shows that both the cloud shade $\kappa(t,\ \Delta t)$ and the cloud cover amount $C(\Delta t)$ are 1D quantities. However, the cloud cover amount is used as a measure for a 2D quantity, i.e., the cloud cover surface area.

Measurements performed in the Romanian town of Timisoara (see Table 3.1 and Fig. 3.1) are used in this section. Global and diffuse solar irradiance (G and G_d, respectively) recorded at the Solar Radiation Monitoring Station of the West University of Timisoara are used here (SRMS 2012). Data are measured at 15 s time interval between sunrise and sunset. Time series of global and diffuse solar irradiance were obtained by using measurements performed from January 1 to December 31, 2009.

Series of sunshine number values are derived from the series of measured solar irradiance values by using the World Meteorological Organization sunshine criterion (WMO 2008), given by Eq. (2.5). All measurements associated to $h < 5°$ have been removed from the database. There are two reasons for this: (1) the pyranometer's accuracy around sunrise and sunset is questionable and (2) $\sin(h)$ tends toward zero for small values of h and, consequently, Eq. (2.5) tends to provide $\xi_{m,j}(t) = 1$, whatever the values of G and G_{d} are.

3.2.1 Statistical Properties

A statistical distribution is uniquely specified by its characteristic function which may be expressed as a Taylor expansion of statistical moments. The simplest approximation for the characteristic function involves using the lowest order moments, i.e., the mean and the variance, which provide information on the location and variability (spread, dispersion) of the data set, respectively. Better approximation for the characteristic function requires adding higher order moments. Commonly the third and fourth moments are used. They provide information on the shape of the distribution.

The statistical moments of order $k, M_k(\xi)$, of a random Boolean variable ξ may be defined by using the following relations (Papoulis 1984):

$$M_k(\xi) = \sum_{\xi=0,1} \xi^k p(\xi), \quad k = 1, 2, \ldots \tag{3.29}$$

In the case that the random Boolean variable is the sunshine number, use of Eqs. (3.25), (3.28a,b), and (3.29) allows writing the mean M:

$$M \equiv M_1(\xi, t, \Delta t) = \sigma(t, \Delta t) = 1 - \kappa(t, \Delta t) \cong 1 - C(t, \Delta t) \tag{3.30}$$

The central statistical moments $M_k(\xi - M)$ of a random Boolean variable ξ are defined as usual (Papoulis 1984):

$$M_k(\xi - M) = \sum_{\xi=0,1} (\xi - M)^k p(\xi), \quad k = 1, 2, \ldots \tag{3.31}$$

The standard deviation is defined as $D \equiv \sqrt{M_2(\xi - M)}$.

Equation (3.31) is used now in the case where the random variable is the sunshine number, for which Eqs. (3.24),(3.25), and (3.28a,b) apply. This yields the first four central statistical moments:

$$M_1(\xi - M, t, \Delta t) = 0 \tag{3.32a}$$

$$M_2(\xi - M, t, \Delta t) = D^2(t, \Delta t) = \kappa(t, \Delta t)(1 - \kappa(t, \Delta t)) \tag{3.32b}$$

$$M_3(\xi - M, t, \Delta t) = \kappa(t, \Delta t)(1 - \kappa(t, \Delta t))(2\kappa(t, \Delta t) - 1) \tag{3.32c}$$

$$M_4(\xi - M, t, \Delta t) = \kappa(t, \Delta t)(1 - \kappa(t, \Delta t))[1 - 3\kappa(t, \Delta t)(1 - \kappa(t, \Delta t))] \tag{3.32d}$$

Equations (3.32a–d) keep their form if $\kappa(t, \Delta t)$ is replaced by $C(t, \Delta t)$. However, the resulting relationships in this case are just approximations since $\kappa(t, \Delta t) \approx C(t, \Delta t)$ (see Eq. (3.28a,b)).

In practice, the moments of order three and four are used to define the skewness γ_3 and the kurtosis γ_4, respectively, (Abramowitz and Stegun 1972):

$$\gamma_3 \equiv \frac{M_3(\xi - M)}{D^{3/2}} \tag{3.33a}$$

$$\gamma_4 \equiv \frac{M_4(\xi - M)}{D^4} - 3 \tag{3.33b}$$

In the case that the random variable is the sunshine number, use of Eqs. (3.33a and 3.33b) yields:

$$\gamma_3(t, \Delta t) = \frac{2\kappa(t, \Delta t) - 1}{\sqrt{\kappa(t, \Delta t)(1 - \kappa(t, \Delta t))}} \tag{3.34}$$

$$\gamma_4(t, \Delta t) = \frac{1 - 6\kappa(t, \Delta t)(1 - \kappa(t, \Delta t))}{\kappa(t, \Delta t)(1 - \kappa(t, \Delta t))} \tag{3.35}$$

Equations (3.34) and (3.35) keep their form if $\kappa(t, \Delta t)$ is replaced by $C(t, \Delta t)$. However, in this case the resulting equations are approximations (see Eq. 3.28a,b).

3.2.2 Time Averaged Statistical Measures

One denotes by $x(t,\Delta t)$ any one of the following quantities: $p(\xi = 1, \Delta t)$, $p(\xi = 0, \Delta t)$, $\sigma(t, \Delta t)$ $\kappa(t, \Delta t)$, $C(t, \Delta t)$, and the four statistical moments defined above. One considers a time interval of duration $\Delta t'$ consisting of m non-overlapping time intervals of duration Δt. Of course, $\Delta t' = m\,\Delta t$. Let t_i $(i = 1,\ldots,m)$ be the time moments in the middle of the m time intervals. One defines the average value $x(\Delta t')$ of $x(t, \Delta t)$ on the time interval of duration $\Delta t'$ in the simple way:

$$x(\Delta t') \equiv \frac{1}{m}\sum_{i=1}^{m} x(t_i, \Delta t) \tag{3.36}$$

Equations (3.28a,b) may be used m times, for t_i $(i = 1,\ldots,m)$. Summation over m and use of Eq. (3.29) yields:

$$C(\Delta t') \approx 1 - \sigma(\Delta t') = \kappa(\Delta t') \tag{3.37}$$

The first Eq. (3.37) shows that the approximate Eq. (3.28a,b) keeps its meaning for arbitrary long time intervals.

3.2.3 Comparison with Measurements

The analyzed data set consists of a large number of periods and this makes the average value (i.e., the first statistical moment) to be found out correctly. A detailed correlation analysis of all instantaneous data sets has been performed in Badescu and Paulescu (2011a). None of these data sets exhibited periodical components. However, the series of *daily mean* sunshine number may be affected by a light seasonality (Paulescu and Badescu 2011; Badescu and Paulescu 2011a). Since here we refer to instantaneous values no filtration of the periodical components has been performed to the data before computing the statistical moments.

3.2.3.1 Classifications Based on Cloud Shade and Observed Total Cloud Cover Amount

It is a common paradigm to stratify the actinometric data according to various classes of observed total cloud cover amount (see Sect. 3.1.1). However, this classification has some weaknesses shown below and a new classification is proposed in Badescu and Paulescu (2011b) and will be discussed next.

Time series of sunshine number values are associated to the series of solar irradiances in the database by using Eq. (2.5). These series may be separated into adjacent equal-duration sequences, if a time interval of duration Δt is adopted. The

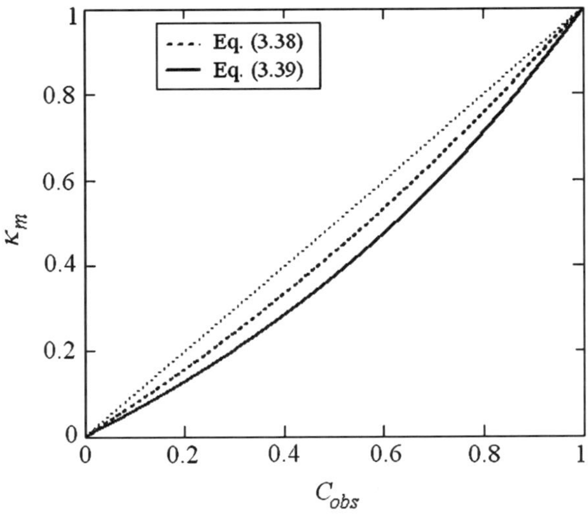

Fig. 3.26 Dependence of *measured* cloud shade $\kappa_m(t, \Delta t)$ on the observed cloud cover amount C_{obs} described by Eqs. (3.38) and (3.39). The diagonal corresponds to the usual assumption $\kappa \approx C$ (see Eq. 3.28a,b). After Badescu and Paulescu (2011b), with permission from Elsevier

"measured" relative sunshine fraction $\sigma_m(t,\Delta t)$ is computed by using Eq. (3.25) for all these sequences, from the associated sunshine number values. The values of the cloud shade $\kappa_m(t, \Delta t)$ are then computed for all sequences by using Eqs. (3.28a,b). This way, the solar irradiance values G and G_d for a given moment t are associated to the cloud shade value $\kappa_m(t, \Delta t)$. Therefore, the actinometric data may be stratified according to the state of the sky by using different classes of cloud shade.

Equation (3.28a,b) suggests that classifications based on cloud shade and total cloud cover amount, respectively, are equivalent in first approximation. However, this is not the case when classes of *observed* values of total cloud cover amount, C_{obs}, are used. Indeed, the estimation (by eye or by cameras) of the total cloud cover amount is subject to well-known errors (Harrison and Coombes 1986, Badescu 1991). The errors affecting the measured values of cloud shade κ_m are smaller in number and of smaller importance than those affecting C_{obs} (Badescu 1991). Many authors reported on the relationship between C_{obs} and κ_m. Few results are reviewed now. Reddy (1974) obtained for the Indian latitudes (8°N–36°N) a yearly variation of the difference $C_{\text{obs}} - \kappa_m$ between 0.02 in March and 0.17 in August. At the same latitudes, Raju and Karuna Kumar 1982) found $C_{\text{obs}} - \kappa_m$ to have maximum values of 0.25 and 0.2, respectively, for C_{obs} in the range 0.4–0.7. Harrison and Coombes (1986) found for the latitudes of Canada (42°N–74°N) that $C_{\text{obs}} - \kappa_m$ can be as high as 0.3. For the Romanian climates and latitudes (44.1°N–47.8°N) we found that C_{obs} is greater than κ_m by as much as 0.2 and the difference $C_{\text{obs}} - \kappa_m$ is a maximum for $C_{\text{obs}} = 0.3\ldots0.7$ (Badescu 1990).

The classification based on cloud shade κ_m has the advantage that avoids eye (or camera) estimation of the state of the sky. The classification based on the observed total cloud cover amount C_{obs} is subjective by nature but is very often used. This prompted some researchers to derive empirical relationships between κ_m and C_{obs}. Two relations developed for the Romanian latitudes and climates are:

Table 3.8 The total number of measurements used in the analysis

	Δt = 600 s	Δt = 1 h	Δt = 3 h
$n_{year,\xi=0}(\Delta t)$	507,954	475,125	412,243
$n_{year,\xi=1}(\Delta t)$	541,046	537,435	512,957
$n_{year,\xi=0,1}(\Delta t)$	1,049,000	1,012,560	925,200

$$\kappa_m(t,\Delta t) = 0.73\,C_{obs}(t,\Delta t) + 0.27C_{obs}^2(t,\Delta t) \tag{3.38}$$

$$\kappa_m(t,\Delta t) = 0.582[\exp(C_{obs}(t,\Delta)) - 1] \tag{3.39}$$

Equation (3.38) is in fact another form of Eq. (3.16), extended for arbitrary time intervals Δt. It was first proposed in Badescu (1991) by using monthly average values of sunshine fraction and observed total cloud cover amount in 29 Romanian localities. Equation (3.39) is obtained from Eq. (3.12) of Paulescu and Schlett (2004) with the following restrictions: $C_{obs} = 1$ for $\kappa_m = 1$ and $C_{obs} = 0$ for $\kappa_m = 0$. Both relations were developed by using daily average values (i.e., Δt equals the daylight length).

The dependence of $\kappa_m(t,\Delta t)$ on $C_{obs}(t,\Delta t)$ predicted by Eqs. (3.38) and (3.39) is shown in Fig. 3.26, together with the approximate equivalence between $\kappa(t,\Delta t)$ and $C(t,\Delta t)$ predicted by Eq. (3.28a,b). For a given value $C = C_{obs}$, κ_m underestimates κ, as expected. The underestimation is smaller in case of Eq. (3.38) than in case of Eq. (3.39). However, Eq. (3.39) has been derived from measurements performed at Timisoara and will be used in the next classification of the sunshine number data.

The classification procedure is applied for all the days during the year as follows. A time duration Δt is first chosen. The maximum number of time subintervals of duration Δt during a day of length Δt_{day} is $n_{\Delta t} = \mathrm{int}(\Delta t_{day}/\Delta t)$, where int( ) denotes the integer part. For each such subinterval a central time moment t may be defined. The sequence of subintervals is placed symmetrically around the noon. All the sunshine number values within a time subinterval are assigned to a cloud shade value $\kappa_m(t,\Delta t)$ and to a simulated observed cloudiness value $C_{obs}(t,\Delta t)$. Three different values of the time duration Δt are used now, i.e., 10 min, one hour, and three hours, respectively. The total number of sunshine number values, for all classes and all the days, is denoted as $n_{year,\xi=0,1}(\Delta t)$ while the total number of sunshine number values $\xi - 0$ and $\xi = 1$ is denoted as $n_{year,\xi=0}(\Delta t)$ and $n_{year,\xi=1}(\Delta t)$, respectively. These quantities are listed in Table 3.8.

A number n_{class} of equal-size cloud shade classes are defined now (the width and the center of the ith class are $\Delta\kappa \equiv 1/n_{class}$ and $\kappa_i \equiv (i-0.5)/n_{class}$, respectively). The available sunshine number database has been stratified by using these cloud shade classes, as follows. Each value $\kappa_m(t,\Delta t)$ obtained during the procedure described above is assigned to one particular cloud shade class. The number of all sunshine number values (i.e., counting both $\xi = 0$ and $\xi = 1$ values) associated to $\kappa_m(t,\Delta t)$ may be associated to that particular cloud shade class, too.

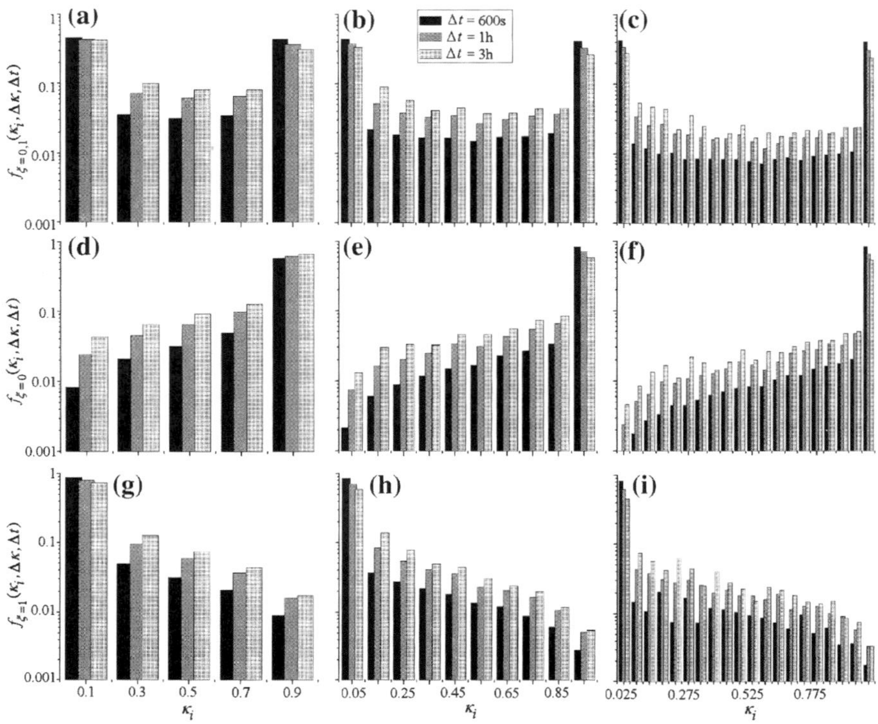

Fig. 3.27 Probability distributions: **a**, **b**, **c** $f_{\xi=0,1}(\kappa_i, \Delta\kappa, \Delta t)$, **d**, **e**, **f** $f_{\xi=0}(\kappa_i, \Delta\kappa, \Delta t)$ and **g**, **h**, **i** $f_{\xi=1}(\kappa_i, \Delta\kappa, \Delta t)$, for three different values Δt: **a**, **d**, **g** $n_{\text{class}} = 5$ cloud shade classes; **b**, **e**, **h** 10 classes; **c**, **f**, **i** 20 classes. On the abscissa κ_i is the center of the i cloud shade class. After Badescu and Paulescu (2011b), with permission from Elsevier

When the procedure is completed for all the days within the year, a histogram of the number of all sunshine number values for all the cloud shade classes is obtained. Probability distributions denoted as $f_{\xi=0,1}(\kappa_i, \Delta\kappa, \Delta t)$ may be built by dividing these histograms through the appropriate values $n_{\text{year},\xi=0,1}(\Delta t)$. Because both values $\xi = 0$ and $\xi = 1$ are involved in calculations, $f_{\xi=0,1}(\kappa_i, \Delta\kappa, \Delta t)$ simply gives the probability to find during the year a time interval of duration Δt with cloud shade in the range $[\kappa_i - \Delta\kappa/2,\ \kappa_i + \Delta\kappa/2)$. This probability distribution is bimodal (see Fig. 3.27a,b,c). A few explanatory comments follow. When very short time intervals Δt are considered (say, slightly longer than the 15 s lag of the measurement series), the associated κ_m value is either 0 or 1 and the probability distribution $f_{\xi=0,1}(\kappa_i, \Delta\kappa, \Delta t)$ is of course strictly bimodal (i.e., it has non-vanishing values only for the cloud shade classes centered on $\kappa_1 = 0.5/n_{n\text{class}}$ and $\kappa_{n\text{class}} = 1 - 0.5/n_{n\text{class}}$, respectively). When the duration Δt increases, it is possible to find during the day some time intervals with values κ_m between 0 and 1. They will bring contributions to the cloud shade classes whose centers κ_i are between κ_1 and κ_{nclass}. The bimodality of the probability distribution maintains for longer time intervals Δt.

Both values $f_{\xi=0,1}(\kappa_1, \Delta\kappa, \Delta t)$ and $f_{\xi=0,1}(\kappa_{n\text{class}}, \Delta\kappa, \Delta t)$ slightly decrease by increasing Δt (see Fig. 3.27a,b,c). However, for classes i between 1 and n_{class} the values of the probability distribution $f_{\xi=0,1}(\kappa_i, \Delta\kappa, \Delta t)$ increase by increasing Δt. This is easy to explain. Cloud cover persistence makes the time intervals of short duration Δt to be rare during the day for given intermediate values of κ_m (consider the case $\Delta t = 60$ s and $\kappa_m = 0.5$, for instance). By increasing Δt but keeping the same value of κ_m, the chance increases to find during the day a time interval whose cloud shade value is κ_m. This time interval will bring additional contribution to one of the cloud shade classes between 1 and n_{class}.

Generally, changing the number n_{class} of cloud shade classes does not change the qualitative properties of $f_{\xi=0,1}(\kappa_i, \Delta\kappa, \Delta t)$ (Fig. 3.27a,b,c). By decreasing the number of classes (or, in other words, by increasing $\Delta\kappa$), the probability of the intermediate values κ_i increases, as expected (compare Fig. 3.27a, b, respectively). The opposite happens in case of increasing the number of classes (compare Fig. 3.27b,c, respectively). Further increasing the number of classes makes the probability distribution to loose its smoothness. This is already visible in Fig. 3.27c, the case $\Delta t = 3$ h.

The above procedure may be repeated to build the histograms for the cases when the sun is covered by clouds (i.e., $\xi = 0$) or the sun is shining (i.e. $\xi = 1$). The appropriate probability distributions $f_{\xi=0}(\kappa_i, \Delta\kappa, \Delta t)$ and $f_{\xi=1}(\kappa_i, \Delta\kappa, \Delta t)$ are obtained by dividing these histograms through the appropriate value $n_{\text{year},\xi=0}(\Delta t)$ and $n_{\text{year},\xi=1}(\Delta t)$, respectively. These distributions are unimodal (see Fig. 3.27d–i).

$f_{\xi=0}(\kappa_i, \Delta\kappa, \Delta t)$ provides the probability for a given moment during the year, with the sun covered by clouds, to belong to a time interval Δt with cloud shade value in the range $[\kappa_i - \Delta\kappa/2, \kappa_i + \Delta\kappa/2)$. A few comments are useful. Note that Eq. (3.27) gives the probability for the sun to be covered by clouds for a particular moment t during a time interval of length Δt during a particular day. Usage of Eqs. (3.27) and (3.28a,b) allows writing this probability as $p(\xi = 0, t, \Delta t) \approx \kappa_m(t, \Delta t)$. This value κ_m belongs to some cloud shade class i, centered on κ_i. It is obvious that the probability $p(\xi = 0, t, \Delta t)$ differs from the probability distribution $f_{\xi=0}(\kappa_i, \Delta\kappa, \Delta t)$, which refers to an arbitrary moment during the year. The probability distribution $f_{\xi=0}(\kappa_i, \Delta\kappa, \Delta t)$ increases by increasing the class center value κ_i, as expected (Fig. 3.27d-f). When the largest class center value $\kappa_{n\text{class}}$ is considered, $f_{\xi=0}(\kappa_{n\text{class}}, \Delta\kappa, \Delta t)$ decreases by increasing Δt. The opposite happens for smallest class center value κ_1. By increasing the number of classes, the probability associated to the intermediate class center values κ_i decreases (compare Fig. 3.27d,f, respectively).

$f_{\xi=1}(\kappa_i, \Delta\kappa, \Delta t)$ provides the probability for a given moment during the year, with the sun shining, to belong to a time interval Δt with cloud shade value in the range $[\kappa_i - \Delta\kappa/2, \kappa_i + \Delta\kappa/2)$. This distribution decreases by increasing the class center value κ_i (Fig. 3.27g–i). When the smallest class center value κ_1 is considered, $f_{\xi=1}(\kappa_1, \Delta\kappa, \Delta t)$ decreases by increasing Δt. The opposite happens for larger class center value κ_i. Other comments made when discussing the features of $f_{\xi=0}(\kappa_i, \Delta\kappa, \Delta t)$ apply here, too.

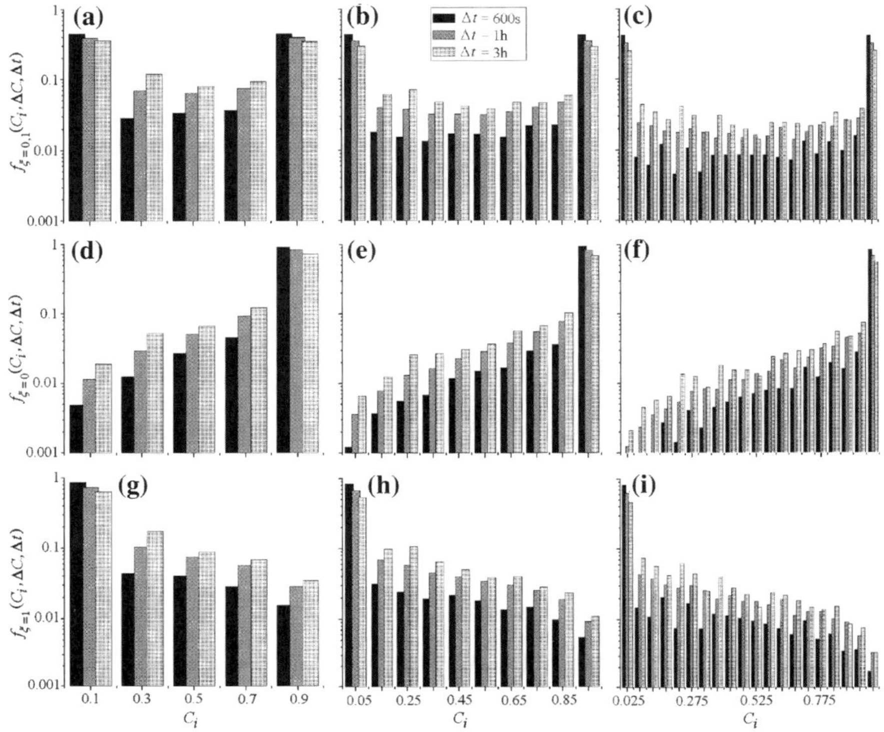

Fig. 3.28 Probability distributions: **a**, **b**, **c** $f_{\xi=0,1}(C_i, \Delta C, \Delta t)$, **d**, **e**, **f** $f_{\xi=0}(C_i, \Delta C, \Delta t)$ and **g**, **h**, **i** $f_{\xi=1}(C_i, \Delta C, \Delta t)$, for three different values Δt. **a**, **d**, **g** $n_{\text{class}} = 5$ classes of cloud cover amount; **b**, **e**, **h** 10 classes; **c**, **f**, **i** 20 classes. The class center values C_i have been obtained by using Eq. (3.39). On the abscissa C_i is the center of the i cloud cover amount class. After Badescu and Paulescu (2011b), with permission from Elsevier

The above procedure may be used to build the probability distributions related to the total cloud cover amount, i.e., $f_{\xi=0,1}(C_i, \Delta C, \Delta t)$, $f_{\xi=0}(C_i, \Delta C, \Delta t)$ and $f_{\xi=1}(C_i, \Delta C, \Delta t)$, respectively. Figure 3.28 shows the results. Most comments made when discussing the probability distributions related to cloud shade apply in this case, too. Note, however, the non-smooth character of the distributions based on C when $n_{class} = 20$ classes of cloud cover are considered (see Figs. 3.28f,i).

Figures 3.27 and 3.28 show that for practical reasons using 10 classes (of cloud shade or total cloud cover amount) provides a robust graphical representation. Increasing the number of classes decreases the smoothness of the probability distributions since some particular classes contain a small (or even null) number of values.

Different functions were tested to represent analytically the eighteen probability distributions shown in Figs. 3.27b,e,h and Figs. 3.28b,e,h for $n_{\text{class}} = 10$ and several values of the time duration Δt. The variables in these functions are the centers of cloud shade and cloud amount classes, κ_i and C_i, respectively, which all

Table 3.9 Fitting coefficients for Eqs. (3.40a, b, c)

ξ	Quantity	Δt	a	b	c	d	e
0	cloud shade	600 s	−3.6621032	−1.6939487	0.8522411	−0.2558878	–
		1 h	−3.1462143	−0.7634130	0.5962169	−0.1777876	–
		3 h	−2.8743943	−0.5397975	0.4736626	−0.1445405	–
	cloud amount	600 s	−3.9940869	−0.6505785	0.9376789	−0.2280056	–
		1 h	−3.3763827	0.2645428	0.7820757	−0.1471312	–
		3 h	−3.0105270	0.1492078	0.6885577	−0.1261044	–
1	cloud shade	600 s	21.594854	2449.5905	−9094.6142	−979.37867	–
		1 h	1.1395148	208.79107	−666.66440	−91.583357	–
		3 h	−0.0315341	75.828760	−233.0666	−35.113937	–
	cloud amount	600 s	6538.1510	755285.77	−2751562.8	−211873.21	–
		1 h	4233.2072	673549.22	−1966777.8	−270052.85	–
		3 h	0.63016984	246.67734	−660.67908	−109.99233	–
0,1	cloud shade	600 s	−4.7656171	1.4822365	−2.9494371	−0.2675414	0.0432186
		1 h	−3.4337544	−0.2140939	−0.9368182	−0.1781281	0.0276758
		3 h	−2.5575068	−1.7274350	0.5441041	−0.1277992	0.0174959
	cloud amount	600 s	−5.1565402	2.1664819	−2.8646944	−0.2506296	0.0470826
		1 h	−3.7641416	0.5237350	−0.9091117	−0.1596641	0.0301526
		3 h	−2.8202938	−0.9695702	0.2894772	−0.1202480	0.0183976

belong to the interval 0.05–0.95. After some trials the following fitting functions have been selected by using the TableCurve 2D software (Systat 2010):

$$\ln[f_{\xi=0}(x, \Delta x, \Delta t)] = a + bx^3 + c \ln x + d \ln^{-1} x \tag{3.40a}$$

$$\ln[f_{\xi=1}(x, \Delta x, \Delta t)] = \frac{a + cx^2}{1 + bx^2 + dx^4} \tag{3.40b}$$

$$\ln\left[f_{\xi=0,1}(x, \Delta x, \Delta t)\right] = a + bx + cx^3 + dx \ln^{-1} x + ex^{-1.5} \tag{3.40c}$$

Here $x \equiv \kappa_i$ and $\Delta x \equiv \Delta\kappa$ (or $x \equiv C_i$ and $\Delta C \equiv \Delta\kappa$ respectively) while the range of variation for the variable x is [0.05, 0.95]. Also, a to e are regression coefficients to be obtained by fitting the functions Eqs. (3.40a, b, c) to the data in Figs. 3.27b,e,h and 3.28b,e,h.

The quite complex functions Eqs. (3.40a, b, c) are mainly defined to provide accurate approximations for the three probability distributions. They are of lesser help in the physical interpretation of the results and do not allow obtaining simple analytical expressions when used in conjunction with the general definition Eq. (3.30) of the statistical moments.

Table 3.9 shows the values of the regression coefficients a to e obtained for all 18 curves in Figs. 3.27b,e,h and 3.28b,e,h. These coefficients depend on the time duration Δt.

The regression accuracy is good since the determination coefficient always exceeds 0.97. Generally, the probability distributions based on κ_i have slightly higher regression accuracy than those based on C_i. However, exceptions exist. There is no obvious dependence of the regression accuracy on the time duration Δt.

Further work is necessary to study the dependence of the coefficients a to e in Eqs. (3.40a, b, c) on the geographical location.

3.2.3.2 Statistical Measures

Comparison between the theoretical results presented in Sect. 3.2.1 and results derived by using measurements is briefly reported in this section. Theory predicts that the variance $D^2(t, \Delta t)$ is given by Eq. (3.32b) while the skewness and kurtosis may be computed by using Eqs. (3.34) and (3.35), respectively. Measurement-based results were obtained for a given time interval of duration Δt centered on the time moment t as follows. First, the mean $M \equiv \sigma(t, \Delta t)$ of the sunshine number values within that interval was computed. Next, the centered statistical moments $M_k(\xi - M)$ of order $k = 2$, 3, and 4 were computed by using the definition Eq. (3.31). The variance D^2 has been subsequently evaluated from Eq. (3.32b) while the skewness and kurtosis were computed by using Eqs. (3.33a) and (3.33b), respectively.

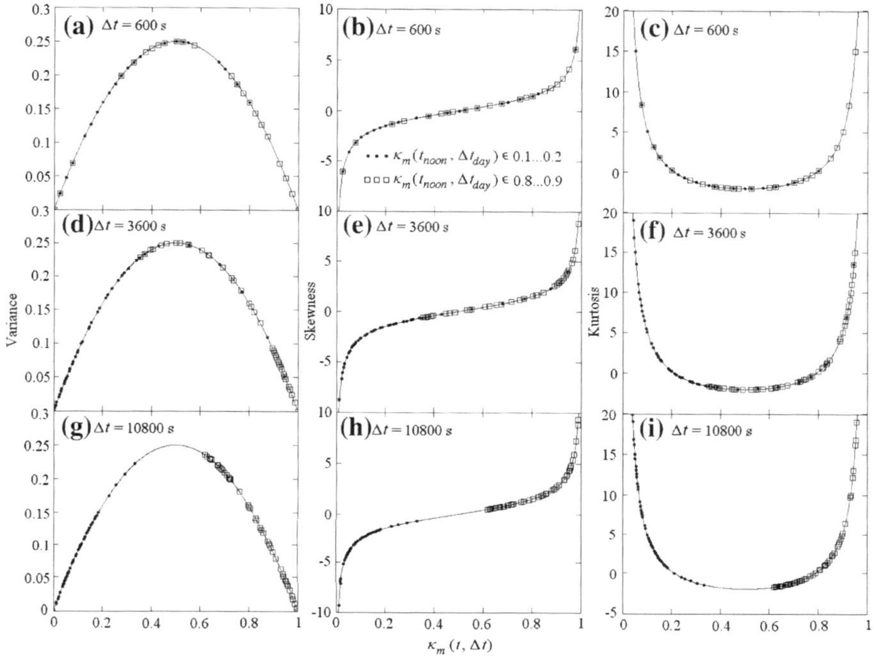

Fig. 3.29 Statistical indicators as a function of the cloud shade $\kappa_m(t, \Delta t)$. Theoretical curves predicted by Eqs. (3.32b), (3.34), and (3.35) (*solid lines*) and values derived from measurements (*symbols*) are shown:. **a**, **d**, **g**—Variance $D^2(t, \Delta t)$; **b**, **e**, **h**—Skewness $\gamma_3(t, \Delta t)$; **c**, **f**, **i**—Kurtosis $\gamma_4(t, \Delta t)$. Data from the days June 12, 2009 (daily cloud shade class 0.1–0.199) and June 30, 2009 (daily cloud shade class 0.8–0.899), respectively, and have been used. After Badescu and Paulescu (2011b), with permission from Elsevier

The dependence of the statistical measures D^2, γ_3, and γ_4 for all these time intervals on the associated cloud shade value $\kappa_m(t, \Delta t)$ and simulated observed total cloud cover amount value $C_{obs,\ s}(t, \Delta t)$ is shown in Figs. 3.29 and 3.30, respectively. Data from two summer days with significantly different state of the sky were used in these figures. The days were selected as follows. First, we computed for all the days during the year 2009 the daily cloud shade values $\kappa_m(t_{noon}, \Delta t_{day})$ (where t_{noon} is the noon time while Δt_{day} is the day-length). Ten classes of days have been created according to their $\kappa_m(t_{noon}, \Delta t_{day})$ values: 0.0–0.099; 0.1–0.199; ..., 0.9–1. The first day (12 June 2009, 3712 sunshine number recordings) has a nearly clear sky and belongs to the cloud shade class 0.1–0.199. The second day (30 June 2009, 3716 sunshine number recordings) has a nearly overcast sky and belongs to the class 0.8–0.899.

Figure 3.29 shows the dependence of the statistical indicators D^2, γ_3 and γ_4 on the cloud shade value $\kappa_m(t, \Delta t)$. Results predicted by theory and by using measurements, respectively, are given. There is very good agreement between the two approaches, for all statistical indicators, for both daily classes of cloud shade and for all values of the time interval duration Δt.

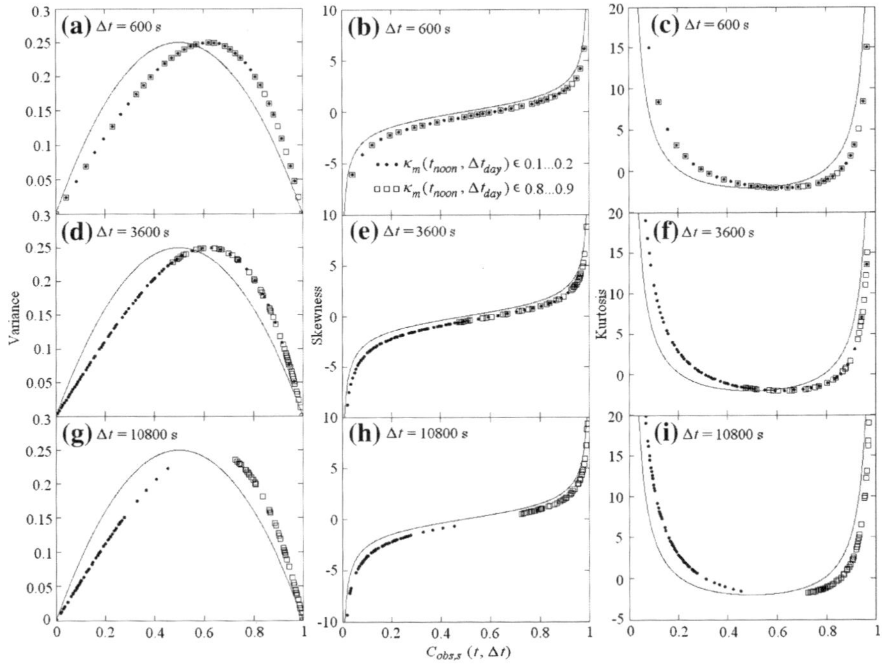

Fig. 3.30 Statistical indicators as a function of the simulated observed cloud amount $C_{\mathrm{obs},s}(t, \Delta t)$. Theoretical curves predicted by Eqs. (3.32b), (3.34), and (3.35) (*solid lines*) and values derived from measurements (*symbols*) are shown: **a**, **d**, **g**—Variance $D^2(t, \Delta t)$; **b**, **e**, **h**—Skewness $\gamma_3(t, \Delta t)$; **c**, **f**, **i**—Kurtosis $\gamma_4(t, \Delta t)$. Data from the days 12 June 2009 (class 0.1–0.199 of daily cloud shade) and 30 June 2009 (class 0.8–0.899 of daily cloud shade) have been used. After Badescu and Paulescu (2011b), with permission from Elsevier

Figure 3.30 shows the dependence of the statistical indicators D^2, γ_3, and γ_4 on the simulated observed total cloud cover amount $C_{\mathrm{obs},s}(t, \Delta t)$. There is a good concordance between theory and measurements when the skewness is considered. Theory slightly overestimates measurements. This applies to both daily classes of cloud shade and to all values of the time interval duration Δt.

With respect to the variance and the kurtosis, the curves derived from measurements are shifted to the right of the theoretical curves but the concordance is still reasonable (Figs. 3.30a,d,g and 3.30c,f,i). The shifting is caused by the fact that Eq. (3.39) is an approximation, which may not be the best fit between cloud shade and observed total cloud cover amount for the data measured in 2009 at Timisoara. For larger values of the time interval duration Δt (Fig. 3.30g–i), the curves obtained from processing the measurement data are not defined for all possible values of $C_{\mathrm{obs},s}(t, \Delta t)$. For example, in case of the sunny day of 12 June the curve is limited to values of $C_{\mathrm{obs},s}(t, \Delta t)$ smaller than 0.5 while for the cloudy day of 30 June the curve is defined only for values of $C_{\mathrm{obs},s}(t, \Delta t)$ larger than 0.7. A short explanation follows. $C_{\mathrm{obs},s}(t, \Delta t)$ may reach one of its extreme values (i.e., 0 or 1) only if during the time interval of duration Δt the sequence of sunshine number recordings keeps a constant value. By increasing the duration Δt, the chance to find such a particular sequence during the day decreases and the chance for $C_{\mathrm{obs},s}(t, \Delta t)$ to reach its extreme value decreases, too. Also, $C_{\mathrm{obs},s}(t, \Delta t)$ tends toward the daily average cloud cover amount when Δt increases and this makes the points to become denser on some portions of the curve.

Summer days with intermediate value of the daily cloud shade have been also analyzed. The day 6 June 2009 (daily cloud shade class 0.4–0.499, 3688 sunshine number recordings) is an example. The agreement between theory and experiments is very good when the dependence of the statistical indicators on the cloud shade $\kappa_m(t, \Delta t)$ is considered, for all values of the time interval duration Δt. Generally, the dependence of the statistical indicators on the simulated observed total cloud cover amount $C_{\mathrm{obs},s}(t, \Delta t)$ for this day has similar features to those already discussed with Fig. 3.30. However, for the largest value of Δt, the curves obtained for the statistical indicators by using measurements in the day with intermediate cloud shade 6 June cover all values of $C_{\mathrm{obs},s}(t, \Delta t)$ between 0 and 1. This has to be compared with the same curves for the days with almost clear sky (12 June) or almost covered sky (30 June), which are limited to values of $C_{\mathrm{obs},s}(t, \Delta t)$ smaller than 0.5 in the first case and larger than 0.7 in the second case (see Fig. 3.30g,h,i).

The dependence of the statistical measures D^2, γ_3, and γ_4 on the associated cloud shade value $\kappa_m(t, \Delta t)$ and simulated observed total cloud cover amount value $C_{obs,s}(t, \Delta t)$ has been analyzed for winter days, too. The very cloudy day 24 December 2009 (class 0.8–0.899 of daily cloud shade), the day with intermediate cloudy sky 28 December 28, 2009 (class 0.4-0.499) and the sunny day 29 December 2009 (class 0.2–0.299) have been studied. Results are shown in Figs. E20-E23 of Badescu and Paulescu (2011b).

The agreement between theory and experiments is always very good when the dependence of the statistical indicators on the cloud shade $\kappa_m(t, \Delta t)$ is considered, for all values of the time interval duration Δt. The dependence of the statistical indicators on the simulated observed total cloud cover amount $C_{\mathrm{obs},s}(t, \Delta t)$ has similar features to those already discussed with Fig. 3.30.

3.2.4 Summary and Discussion

The sunshine number database is stratified into a given number of classes of cloud shade and total cloud cover amount, respectively. These classifications may be seen as equivalent in first approximation. It is shown, however, that this is not true when the classification is made upon *observed* total cloud cover amount, which is subjected to various well-known errors.

The probability distribution for a time moment to belong to a time interval of given duration Δt and given cloud shade class is bimodal. Generally, the probability increases by increasing Δt and by decreasing the number of cloud shade classes. However, the opposite happens in case of the smallest and largest cloud shade classes. The probability distribution for the sun to be covered by clouds, for a time moment within a time interval of given duration Δt and given cloud shade class is unimodal. The probability increases by increasing Δt and by decreasing the number of cloud shade classes, but not for the largest cloud shade class. The probability distribution for the sun shining, for a time moment within a time interval of given duration Δt and given cloud shade class is unimodal. The probability increases by increasing Δt and by decreasing the number of cloud shade classes, but not for the smallest cloud shade class. Similar results were obtained in case of probability distributions based on classes of total cloud cover amount.

The probability distributions were represented analytically in the case of 10 classes of cloud shade and total cloud cover amount, respectively.

Statistical measures for the sunshine number are defined. They are the first four statistical moments, i.e., the mean, the variance, the skewness and the kurtosis. The theory allows expressing these measures as a function of a single parameter very often measured by the meteorological stations, namely the cloud shade. Alternately, this single parameter may be the total cloud cover amount.

The dependence of the four statistical indicators on the cloud shade value has been evaluated by theory and by using measurements, respectively. There is very good agreement between the two approaches for all statistical indicators, for all summer and winter days, which belong to various classes of cloud shade.

The dependence of the four statistical indicators on the simulated observed cloud cover amount has been analyzed, too. There is a good concordance between theory and measurements when the third statistical moment (i.e., the skewness) is considered. In case of the second and fourth moments (i.e., the variance and the kurtosis, respectively) the curves derived by measurements are shifted to the right of the theoretical curves but the concordance is still reasonable.

The results are useful for those applications where the fluctuating nature of solar radiation has to be taken into account, mainly when the systems involved have a nonlinear response and are very sensitive to the instantaneous values.

References

Abramowitz M, Stegun IA (eds) (1972) Handbook of Mathematical Functions with Formulas, Graphs, and Mathematical Tables, 9th edn. Dover, New York, p 928

Aguilar RJ, Collares-Pereira M, Conde JP (1988) Simple procedure for generating sequences of daily radiation values using a library of Markov transition matrices. Sol Energy 40:269–279

Aquilar R, Collares-Pereira (1992) Tag : a time—dependent, autoregressive, Gaussian model for generating synthetic hourly radiation. Sol Energy 49:167–174

Akinoglu BG, Ecevit A (1990) Construction of a quadratic model using modified Angstrom coefficients to estimate global solar radiation. Sol Energy 45:85–92

Alnaser WE (1989) Empirical correlation for total and diffuse radiation in Bahrain. Energy 14:409–414

Alnaser WE (1993) New model to estimate the solar global irradiation using astronomical and meteorological parameters. Renew Energy 3:175–177

Badescu V (1990) Observations concerning the empirical relationship of cloud shade to point cloudiness. J Appl Meteorol 29:1358–1360
Badescu V (1991) Studies concerning the empirical relationship of cloud shade to point cloudiness (Romania). Theor Appl Climatol 44:187–200
Badescu V (1992) Over and under estimation of cloud amount: theory and Romanian observations. Int J Sol Energy 11:201–209
Badescu V (1997) Use of autoregressive models to generate series of daily averaged point cloudiness values. Renew Energy 12(1):71–82
Badescu V (2000) Simple model to generate daily averaged point cloudiness data. Int J Sol Energy 20(2):129–148
Badescu V (2008) Use of sunshine number for solar irradiance time series generation. In: Badescu V (ed) Modeling solar radiation at the Earth surface. Springer, Berlin, p 327
Badescu V, Zamfir E (1996) Different cloud cover classifications of the day and implications on the estimation of collected solar energy. World Renewable Energy Congress, June 15-21, Denver, Colorado, vol. 3, pp 2066–2069
Badescu V, Paulescu M (2011a) Autocorrelation properties of the sunshine number and sunshine stability number. Meteorol Atmos Phys 112:139–154
Badescu V, Paulescu M (2011b) Statistical properties of the sunshine number illustrated with measurements from Timisoara (Romania). Atmos Res 101(1–2):194–204
Bahel V, Srinivasan R, Baksh H (1986) Solar radiation for Dhahran; Saudi Arabia. Energy 11:985–989
Boland J (1995) Time-series analysis of climatic variables. Sol Energy 55(5):377–388
Danescu A, Bucurenciu S, Petrescu, S (1980) Utilizarea energiei solare. Ed. Tehnica, Bucuresti, pp 31–34
Davies JA, McKay DC (1982) Estimating solar irradiance and components. Sol Energy 29:55–64
Davies JA, McKay DC, Luciani G, Abdel-Wahab M (1988) Validation of models for estimating solar radiation on horizontal surfaces. IEA Task IX, Final Report, vol. 1, Atmospheric Environment Service of Canada, Downsview, Ontario, Canada
Drake F, Mulugetta Y (1996) Assessment of solar and wind energy resources in Ethiopia I. Sol Energy 57:205–217
ENEC Ethiopian National Energy Commission (1986 Co-operation agreement in the energy sector between ENEC and CESEN-ANSALDO/FINMECCANICA group, Main Report, ENEC, Addis Abeba
Erhan E (1979) Clima si microclimatele din zona orasului Iasi. Junimea, Iasi, pp 27–29
Festa R, Ratto CF (1993) Solar radiation statistical properties, Report No. IEA-SCHP-9E – 4, IEA Task 9 Solar radiation and Pyranometry studies, University of Genova, Genova
Gansler RA, Klein SA, Beckman WA (1994) Assessment of the accuracy of generated meteorological data for use in solar energy simulation studies. Sol Energy 53(3):279–287
Gariepy J (1980) Estimation du rayonnement solaire global. Internal Report, Service of Meteorology, Government of Quebec, Canada
Gordon JM, Reddy TA (1988) Time series analysis of daily horizontal solar radiation. Sol Energy 41:215–226
Halouani N, Nguyen CT, Vo-Ngoc D (1993) Calculation of monthly average global solar radiation on horizontal surfaces using daily hours of bright sunshine. Sol Energy 50:247–258
Harrison AW, Coombes CA (1986) Empirical relationship of cloud shade to point cloudiness (Canada). Sol Energy 37:417–421
IEA (1984) Handbook of methods of estimating solar radiation. IEA Task V, Subtask B, Swedish Meteorological and Hydrological Institute, Norrkoping, pp 109
Knight KM, Klein SA, Duffie JA (1991) A methodology for the synthesis of hourly weather data. Sol Energy 46(2):109–120
Nakada Y, Takahashi H, Ichida K, Minemoto T, Takakura H (2010) Influence of clearness index and air mass on sunlight and outdoor performance of photovoltaic modules. Curr Appl Phys 10(2):S261–S264

Ogelman H, Ecevit A, Tasdemiroglu A (1984) A new method for estimating solar radiation from bright sunshine data. Sol Energy 33:619–625

Olseth JA (1984) A probability density function for daily insolation within the temperate storm belts. Sol Energy 33:533 (Erratum in Solar Energy (1986) 36, 479)

Olseth JA, Skartveit A (1987) A probability density model for hourly total and beam irradiance on arbitrary oriented planes. Sol Energy 39:343–351

Papoulis A (1984) Probability, random variables, and stochastic processes, 2nd edn. McGraw-Hill, New York, pp 145–149

Paulescu M, Schlett Z (2004) Performance assessment of global solar irradiation models under Romanian climate. Renew Energy 29:767–777

Paulescu M, Badescu V (2011) New approach to measure the stability of the solar radiative regime. Theor Appl Climatol 103:459–470

Polo J, Zarzalejo LF, Marchante R, Navarro AA (2011) A simple approach to the synthetic generation of solar irradiance time series with high temporal resolution. Sol Energy 85(5):1164–1170

Raju ASN, Karuna Kumar K (1982) Comparison of point cloudiness and sunshine derived cloud cover in India. Pure Appl Geophys 120:495–502

Reddy SJ (1974) An empirical method for estimating sunshine from total cloud amount. Sol Energy 15:281–284

Rehman S (1998) Solar radiation over Saudi Arabia and comparisons with empirical models. Energy 23:1077–1082

Rietvelt MR (1978) A new method for estimating the regression coefficients in the formula relating solar radiation to sunshine. Agric Meteorol 19:243–252

RMHI (1974) Basis of meteorological data. Romanian Meteorological and Hydrological Institute, Bucharest

Skartveit A, Olseth JA (1992) The probability density and autocorrelation of short-term global and beam irradiance. Sol Energy 49(6):477–487

SRMS Solar Radiation Monitoring Station (2012) Physics Department, West University of Timisoara, Romania. http://solar.physics.uvt.ro/srms

Systat (2010) TableCurve 2D. Systat Software Inc., Chicago, IL, USA http://www.systat.com/

WMO (1981) Meteorological aspects of the utilization of solar radiation as an energy source. World Meteorological Organisation, Technical Note 172, Geneva

WMO (2008) Guide to Meteorological Instruments and Methods of Observation, WMO-No.8, 2008, Ch. 8 – Measurement of sunshine duration, pp I8–4. http://www.wmo.int/ pages/prog/www/IMOP/publications/CIMO-Guide/CIMO_ Guide-7th_Edition-2008.html.

Young KL, Woo MK, Munro DS (1995) Simple approaches to modelling solar radiation in the Arctic. Sol Energy 54(1):33–40

Zabara K, Yianoulis P (1992) Conditional probabilities of daily relative sunshine data and the dependence on the weather of the previous day. Sol Energy 48(6):421–427

Zamfir E, Oancea C, Badescu V (1994) Cloud cover influence on long-term performance of flat plate solar collectors. Renew Energy 4(1):339–347

Chapter 4
Stability of the Radiative Regime

The stability (or fluctuation) of the daily radiative regime might be equally important in some practical cases, such as the management of power grid (see Chap. 1). Therefore, a day classification according to the stability of the radiative regime is useful. A study on this line has been performed by Tomson (2010). The increment of solar global irradiance ΔG has been defined by this author as the difference between the irradiance values in a sequence of recordings $G(t)$. ΔG is used as an indicator of the radiative regime fluctuation. The series of irradiance values $G(t)$ is characterized as having a small-scale fluctuation, if $\Delta G < 50$ W/m^2/s for clear sky condition and $\Delta G < 150$ W/m^2/s in overcast sky situation. A Boolean variable is used to describe the irradiance time series, which equals 1 when $G(t) > G_{\text{aver}}$ and equals 0 when $G(t) < G_{\text{aver}}$ (here G_{aver} is the moving average value of solar irradiance time series). Tomson (2010) concluded that the changing speed of solar irradiance has an exponential density of probability and small-scale and large-scale fluctuations always coexist.

4.1 Measures for Day Classification (Cloud Shade, Clearness Index, Fractal Dimension)

In this section, Δt equals the length of daylight (i.e., the time interval between sunrise and sunset) and its value depends on day.

4.1.1 Classes of Cloud Shade

Classes based on cloud shade have been used in Sect. 3.2.3.1. Time series of sunshine number values is associated to the solar irradiance series during that time interval, by using Eq. (3.29). The relative sunshine fraction $\Delta\sigma_m(\Delta t)$ ("m" comes

M. Paulescu et al., *Weather Modeling and Forecasting of PV Systems Operation*, Green Energy and Technology, DOI: 10.1007/978-1-4471-4649-0_4,

from "measurement") is computed by using Eq. (2.5), from the associated sunshine number values. The cloud shade value $\kappa_m(\Delta t)$ is then computed using Eq. (3.26). Any actinometric or meteorological parameter characterizing the given day may be associated to that cloud shade value $\kappa_m(\Delta t)$. One denotes by X that parameter. (Examples of X are the daily averaged solar global irradiance or the daily averaged air temperature, for instance). The procedure may be repeated for X for all the days in a year. For each day, an average value of the parameter X is computed. For the same day $\kappa_m(\Delta t)$ is also computed. The resulting values of the cloud shade $\kappa_m(\Delta t)$ may then be grouped into classes. Consequently, the associated values of the (radiometric or meteorological) parameter X may be stratified according to these classes of cloud shade.

4.1.2 Classes of Observed Total Cloud Cover Amount

Classes based on total cloud cover amount have been used in Sect. 3.2.3.1. Using Eq. (3.28a) yields $C(\Delta t) \approx \kappa_m(\Delta t)$. This suggests that classifications based on cloud shade and cloud cover are equivalent. However, this is not the case when classes of observed values of cloud cover amount, C_{obs}, are used. Indeed, the estimation by eye of the cloud cover amount is subjected to well-known errors, as discussed in Sect. 3.2.3.1. The errors affecting the measured values of cloud shade, κ_m, are smaller in number and of smaller importance than those affecting C_{obs} (Badescu 1991).

The classification proposed in this section, based on cloud shade, has the advantage that avoids the problems risen by the sky state estimation. However, the classification based on the observed total cloud cover amount is subjective by nature but is very often used. Equation (3.39) may be used to stratify the irradiance database in various classes of *simulated* observed cloud cover amount $C_{obs,s}$. The procedure is as follows. For a given value of Δt, the value κ_m is first computed, as shown in Sect. 3.2.3.1. This value is next used as an entry in Eq. (3.39), allowing to obtain $C_{obs,s}$.

4.1.3 Classes of Clearness Index

The actual level of solar irradiance at Earth surface results by superposition of deterministic and stochastic factors. Among the deterministic factors one quotes the seasonal and diurnal variations of the astronomical parameters. One denotes by G_{ext} the horizontal extraterrestrial solar irradiance. The stochastic factors are sometimes quantified by means of the instantaneous clearness index (Liu and Jordan 1960):

$$k_{t,\text{instant}} = G/G_{\text{ext}} \tag{4.1}$$

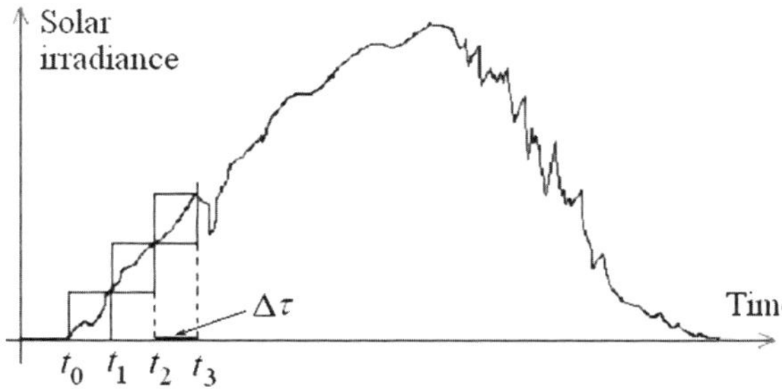

Fig. 4.1 Covering the global solar irradiance signal with rectangles

Thus, the clearness index accounts for all random meteorological influences being a measure of the atmospheric transparency. The clearness index can also be defined for global solar irradiation over a given time interval (hour, day, month) as already shown in Sect. 3.1.3. For a review of statistical behavior of solar radiation components based on clearness index see Tovar-Pescador (2008).

One denotes by H and H_{ext} the daily horizontal global solar irradiation on the ground and the daily solar irradiation at the top of the atmosphere, respectively. In this section, we are using the daily clearness index for global solar irradiation, k_{t}, defined as explained in Sect. 3.1.3 by:

$$k_t \equiv H/H_{\text{ext}} \tag{4.2}$$

4.1.4 Classes Based on Fractal Dimension

The fractal dimension D of a curve lies between 1 and 2, depending on how much surface area the curve "fills" (Dubuc et al. 1989). A fractal dimension $D(\Delta t)$ may be assigned to solar irradiance $G(t)$ during the time interval of duration Δt. Here we use the method proposed in Harrouni (2008), where the curve described by the solar irradiance is covered by rectangles (Fig. 4.1).

For $n + 1$ different time moments t_j separated by the same interval $\Delta\tau$, the area $S(\Delta\tau, \Delta\text{t})$ of this coverage is given by:

$$S(\Delta\tau, \Delta t) = \sum_{j=0}^{n-1} \left|G(t_j + \Delta\tau) - G(t_j)\right|\Delta\tau \tag{4.3}$$

The fractal dimension $D(\Delta t)$ represents the slope of the log–log plot $\ln\left(\frac{S(\Delta\tau)}{\Delta\tau^2}\right) = f\left(\ln\frac{1}{\Delta\tau}\right)$ fitted by the least square method:

$$\ln\left(\frac{S(\Delta\tau, \Delta t)}{\Delta\tau^2}\right) = D(\Delta t)\ln\left(\frac{1}{\Delta\tau}\right) + C, \quad \Delta\tau \to 0 \tag{4.4}$$

Usage of Eq. (4.4) requires choosing different values $\Delta\tau$ and computation of the associated surface area $S(\Delta\tau, \Delta t)$.

4.1.5 Application

All days in the database described in Sect. 3.2 are classified according their radiative regime. They refer to measurements performed from January 1 to December 31, 2009 in the Romanian town of Timisoara. Data are recorded at 15 s time interval between sunrise and sunset. All measures described in Sects. 4.1.2–4.1.5 are used for days classification.

Figure 4.2 shows that the daily averaged sunshine number series exhibits a random variation. Visual inspection shows, however, that during the days of the warm season (i.e., day 90–day 270) the average sunshine number is larger than during the rest of the year. This agrees with common experience.

Figure 4.3 shows the frequency distributions for 10 classes of daily averaged cloud shade, daily simulated cloud cover amount, fractal dimension of daily global solar irradiation, and daily clearness index. The classifications based on cloud shade and observed cloud cover amount show similar features. They are both bimodal, with larger frequencies at the largest classes (Fig. 4.3a, b). The bimodality is more obvious in case of cloud shade classification.

The frequency distributions associated to clearness index and fractal dimension are both unimodal (Fig. 4.3c, d). However, these two measures have different physical significance. The fractal dimension is easier to be related to the sunshine number. Indeed, it is a measure of the deviation between the time distribution of solar radiation during a real day and a certain constant value associated to the solar radiation distribution during a clear sky day (or during an overcast sky day). For a given day, the more the sunshine number is changing in time, the larger the fractal dimension is. The clearness index is a measure of solar radiation extinction in the atmosphere, which includes effects due to clouds but also effects due to radiation interaction with other atmospheric constituents. For a given cloud cover amount, different values of the clearness index may be conceived, due to different atmospheric contents of water vapor and aerosols, for instance. Note that some of the larger classes are empty for both kinds of measures (i.e., fractal dimension and clearness index) (Fig. 4.3c, d). This is different from the case of the classifications based on cloud cover amount and cloud shade, where all the 10 classes are populated.

A few more details are given by using the (sunshine number based) cloud shade classification. Figure 4.4 shows the daily variation of the global and diffuse irradiance on horizontal surface during the days with daily average cloud shade between 0.8 and 0.9. This is the least populated class, containing 16 days only.

Figure 4.5 shows the daily variation of the sunshine number during the same days considered in Fig. 4.4. An interesting aspect deserves attention. It will be discussed now by using the data from the days 28 and 29 October 2009. Figure 4.4 shows that, despite these 2 days belong to the same daily cloud shade class, the stability of their radiative regime is different: 28 October has a significantly larger fluctuation of solar radiation than 29 October. The sunshine number variation during the 2 days confirms this observation (see Fig. 4.5).

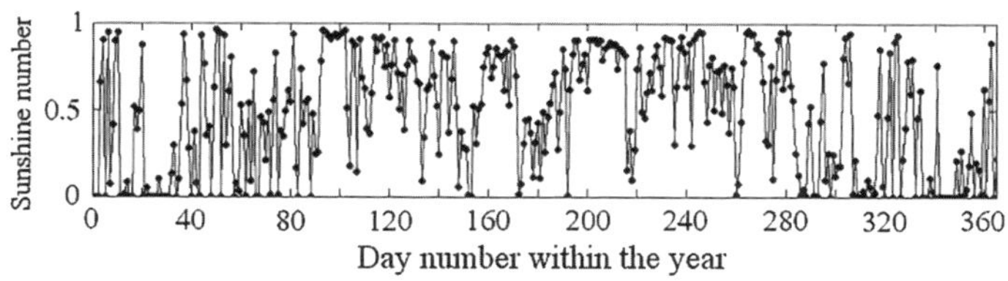

Fig. 4.2 Variation of the daily averaged sunshine number as a function of day number during the year 2009 at Timisoara. From Paulescu and Badescu (2011) with permission from Springer

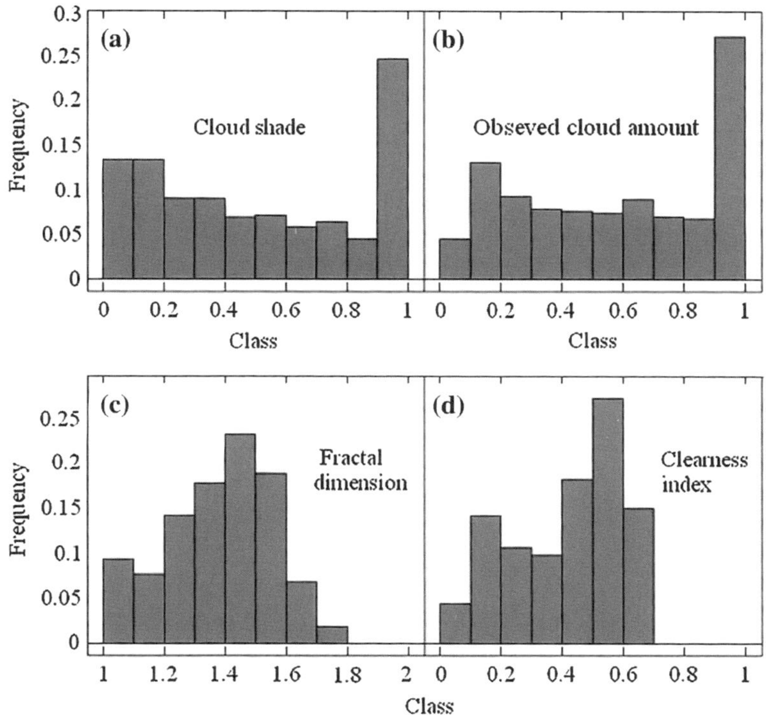

Fig. 4.3 Frequency distributions for 10 classes of **a** daily averaged cloud shade, **b** daily averaged simulated observed cloud cover amount, **c** the fractal dimension of daily global solar irradiation signal and **d** daily clearness index. Data from Timisoara during January–December 2009 have been used. From Paulescu and Badescu (2011), with permission from Springer

The overall conclusion obtained from the visual inspection of Figs. 4.4 and 4.5 is that the solar radiation fluctuation in days belonging to the same cloud shade class may be quite different. This conclusion also applies for the other three classifications (based on cloud cover amount, fractal dimension, and clearness index, respectively). Note, however, that the fractal dimension combined with the clearness index can be used to describe the fluctuation of solar radiation during a

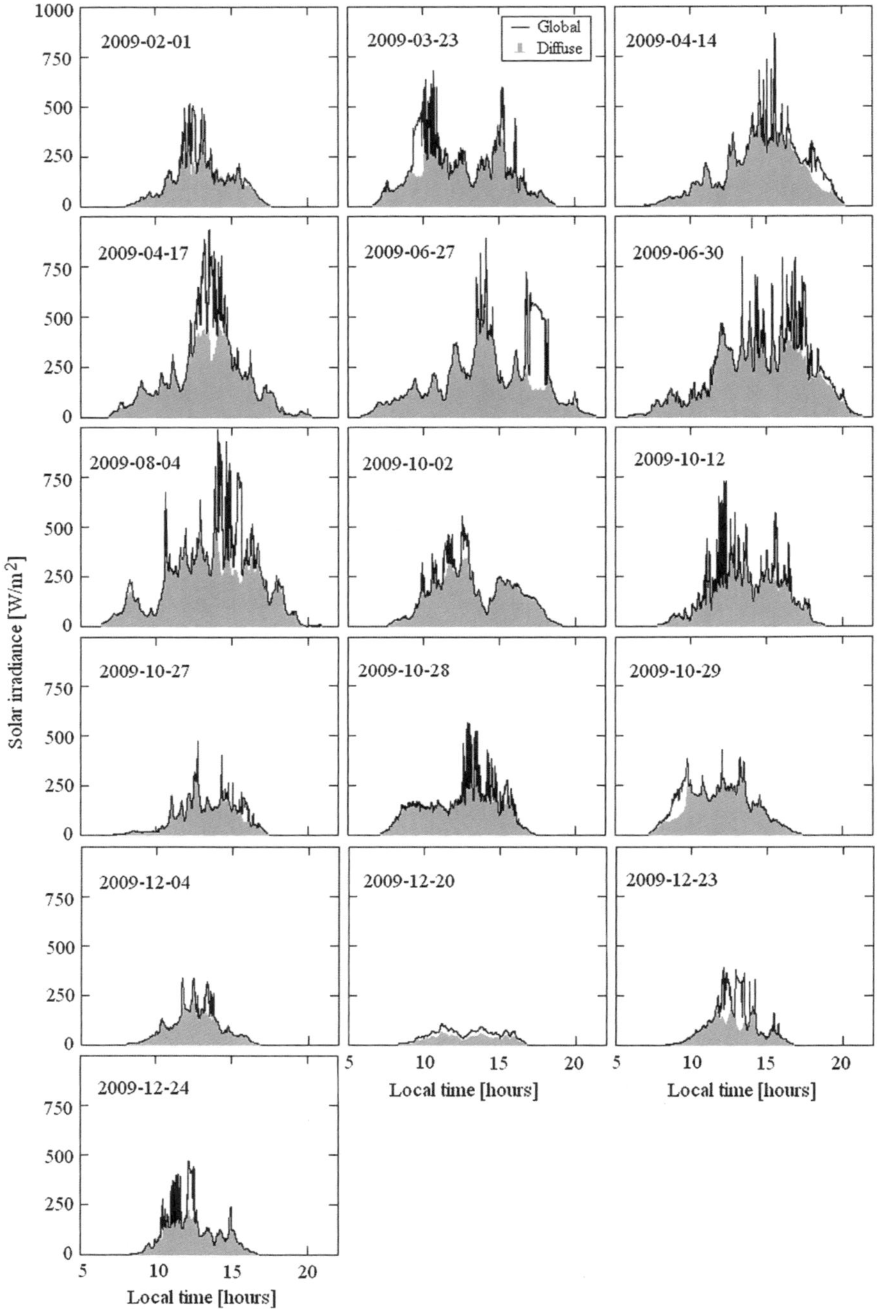

Fig. 4.4 Global and diffuse solar irradiance on horizontal surface during the days with daily average cloud shade between 0.8 and 0.9. Data from Timisoara during January–December 2009 have been used. After Paulescu and Badescu (2011), with permission from Springer

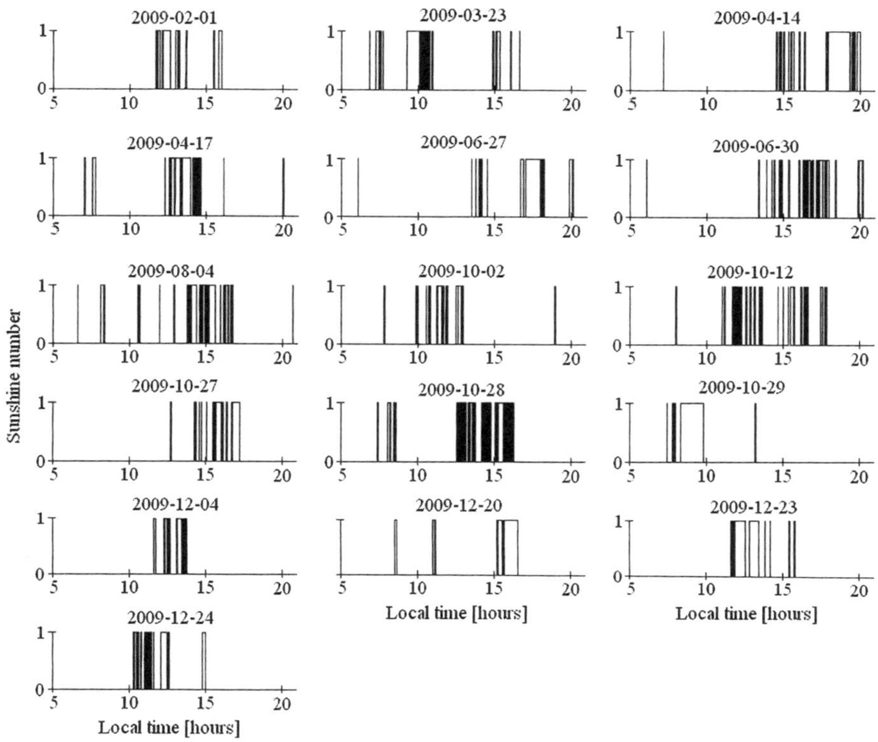

Fig. 4.5 Sunshine number variation during the days with daily average cloud shade between 0.8 and 0.9. Data from Timisoara during January–December 2009 have been used. After Paulescu and Badescu (2011), with permission from Springer

given day (Harrouni 2008). Other ways to quantify the radiation fluctuation by using combinations of the above four classifications could be equally well imagined. A new, single parameter to describe the radiation fluctuation during a day is defined in Sect. 4.2.

The above study demonstrates that the days may be stratified into classes of cloud shade and total cloud cover amount, respectively. These classifications may be seen as equivalent in first approximation. It is shown, however, that this is not true when the classification is made upon observed total cloud cover amount, which is subjected to various well-known errors. Other classification criteria were the daily averaged clearness index and the fractal dimension of the solar global irradiance signal during the day. Classifications based on cloud shade and observed cloud cover amount are bimodal, with larger frequencies at the largest classes. The frequency distributions associated to clearness index and fractal dimension, respectively, are both unimodal.

4.2 The Sunshine Stability Number

The statistical measures associated to the Boolean random variable $\xi(t)$ give useful information when dynamical properties are not the main concern. However, in Sect. 4.1.5, it has been shown that the same cloud shade value $\kappa(\Delta t)$ may be obtained for rather different distributions of the sunshine number $\xi(t)$ during the time interval of duration Δt.

The number of changes that ξ exhibits during a time interval of duration Δt may be used to characterize the stability of the radiative regime. For example, let us consider two hypothetical days having the same value of the daily averaged sunshine number. In the first case, the sky is clear during the first part of the day and fully covered by clouds during the last part. The number of changes *of* ξ during this first day is 1. In the second case, time periods with clear sky and cloudy sky alternate during the day. The number of changes of ξ during this second day is larger than 1. The radiative regime in the first day is said to be more stable than in the second day.

Let us assume an equidistant time moment series t_j ($j = 1,2,\ldots,n$) during the time interval of duration Δt. One denotes $\Delta \equiv t_{j+1} - t_j$ ($j = 1,3,\ldots,n - 1$). The fluctuations of the sunshine number ξ during Δt may be described by using the *sunshine stability number* $\zeta(t_j, \Delta t)$ ($j = 2,3,\ldots,n$), which is a random Boolean variable defined by:

$$\zeta(t_{j\geq 2}, \Delta\tau) \equiv \begin{cases} 1 & \text{if} \begin{cases} \xi(t_j) < \xi(t_{j-1}) & (\text{when} \quad \xi(t_1) = 1) \quad \text{or} \\ \xi(t_j) > \xi(t_{j-1}) & (\text{when} \quad \xi(t_1) = 0) \end{cases} \\ 0 & \text{otherwise} \end{cases} \tag{4.5}$$

A few comments about the definition Eq. (4.5) are useful. First, one assumes that in the very morning the sun is not covered by clouds (i.e. $\xi(t_1) = 1$). According to Eq. (4.5), $\zeta = 1$ only for those moments when the sun is just covered by clouds. Thus, counting the nonnull values *of* ζ provides a measure for the phenomenon of sun's disappearance from the sky. Second, one assumes that in the very morning the sun is covered by clouds (i.e. $\xi(t_1) = 0$). Then, according to Eq. (4.5), $\zeta = 1$ only for those moments when the sun is just released from the clouds. This time, counting the nonnull values of ζ provides a measure for the phenomenon of sun's apparition on the sky. To conclude, depending on the initial value $\xi(t_1)$, Eq. (4.5) quantifies just one of the two different phenomena: sun appearance and sun disappearance on/from the sky, respectively. Of course, various parameters may be defined to quantify both phenomena in the same time. However, in this chapter Eq. (4.5) is used.

The probability that the sunshine number will change its initial value $\xi(t_1)$ during Δt is denoted $p(\zeta = 1, \Delta t)$ and the probability that the sunshine number will keep its initial value is denoted $p(\zeta = 0, \Delta t)$. The two probabilities are related by the following normalization condition:

$$p(\zeta = 0, \Delta t) + p(\zeta = 1, \Delta t) = 1 \tag{4.6}$$

The average value of the sunshine stability number during the interval Δt is denoted $\bar{\zeta}(\Delta\tau, \Delta t)$. Note that $\bar{\zeta}(\Delta\tau, \Delta t)$ is not a Boolean variable. It ranges between

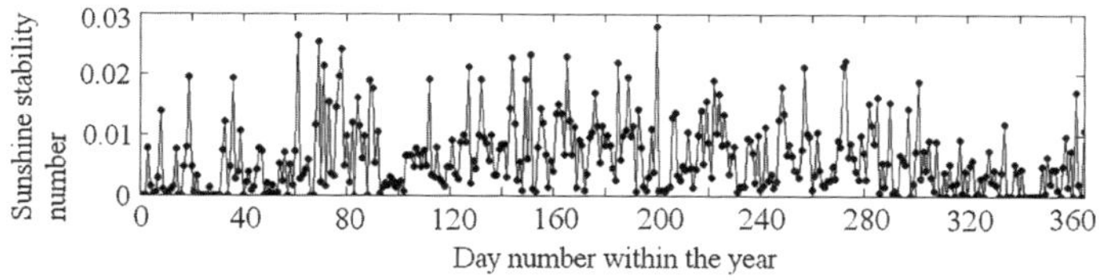

Fig. 4.6 Variation of the daily averaged sunshine stability number $\bar{\zeta}(\Delta\tau, \Delta t)$ as a function of day number during the year 2009 at Timisoara. In all days $\Delta\tau = 15$ s and Δt is the daylight length. From Paulescu and Badescu (2011), with permission from Springer

0 (in the extreme case when the instantaneous values of the sunshine number ξ are all 0 or 1, respectively, for all time moments t_j ($j = 1,2,\ldots,n$) during Δt) and 1 (in the extreme case when the instantaneous values of the sunshine number ξ change for every two consecutive moments t_{j-1} and t_j during Δt). The radiative regime is *fully stable* in the first case and *fully unstable* in the last case.

Figure 4.6 shows the variation of the daily averaged sunshine stability number $\bar{\zeta}(\Delta\tau, \Delta t)$ as a function of day number during the year 2009 at Timisoara (for this particular case, $\Delta = 15$ s and Δt equals the daylight length). $\bar{\zeta}(\Delta\tau, \Delta t)$ ranges between 0 and 0.028. The radiative regime is rather stable.

Figure 4.7 shows the cumulative frequency curves for 10 classes of daily averaged sunshine stability number, daily averaged cloud shade, fractal dimension of daily global solar irradiation signal, and daily clearness index, respectively. The classifications based on sunshine stability number, clearness index, and fractal dimension show similar features in the sense that the cumulative contribution of the larger classes is very small (Fig. 4.7a, c, d). This corresponds to unimodal frequency distributions (see also Fig. 4.3c, d). The slope of the cumulative distribution is large for the smallest and largest classes of the cloud shade classification (Fig. 4.7b). This corresponds to a bimodal frequency distribution, in agreement with Fig. 4.3a.

Figure 4.8 shows the frequency distribution of the days classified according to their daily averaged sunshine stability number $\bar{\zeta}(\Delta\tau, \Delta t)$. This is a unimodal distribution, as expected. Classes of smaller values of $\bar{\zeta}(\Delta\tau, \Delta t)$ are the most populated and this agrees with the perception of a rather stable radiative regime emphasized when discussed Fig. 4.6.

Data analysis show that the solar radiation fluctuation in days belonging to the same class (of cloud shade, cloud cover amount, clearness index, or fractal dimension) may be quite different. Therefore, other ways of properly quantifying the fluctuations of solar radiation during a given day should be found. A new parameter is defined to quantify the stability of the radiative regime, namely the sunshine stability number ζ. The average value $\bar{\zeta}$ of the sunshine stability number during any time interval ranges between 0 and 1. In practice, when applied to the 2009 days in Timisoara, the daily averaged sunshine stability number ranges between 0 and 0.028. The frequency distribution of the days classified according to their average sunshine stability number is unimodal.

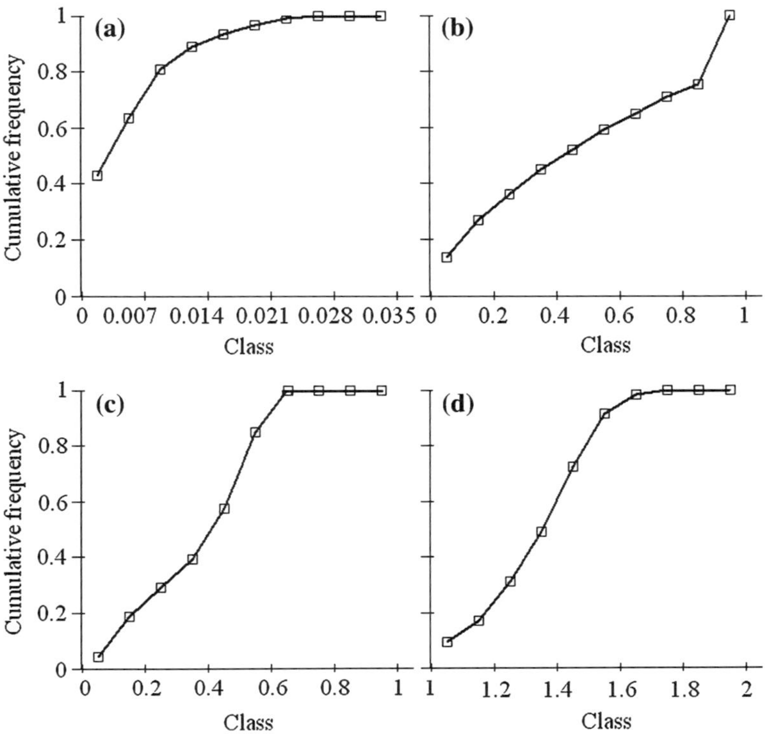

Fig. 4.7 Cumulative frequency for 10 classes of daily average values of **a** the sunshine stability number $\bar{\zeta}(\Delta\tau, \Delta t)$, **b** cloud shade, **c** clearness index, and **d** fractal dimension. Data from Timisoara during January–December 2009 have been used. In all cases, $\Delta\tau = 15$ s and Δt is the daylight length. From Paulescu and Badescu (2011), with permission from Springer

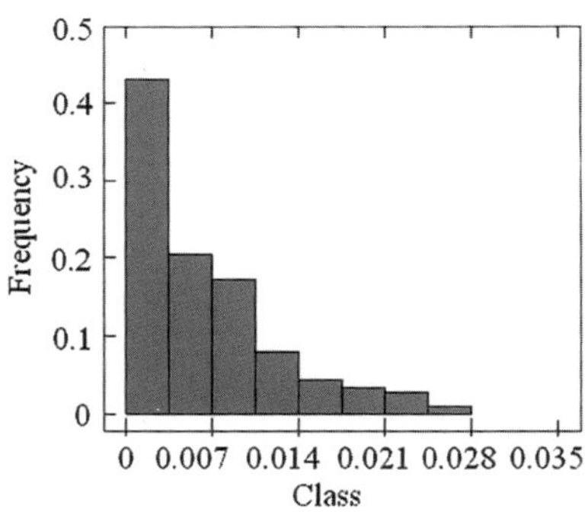

Fig. 4.8 Frequency distribution for 10 classes of daily average sunshine stability number $\bar{\zeta}(\Delta\tau, \Delta t)$. Data from Timisoara during January–December 2009 have been used. In all cases, $\Delta\tau = 15$ s and Δt is the daylight length

4.3 The Radiative Regime. Disorder and Complexity

The informational entropy is very often used as a measure of disorder. We could define the entropy $S_x(\Delta\tau, \Delta t)$ associated to a random Boolean variable x (here x stands for either the sunshine number ξ or the sunshine stability number ζ):

$$S_x(\Delta\tau, \Delta t) \equiv -\sum_{x=0,1} p(x, \Delta t) \ln p(x, \Delta t) \tag{4.7}$$

However, this tacitly assumes that the size of the system, as measured by the number of states available to it, does not change. In fact, if the number of states of the system increases then the entropy, and therefore the disorder of the system will also increase for no other reason than the increase in the number of states. To circumvent this problem, the "disorder" $\Delta_x(\Delta\tau, \Delta t)$ can be defined as (Landsberg 1984; Davison and Shiner 2005):

$$\Delta_x(\Delta\tau, \Delta t) \equiv \frac{S_x(\Delta\tau, \Delta t)}{S_{x,\max}(\Delta\tau, \Delta t)} \tag{4.8}$$

where $S_{x,\max}(\Delta\tau, \Delta t)$ is the maximum entropy which occurs in the simplest case at the equiprobable distribution $p(x = 1, \Delta t) = p(x = 0, \Delta t) = 1/2$. The order $\Omega_x(\Delta\tau, \Delta t)$ is defined as (Shiner et al. 1999):

$$\Omega_x(\Delta\tau, \Delta t) \equiv 1 - \Delta_x(\Delta\tau, \Delta t) \tag{4.9}$$

Various complexity measures are defined in the literature. Here we are using the simple complexity $\Gamma_x^{ab}(\Delta\tau, \Delta t)$ of disorder strength a and order strength b, which is defined by (Shiner et al. 1999):

$$\Gamma_x^{ab}(\Delta\tau, \Delta t) = \Delta_x^a(\Delta\tau, \Delta t)\Omega_x^b(\Delta\tau, \Delta t) \tag{4.10}$$

This general expression covers the three usual categories of complexity measures described in Shiner et al. (1999). However, results presented in the following will show which particular values of the strength a and b are appropriate in particular cases.

The statistical measures defined in this section have been used to classify the days in the database. Table 4.1 shows results obtained for selected days in the class of daily averaged cloud shade 0.4–0.5. Some of these days are characterized by small values of the daily averaged sunshine stability number ζ (i.e., these days have small solar radiation fluctuations), while the others days have larger values of ζ (i.e., larger fluctuations of radiation).

Table 4.1 shows that the disorder and the complexity based on sunshine number ξ (i.e., $\Delta_\xi(\Delta\tau, \Delta t)$ and $\Gamma_\xi^{11}(\Delta\tau, \Delta t)$, respectively) do not differentiate the days with small and large number of sunshine number fluctuations. On the contrary, both the disorder and the complexity based on the sunshine stability number ζ (i.e., $\Delta_\zeta(\Delta\tau, \Delta t)$ and $\Gamma_\zeta^{11}(\Delta\tau, \Delta t)$,) are obviously smaller in the days with smaller values

Table 4.1 Statistical measures for selected days in the cloud shade class 0.4–0.5

Daily averaged sunshine stability number $\bar{\zeta}$	Day symbol	Disorder $\Delta_{\xi}(\Delta\tau, \Delta t)$	Complexity $\Gamma_{\xi}^{11}(\Delta\tau, \Delta t)$	Disorder $\Delta_{\zeta}(\Delta\tau, \Delta t)$	Complexity $\Gamma_{\zeta}^{11}(\Delta\tau, \Delta t)$
0.0033	090304	0.997	0.0034	0.033	0.032
0.0046	090117	0.999	0.0004	0.043	0.041
0.0078	090603	0.999	0.0016	0.066	0.062
0.0087	090504	0.999	0.0000	0.073	0.067
0.015	090314	0.991	0.0085	0.114	0.101
0.019	090205	0.997	0.0028	0.138	0.118

In all cases $\Delta\tau = 15$ s and Δt is the daylight length. The day symbol has the general form *yymmdd* where *yy* is the year (09 stands for 2009), *mm* is the month number in the year, and *dd* is the day number in the month

of the sunshine stability number than in the other days. Any of these last two measures can be used to classify the days from the point of view of the stability of the radiative regime.

The ξ -based complexity $\Gamma_{\xi}^{11}(\Delta\tau, \Delta t)$ does not scale with the ξ -based disorder $\Delta_{\xi}(\Delta\tau, \Delta t)$ (see Table 4.1). The complexity $\Gamma_{\xi}^{01}(\Delta\tau, \Delta t)$ is more appropriate to be used for the characterization of days from the point of view of the sunshine number. The ζ -based complexity $\Gamma_{\zeta}^{11}(\Delta\tau, \Delta t)$ scales very well with the ζ -based disorder $\Delta_{\zeta}(\Delta\tau, \Delta t)$ (see Table 4.1). Taking into account Eq. (4.10), we conclude that the complexity $\Gamma_{\zeta}^{10}(\Delta t)$ is a simpler measure of complexity than $\Gamma_{\zeta}^{11}(\Delta\tau, \Delta t)$. However, both measures may be used for the characterization of days from the point of view of the fluctuations of solar global radiation.

Other measures were introduced in this section to properly quantify the daily fluctuations of global solar irradiance. They are based on the concepts of disorder and complexity, respectively. Measures based on the sunshine stability number are more appropriate to characterize the fluctuations of the radiative regime than those based on the sunshine number.

4.4 The Radiative Regime. Days Ranking

Ranking the days from the view point of the stability of their radiative regime may be performed by using various criteria, such as the daily average value of the sunshine stability number ζ, and the ζ -based disorder and complexity, respectively. The fact that ζ is a Boolean variable has specific consequences. Tests performed by using the database described in Sect. 3.2 shows that the resulted day hierarchies are similar for all three criteria. The criterion based on the daily average value of ζ is simpler and will be used in the following.

Figure 4.9 shows the ranking of all days belonging to the daily average cloud shade class 0.4–0.5. Three ranking criteria were used there, namely the daily mean value of sunshine stability number, the daily averaged clearness index, and the

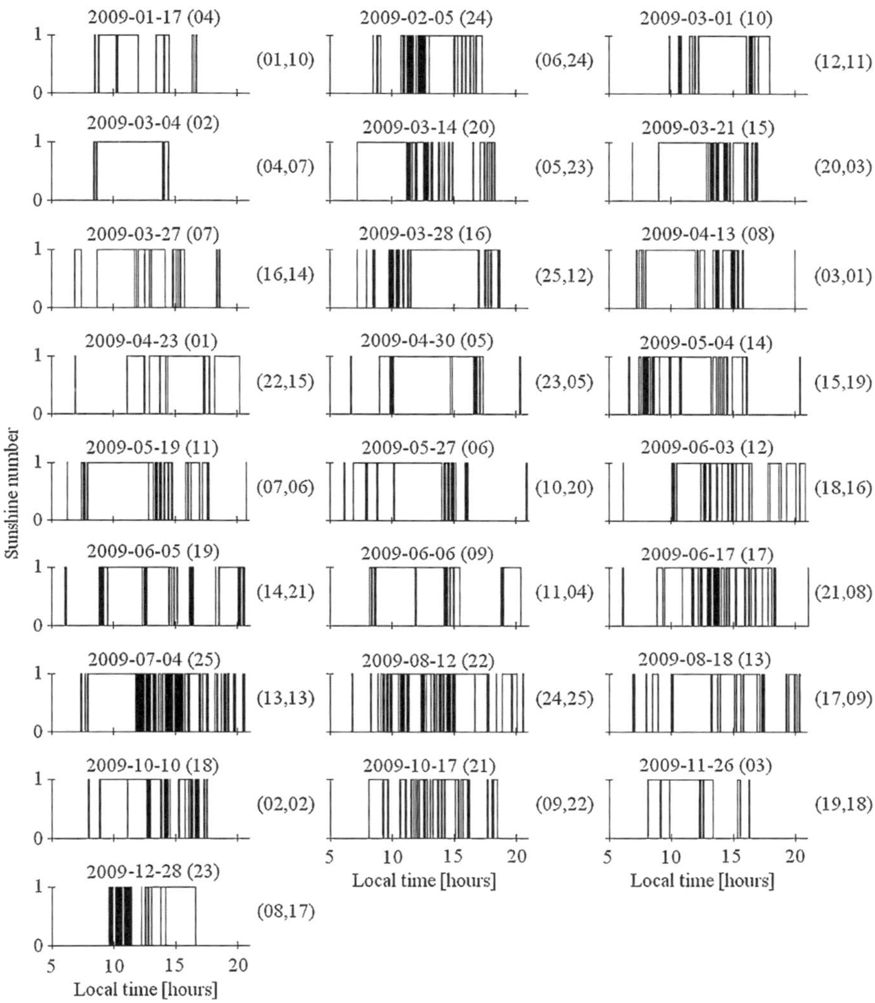

Fig. 4.9 Diurnal variation of the sunshine number ξ in all the days with mean cloud shade 0.4–0.5. In subfigures heading the day rank according to the daily mean value of sunshine stability number $\bar{\zeta}(\Delta\tau, \Delta t)$ is shown between parentheses. To the right of each subfigure, the ranks according to the daily averaged clearness index and fractal dimension, respectively, are also shown. In all cases, $\Delta\tau = 15$ s and Δt is the daylight length. From Paulescu and Badescu (2011), with permission from Springer

fractal dimension, respectively. Visual inspection of the graphs in Fig. 4.9 shows a good agreement between the fluctuations of the sunshine number and the ranking induced by the criterion based on the daily average sunshine stability number $\bar{\zeta}$. This applies to the criteria based on the ζ -based disorder and complexity, too (see comments at the end of Sect. 4.3).

Figure 4.10 shows the daily variation of the sunshine stability number ζ during all days belonging to the daily average cloud shade class 0.4–0.5. These days were

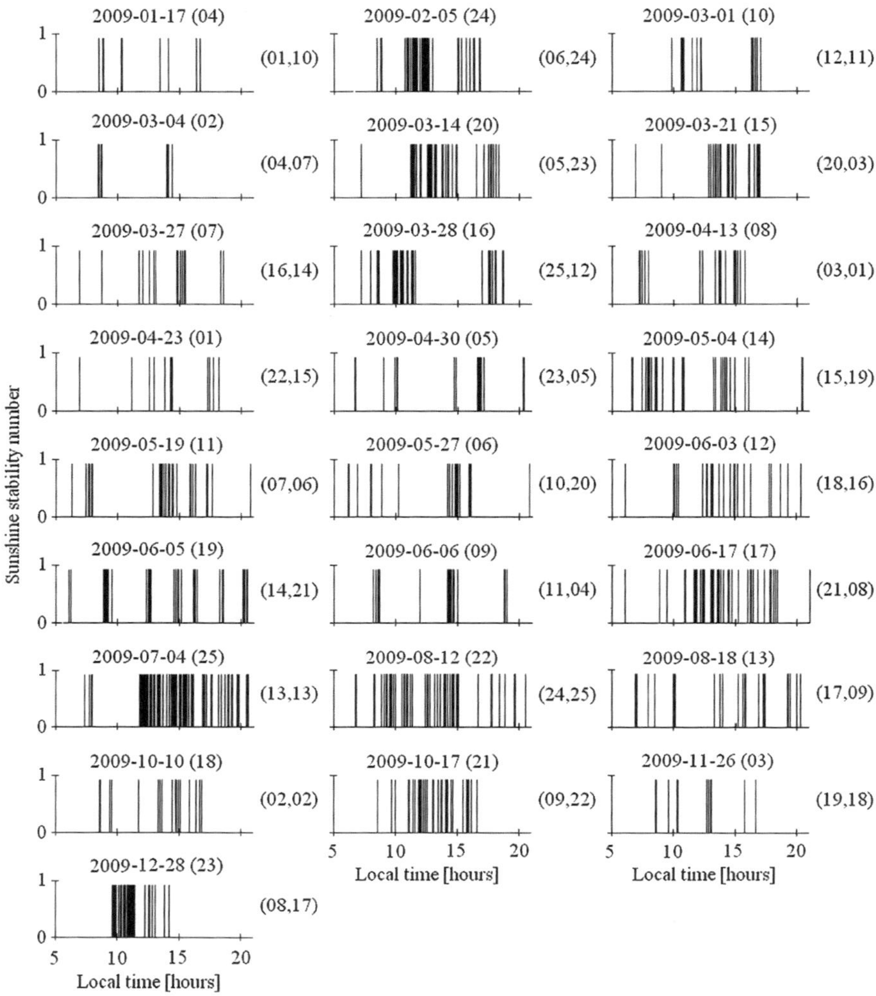

Fig. 4.10 Diurnal variation of the sunshine stability number ζ in all the days with mean cloud shade 0.4–0.5. In subfigures heading the day rank according to the daily mean value of sunshine stability number $\bar{\zeta}(\Delta\tau, \Delta t)$ is shown between parentheses. To the right of each subfigure the ranks according to the daily averaged clearness index and fractal dimension, respectively, are also shown. In all cases, $\Delta\tau = 15$ s and Δt is the daylight length

ranked according to three different criteria, namely the daily mean value of sunshine stability number ζ, the daily averaged clearness index, and the fractal dimension, respectively. Visual inspection of the graphs in Fig. 4.10 confirms conjectures discussed at Fig. 4.9: the ranking induced by the criterion based on daily average sunshine stability number ζ is in good agreement with the radiative regime of the day.

Rankings similar to those shown in Figs. 4.9 and 4.10 were prepared for all days in the daily average cloud shade classes 0.1–0.2 and 0.8–0.9, respectively (Paulescu and Badescu 2011). They confirm previous conclusions.

In conclusion, ranking the days from the view point of the stability of their radiative regime may be performed by using various criteria, such as the daily average value of the sunshine stability number ζ and the ζ-based disorder and complexity, respectively. The resulted day hierarchies are similar for all three criteria. However, the criterion based on the daily average value of ζ is simpler.

4.5 The Radiative Regime. Sequential Characteristics

In this section, we describe in more detail the sequential characteristics of the time series of the sunshine number and the sunshine stability number. Our first objective is to find appropriate transforms to allow us to obtain statistical equilibrium for the datasets. This is a central feature in the development of time series models. A somewhat restricted form of statistical equilibrium is stationarity, which means that the time series is adequately described by its mean, variance, and spectral density (or, equivalently, autocorrelation) function. All these three statistical indicators are used here. Our second objective is to obtain forecast models, which are such that the mean square of the deviations between the actual and forecasted values (i.e., the residuals) is as small as possible. A classical technique developed by Box and Jenkins (1970) for time series modeling will be applied (see Boland (2008) for a detailed presentation). It uses a combination of autoregressive (AR), integration (I) and moving average (MA) terms in the general autoregressive integrated moving average (ARIMA) models . The general ARIMA (p, d, q) allows us to evaluate the variable z (here identifies the sunshine number ξ or sunshine stability number ζ) at the discrete time t as a function of its values at previous time moments. Its general form is (Box and Jenkins 1970, p. 11):

$$w_t = \varphi_1 w_{t-1} + \cdots + \varphi_p w_{t-p} + a_t - \theta_1 a_{t-1} - \cdots - \theta_q a_{t-q} \tag{4.11}$$

where the new variable w_t is obtained by differencing d times the variable z_t:

$$w_t = \nabla^d z_t \tag{4.12}$$

In Eq. (4.11) φ_i ($i = 1, 2, \ldots, p$) are the AR coefficients, θ_i ($i = 1, 2, \ldots, q$) are the MA coefficients and a_t is white noise with zero mean and standard deviation σ_a. Sometimes an adjustment constant θ_0 is included into Eq. (4.11). The coefficients φ_i and θ_i as well as the standard deviation of the noise, σ_a, are obtained in the following by using the maximum likelihood method (Box and Jenkins 1970, Chap. 7). In Eq. (4.12) ∇ stands for the differencing operator $\nabla^d z_t = (1 - B)^d z_t$, where B is the backshift operator $B z_t = z_{t-1}$.

The procedure to find the "appropriate" model for a given time series is as follows. First, the level of differencing the data is gradually increased. According to the parcimony principle, we tried to obtain a sample spectrum with as small systematic tendencies as possible. Second, the significance of the fitted φ_i and θ_i coefficients was estimated. The following criterion was adopted: we tested if the t-statistics greater than 2 in magnitude corresponded to p-values less than 0.05. If they did not, we tried to refit the model with the least significant variable excluded. One has to remind that the t-statistic is just the estimated coefficient divided by its own standard error. Thus, it measures the number of standard deviations that the estimated coefficient is different from zero and it is used to test the hypothesis that the true value of the coefficient is nonzero, in order to confirm that the coefficient really belongs to the model. The p-value is the probability of observing a given value (or a value larger than the given value) of the t-statistic under the null hypothesis that the true coefficient value is zero. If the p-value is greater than 0.05 (say)—which occurs roughly when the t-statistic is less than 2 in absolute value—this means that the coefficient may be only "accidentally" significant. To obey the parsimony principle the number of coefficients was kept as small as possible.

4.5.1 Sunshine Number. Sequential Characteristics

4.5.1.1 Sunshine Number. Series of Daily Averaged Values

The yearly series of the daily averaged values of the sunshine number has been analyzed. A first-order differencing ($d = 1$) is necessary to remove (in part) the trends from the series. We performed several tests which showed that ARIMA models without the adjustment constant θ_0 give better results than models with that constant included.

The series of daily averaged values of the sunshine number during 2009 at Timisoara are best described by the ARIMA(0,1,2) model with $\theta_1 = 0.623365$ (with a standard error of 0.0512469, a t-statistic of 12.1639, and a p-value of 0.000000) and $\theta_2 = 0.2249$ (with a standard error of 0.0515, a t-statistic of 4.3695, and a p-value of 0.000016). The estimated white noise standard deviation is 0.2884 and the estimated white noise variance is 0.0832 (for 362 degrees of freedom). Figure 4.11 shows the time series of the daily averaged values of the sunshine number $\bar{\xi}(\Delta\tau, \Delta t)$ during 2009 as generated by this ARIMA(0,1,2) model. The root mean square error (RMSE) of the residuals is 0.2879. There is a good similarity between the sequential features of the observed time series (Fig. 4.2) and the synthetic time series (Fig. 4.11).

4.5.1.2 Sunshine Number. Series of 15 s Lag Values

The time series of the sunshine number ξ have been analyzed for all days belonging to particular daily cloud shade classes. A first-order differencing ($d = 1$) is necessary to remove (in part) trends from the series. Several tests showed that

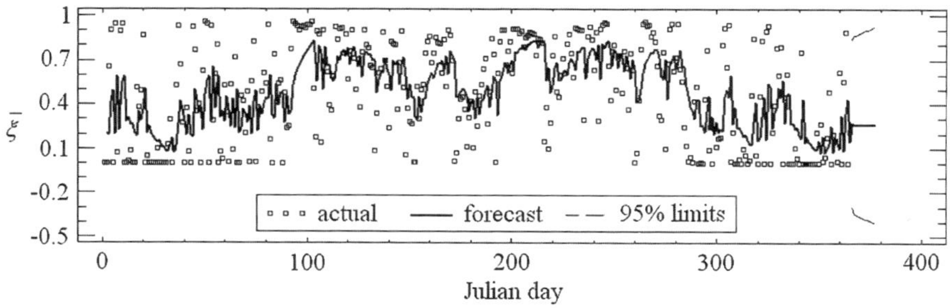

Fig. 4.11 ARIMA(0,1,2) model in the case of a time series of the daily averaged values of the sunshine number $\bar{\xi}(\Delta\tau, \Delta t)$. Data from Timisoara during January–December 2009 have been used. Forecast for the next 12 days are also shown together with the ±95 % confidence interval. From Badescu and Paulescu (2011), with permission from Springer

Table 4.2 Mean absolute error (MAE) and root mean squared error (RMSE) for various models when applied to the sunshine number time series of 13 April 2009 (daily cloud shade class 0.4–0.499)

Model	MAE	RMSE
Random walk	0.0320	0.1151
Linear trend	0.3090	0.3928
Simple MA of three terms	0.0220	0.1308
Simple exponential smoothing	0.0149	0.1135
Brown's linear exponential smoothing	0.0251	0.1229
ARIMA(0,1,0)	**0.0132**	0.1151
ARIMA(1,0,0)	0.0207	0.1147
ARIMA(2,0,0)	0.0202	0.1132
ARIMA(2,1,0)	0.0149	0.1135
ARIMA(2,1,1)	0.0148	0.1135
ARIMA(2,1,2)	0.0257	**0.1126**

Numbers in bold show the best performance

ARIMA models without the adjustment constant θ_0 give better results than models with that constant included. For each day, the most suitable ARIMA model has been selected by using the parsimony principle, as described in the paragraphs following Eq. (4.12).

The process is illustrated by using the sunshine number time series during 13 April 2009. This day belongs to the daily cloud shade class 0.4–0.499, which contains days with almost equal total duration of shade and bright sunshine, respectively. Several models have been applied to this time series. Table 4.2 shows the main indicators of accuracy of these models. Generally, the ARIMA models have a better performance than the other models. The lowest bias error is associated with the ARIMA(0,1,0) model, while the lowest RMSE value corresponds to the ARIMA(2,1,2) model.

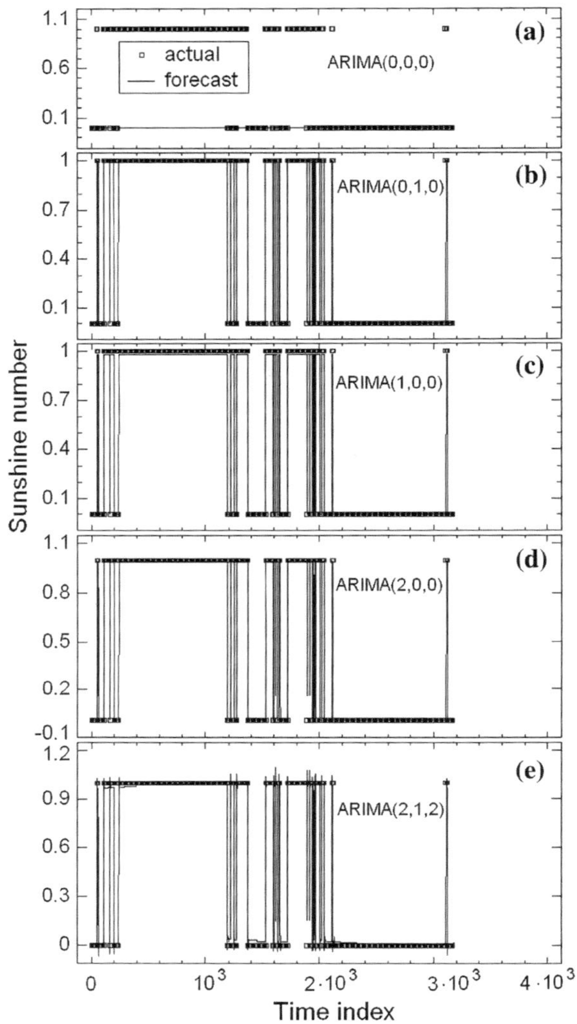

Fig. 4.12 Various ARIMA models applied to the sunshine number time series for 13 April 2009. **a** ARIMA(0,0,0); **b** ARIMA(0,1,0); **c** ARIMA(1,0,0); **d** ARIMA(2,0,0); **e** ARIMA(2,1,2). The time index refers to intervals of 15 s. From Badescu and Paulescu (2011), with permission from Springer

Forecasts provided by some ARIMA models are shown in Fig. 4.12. The diagrams of the residual autocorrelation coefficients of these models are shown in Fig. 4.13. The model ARIMA(0,0,0) is shown in Fig. 4.12a. The time series has positive autocorrelations out to a high number of lags (Fig. 4.13a). The usual procedure to obtain stationarity is by differencing. Here first-order differencing is used (i.e., $d = 1$). Figure 4.12b shows the forecasts by the ARIMA(0,1,0) model, while Fig. 4.13b shows the associated residual autocorrelation coefficients. The - first autocorrelation coefficient is negative. This means the series do not need a higher order of differencing. Since the lag-1 autocorrelation is higher than -0.5, the series is not over differenced. Using one autoregressive term (i.e., $p = 1$) instead of differencing yields an almost similar diagram of autocorrelation

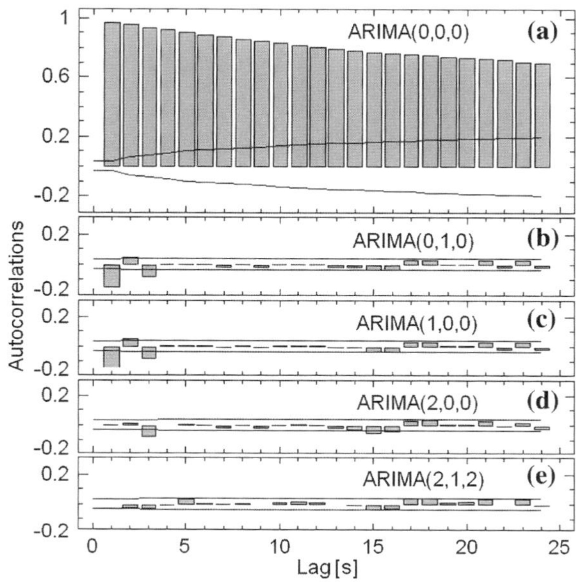

Fig. 4.13 Residual autocorrelation coefficients for the ARIMA models of Fig. 4.12. The lags refer to intervals of 15 s. The *solid lines* show the ±95 % confidence interval. From Badescu and Paulescu (2011), with permission from Springer

coefficients (compare Fig. 4.13c with b). The forecasts are also similar (compare Fig. 4.12c with b). This similarity is supported by the large value of the first AR coefficient ($\varphi_1 = 0.98$). Better performance is obtained by increasing the number of the AR terms (i.e., $p = 2$). Figure 4.13d shows that the third residual autocorrelation coefficient of the ARIMA(2,0,0) model is still negative. Further adding differencing (i.e., $d = 1$) and one MA term (i.e., $q = 1$) do not significantly improve the performance. The best performance is obtained by using two MA terms ($q = 2$). Figure 4.13e shows that all autocorrelation coefficients of the ARIMA(2,1,2) model are confined to the ±95 % confidence interval.

Table 4.3 shows as an example the best ARIMA models for all days belonging to the daily cloud shade class 0.5–0.599. Most models require just a single autocorrelation coefficient and a single MA coefficient. The models requiring more than one autocorrelation coefficient are associated with days in the cold season. There is no obvious relation between the white noise standard deviation and the type of the ARIMA model or the season. Tables similar to Table 4.3 can be found for all the cloud shade classes in Badescu and Paulescu (2011).

4.5.1.3 Sunshine Number. Forecasting Time Series

The results listed in Table 4.3 recommend ARIMA(p,1,q) to be used for forecasting or synthesis of the sunshine number time series. However, this model, as well as the ARIMA(p,0,q) model, raises difficult problems when used in practice. This is obvious from Fig. 4.12e and (less obvious) from Fig. 4.12d. The forecasted time series contains noninteger numbers rather than a sequence of 0 and 1 as would be expected for a Boolean variable such as the sunshine number. A brief

Table 4.3 Most suitable ARIMA(p,d,q) models without constant for time series of sunshine number during all the days in the daily cloud shade class 0.5–0.599

Date	ARIMA model	φ_1	φ_2	φ_3	θ_1	θ_2	θ_3	White noise standard deviation, σ_a	White noise standard deviation σ_a for the ARIMA(0,1,0) model
2009-01-08	(1,1,1)	0.64494	–	–	0.88073	–	–	0.15966	0.16702
2009-01-19	(1,1,1)	0.59682	–	–	0.89425	–	–	0.18519	0.19743
2009-03-09	(0,1,0)	–	–	–	–	–	–	0.15342	0.15342
2009-03-10	(1,1,1)	0.65571	–	–	0.87019	–	–	0.21629	0.22479
2009-03-12	(1,1,1)	0.74387	–	–	0.93369	–	–	0.19900	0.20680
2009-03-19	(1,1,1)	0.61375	–	–	0.84709	–	–	0.21104	0.22000
2009-03-26	(1,1,1)	0.19205	–	–	0.44087	–	–	0.14740	0.15200
2009-03-30	(1,1,1)	0.72216	–	–	0.94153	–	–	0.18604	0.19509
2009-06-24	(0,1,0)	–	–	–	–	–	–	0.14474	0.14474
2009-06-25	(2,1,1)	0.60879	−0.07870	–	0.78664	–	–	0.17712	0.18396
2009-06-29	(0,1,0)	–	–	–	–	–	–	0.12911	0.12911
2009-07-01	(0,1,0)	–	–	–	–	–	–	0.12925	0.12925
2009-07-03	(1,1,1)	0.27822	–	–	0.52705	–	–	0.06746	0.06968
2009-07-08	(1,1,1)	0.69633	–	–	0.93131	–	–	0.18795	0.19773
2009-08-10	(1,1,1)	0.62866	–	–	0.88768	–	–	0.18505	0.19499
2009-08-11	(1,1,1)	0.75392	–	–	0.90780	–	–	0.13950	0.14325
2009-09-05	(1,1,1)	0.66196	–	–	0.84820	–	–	0.18405	0.18961
2009-09-08	(1,1,1)	0.69271	–	–	0.87595	–	–	0.12605	0.13004
2009-09-11	(3,1,3)	−0.55869	0.32929	0.52016	−0.46663	0.44992	0.66220	0.08713	0.08880
2009-09-19	(1,1,1)	0.57956	–	–	0.74780	–	–	0.14226	0.14524
2009-10-16	(0,1,1)	–	–	–	0.22460	–	–	0.10131	0.10387
2009-10-21	(3,1,3)	0.08971	0.15581	0.45858	0.36496	0.20280	0.34730	0.11011	0.11590
2009-11-13	(2,1,2)	0.13201	0.44768	–	0.37699	0.48743	–	0.13073	0.13623
2009-11-17	(2,1,3)	−0.39473	−0.71271	–	−0.03221	−0.39426	0.16652	0.08970	0.09948
2009-11-29	(2,1,2)	0.01927	0.46938	–	0.18002	0.63789	–	0.08392	0.08681
2009-12-21	(0,1,0)	–	–	–	–	–	–	0.04437	0.04437

The correlation and MA coefficients are shown together with the white noise standard deviations. From Badescu and Paulescu (2011), with permission from Springer

explanation follows. The forecast z_t for the sunshine number at time t is obtained for previous values $z_{t-1}, z_{t-2}, \ldots$ of the sunshine number, which are integer values. However, the AR and MA coefficients entering Eq. (4.11) are not generally integer values and, as a result, the forecast z_t is generally noninteger.

A solution to partially avoid this obstacle is to use ARIMA(0,d,0) models. From Eq. (4.11) one sees that these models do not contain AR and MA coefficients. Practically, the use of Eq. (4.11) means developing a model for the white noise a_t of the time series w_t. Note that w_t is not a Boolean variable but it still is an integer-value random variable. Consequently, the white noise a_t is not a Boolean random variable. To obtain a time series consisting of a sequence of 0 and 1 from Eq. (4.11) one needs additional assumptions beyond the Box–Jenkins theory of ARIMA(0,d,0) modeling.

Note that the model ARIMA(0,1,0) is equivalent to an ARIMA(1,0,0) model without differencing ($d = 0$), and one autoregressive terms ($p = 1$) having the auto-regressive coefficient $\varphi_1 = 1$. Table 4.2 shows that the ARIMA(0,1,0) model has reasonable good performance. ARIMA(0,2,0) models may be also used. This is described in the following.

ARIMA(0,*d*,0) Models. White Noise for Integer-Value Time Series

Definitions

Various definitions of the white noise are used in practice. Brown (1983) defines it as "… a stationary random process having a constant spectral density function." Papoulis (1984) defines it differently: "We shall say that a process z_t is white noise if its values z_t and z_s are uncorrelated for every t and s: $\mathrm{Cov}(t, s) = 0, \forall(t,s)$". It has been shown that one cannot conclude that a white noise process has zero mean solely from Papoulis' definition of white noise. However, if one also knows that the power spectral density of the process is constant, then one knows that the mean is zero (Yates 2009).

The following definitions are used in this section. A *covariance stationary time series*, z_t, has the properties: (1) The expectation value $E(z_t)$, denoted μ, is a constant for all t; (2) The variance $\mathrm{Var}(z_t)$, denoted $\sigma^2(<\infty)$ is a constant for all t, and (3) The covariance $\mathrm{Cov}(z_t, z_{t-\tau})$, denoted $\gamma(\tau)$ is a constant for all t and τ. *An identical distribution (i.i.d.) time series* is a covariance stationary time series with $\gamma(0) = \sigma^2$ and $\gamma(i) = 0$ ($i = 1,2,\ldots$). *A white noise* is a covariance stationary time series with $E(z_t) = \mu = 0$ and $\mathrm{Cov}(z_t, z_s) = 0$ if $t \neq s$.

The white noise time series form the basic building blocks for the construction of more complicated time series.

ARIMA(0,d,0) Models

The following relationship is considered:

$$y_t(z_t, z_{t-1}, z_{t-2}, \cdots) = a_t \tag{4.13}$$

where y_t, z_t and a_t are time series. The mean values and the standard deviation values of the time series z_t, y_t, a_t are denoted μ_z, μ_y, μ_a and σ_z, σ_y, σ_a, respectively. Generally, $\mu_z \neq \mu_y$ and $\sigma_z \neq \sigma_y$. If y_t is a covariance stationary series, from Eq. (4.13) one sees that

$$\mu_y = \mu_a \text{ and } \sigma_y = \sigma_a \tag{4.14a, b}$$

ARIMA(0,d,0) modeling starts from time series z_t and by appropriate differencing transformations yields relationships of the form Eq. (4.13) where y_t is a covariance stationary series and a_t is a zero-mean white noise.

In the present section, z is either the sunshine number ξ or the sunshine stability number ζ and the associated (Boolean) time series z_t consists of a sequence of 0 and 1. Three particular cases (associated to three different functions y_t) are of interest here. They are described next.

ARIMA(0,0,0) Models

ARIMA modeling *with constant* is considered and the following particular form of Eq. (4.13) is used:

$$y_t = z_t - \mu_z \tag{4.15}$$

where $\mu_z \in [0,1]$. From Eqs. (4.13) and (4.15) one sees that both y_t and a_t consist of sequences of $d \equiv -\mu_z$ and $b \equiv 1 - \mu_z$. The total number of values, the number of d's and the number of b's in the white noise series is denoted N, N_d and N_b, respectively. Then:

$$N = N_d + N_b \tag{4.16}$$

For a zero-mean white noise one finds after some algebra:

$$\mu_a \equiv \frac{\sum_{i=1}^{N} n_i}{N} = \frac{N_d(-\mu_z) + N_b(1 - \mu_z)}{N} = 0 \tag{4.17}$$

$$\sigma_a = \left[(1/N) \sum_{i=1}^{N} (n_i - \mu_a)^2 \right]^{1/2} = \sqrt{\mu_z(1 - \mu_z)} \tag{4.18}$$

Equation (4.18) shows that knowing the mean value μ_z of the time series z_t means knowing the white noise standard deviation σ_a. The reciprocal holds. A simple procedure to generate the white noise values is as follows. A random number (say α) distributed uniformly between 0 and 1 is generated. Then, the white noise value a_t is;

$$a_t = \begin{cases} 1 & \text{if } \ 0 \leq \alpha \leq \mu_z \\ 0 & \text{otherwise} \end{cases} \tag{4.19}$$

The input for the generation procedure is σ_a (or μ_z).

ARIMA(0,1,0) Models

ARIMA modeling *without constant* is considered and the following particular form of Eq. (4.13) is used:

$$y_t = z_t - z_{t-1} \tag{4.20}$$

The time series y_t consists of a sequence of -1, 0, and 1. Equation (4.13) shows that this applies for the zero-mean white noise a_t, too. The total number of white noise series values is N and the number of -1, 0, and 1 in the white noise series is denoted N_{-1}, N_0, and N_1, respectively. Then:

$$N = N_{-1} + N_0 + N_1 \tag{4.21}$$

The definition of white noise's mean and standard deviation gives:

$$\mu_a \equiv \frac{\sum_{i=1}^{N} n_i}{N} = \frac{N_1 - N_{-1}}{N} = 0 \tag{4.22}$$

$$\sigma_a^2 = \frac{\sum_{i=1}^{N} (n_i - \mu_a)^2}{N} = \frac{N_{-1} + N_1}{N} \tag{4.23}$$

By using Eqs. (4.21)–(4.23) one finds after some algebra:

$$N_0/N = 1 - \sigma_a^2 \tag{4.24}$$

$$N_{-1}/N = N_1/N = \sigma_a^2/2 \tag{4.25}$$

To generate a series of white noise values of standard deviation σ_a, the following procedure is adopted. A random number (say α) distributed uniformly between 0 and 1 is generated. One denotes:

$$p_0 = 1 - \sigma_a^2 \tag{4.26}$$

Then, the white noise value a_t is:

$$a_t = \begin{cases} 0 & \text{if} \quad 0 \le \alpha < p_0 \\ -1 & \text{if} \quad p_0 \le \alpha \le \frac{1}{2}(1 + p_0) \\ 1 & \text{otherwise} \end{cases} \tag{4.27}$$

The input for the generation procedure is σ_a.

ARIMA(0,2,0) Models

ARIMA modeling *without constant* is considered and the following particular form of Eq. (4.13) is used:

$$y_t = z_t - 2z_{t-1} + z_{t-2} \tag{4.28}$$

The time series y_t consists of a sequence of -2, -1, 0, 1, and 2.

Two types of zero-mean white noises may be considered for this series. The first type of white noise consists of a sequence of −1, 0, and 1. Such a white noise may be generated by using the procedure described by Eqs. (4.15)–(4.19). The input for the generation procedure is the white noise standard deviation σ_a.

A second type of white noise consists of a sequence of −2, −1, 0, 1, and 2. The total number of white noise series values is *N* and the number of −2, −1, 0, 1, and 2 in the white noise series is denoted N_{-2}, N_{-1}, N_0, N_1, and N_2, respectively. Then:

$$N = N_{-2} + N_{-1} + N_0 + N_1 + N_2 \tag{4.29}$$

The definition of the mean and standard deviation gives:

$$\mu_a \equiv \frac{-2N_{-2} - N_1 + N_{-1} + 2N_2}{N} \tag{4.30}$$

$$\begin{aligned}\sigma_a^2 &= \frac{\sum_{i=1}^{N}(n_i - \mu_a)^2}{N} \\ &= \frac{N_{-2}(-2-\mu_a)^2 + N_{-1}(-1-\mu_a)^2 + N_0(-\mu_a)^2 + N_1(1-\mu_a)^2 + N_2(2-\mu_a)^2}{N}\end{aligned} \tag{4.31}$$

Also, the definition of the standardized skewness and kurtosis, s_a and k_a, respectively, gives:

$$\begin{aligned}s_a &= \frac{\sum_{i=1}^{N}(n_i - \mu_a)^3}{N\sigma_a^3} \\ &= \frac{N_{-2}(-2-\mu_a)^3 + N_{-1}(-1-\mu_a)^3 + N_0(-\mu_a)^3 + N_1(1-\mu_a)^3 + N_2(2-\mu_a)^3}{N}\end{aligned} \tag{4.32}$$

$$\begin{aligned}k_a &= \frac{\sum_{i=1}^{N}(n_i - \mu_a)^4}{N\sigma_a^4} \\ &= \frac{N_{-2}(-2-\mu_a)^4 + N_{-1}(-1-\mu_a)^4 + N_0(-\mu_a)^4 + N_1(1-\mu_a)^4 + N_2(2-\mu_a)^4}{N}\end{aligned} \tag{4.33}$$

The five Eqs. (4.29)–(4.33) may be solved for the unknown N_{-2}, N_{-1}, N_0, N_1, and N_2. The solution is given in Table 4.4, which shows that generally $N_{-2} \neq N_2$ and $N_{-1} \neq N_1$. The solution in Table 4.4 simplifies considerably in case of a zero-mean white noise ($\mu_a = 0$).

To generate a series of white noise values of standard deviation σ_a, the following procedure is adopted. A random number (say α) distributed uniformly between 0 and 1 is generated. Then, the white noise value a_t is:

Table 4.4 Solution of the Eqs. (4.29)–(4.33)

$$\frac{N_{-2}}{N}=\frac{1}{24}k_a\sigma_a^4-\frac{1}{24}\sigma_a^2-\frac{1}{24}\mu_a^3-\frac{1}{24}\mu_a^2-\frac{1}{12}s_a\sigma_a^3+\frac{1}{12}\mu_a+\frac{1}{4}\mu_a^2\sigma_a^2-\frac{1}{4}\mu_a\sigma_a^2+\frac{1}{24}\mu_a^4+\frac{1}{6}\mu_a s_a\sigma_a^3$$

$$\frac{N_{-1}}{N}=-\frac{1}{6}k_a\sigma_a^4+\frac{2}{3}\sigma_a^2+\frac{1}{6}\mu_a^3+\frac{2}{3}\mu_a^2+\frac{1}{6}s_a\sigma_a^3-\frac{2}{3}\mu_a-\mu_a^2\sigma_a^2+\frac{1}{3}\mu_a\sigma_a^2-\frac{1}{6}\mu_a^4-\frac{2}{3}\mu_a s_a\sigma_a^3$$

$$\frac{N_0}{N}=1-\frac{5}{4}\mu_a^2+\frac{1}{4}k_a\sigma_a^4-\frac{5}{4}\sigma_a^2+\frac{1}{4}\mu_a^4+\frac{3}{2}\mu_a^2\sigma_a^2+\mu_a s_a\sigma_a^3$$

$$\frac{N_1}{N}=\frac{2}{3}\mu_a^2-\frac{1}{6}k_a\sigma_a^4+\frac{2}{3}\sigma_a^2-\frac{1}{6}\mu_a^3-\frac{1}{6}\mu_a^4+\frac{2}{3}\mu_a-\mu_a^2\sigma_a^2-\frac{1}{2}\mu_a\sigma_a^2-\frac{1}{3}\mu_a s_a\sigma_a^3-\frac{1}{6}s_a\sigma_a^3$$

$$\frac{N_2}{N}=-\frac{1}{24}\mu_a^2+\frac{1}{24}k_a\sigma_a^4-\frac{1}{24}\sigma_a^2+\frac{1}{6}\mu_a^3+\frac{1}{24}\mu_a^4-\frac{1}{12}\mu_a+\frac{1}{4}\mu_a^2\sigma_a^2+\frac{1}{2}\mu_a\sigma_a^2+\frac{1}{6}\mu_a s_a\sigma_a^3+\frac{1}{12}s_a\sigma_a^3$$

$$a_t=\begin{cases}0 & \text{if} \quad \alpha\leq\frac{N_0}{N}\\ -2 & \text{if} \quad \alpha\in\left(\frac{N_0}{N},\frac{N_0}{N}+\frac{N_{-2}}{N}\right]\\ -1 & \text{if} \quad \alpha\in\left(\frac{N_0}{N}+\frac{N_{-2}}{N},\frac{N_0}{N}+\frac{N_{-2}}{N}+\frac{N_1}{N}\right]\\ 2 & \text{if} \quad \alpha\in\left(\frac{N_0}{N}+\frac{N_{-2}}{N}+\frac{N_1}{N},\frac{N_0}{N}+\frac{N_{-2}}{N}+\frac{N_1}{N}+\frac{N_2}{N}\right]\\ 1 & \text{otherwise}\end{cases} \tag{4.34}$$

The input for the generation procedure is the white noise standard deviation σ_a and the standardized skewness and kurtosis, s_a and k_a, respectively.

The Analog Principle

Most practical applications are using the *analog principle*, which turns out to have a sound basis in statistical theory. This means that one estimates population moments by the analogous sample moment, i.e., replace expected values with analogous sample moments. Thus, the values of σ_a, s_a, and k_a are obtained from sample analysis.

ARIMA(0,d,0) Models. Synthesis of Boolean Time Series

Defining the Problem

ARIMA(0,d,0) modeling is based on the general relationship Eq. (4.13). There, z_t is a Boolean time series (i.e., it consists of a series of 0 and 1). Time series synthesis means generating z_t values by using previous values z_{t-1}, z_{t-2}, ... (which belong to a Boolean time series) and the present value of the white noise, a_t (which

Table 4.5 Rules for the composition $w_t \equiv z_{t-1} \times z_t \equiv z_t - z_{t-1}$

z_{t-1}	z_t	w_t
0	0	0
1	0	−1
0	1	1
1	1	0

generally does not belong to a Boolean time series). To obtain a final Boolean time series z_t, one needs additional assumptions beyond the Box–Jenkins theory of ARIMA(0,d,0) modeling. These assumptions are called here composition rules. They are shown here for the three ARIMA(0,d,0) models.

ARIMA(0,0,0) Models. Boolean Time Series

From Eqs. (4.13) and (4.15) one finds:

$$z_t = \mu_z + a_t \tag{4.35}$$

where $\mu_z \in [0,1]$ and the white noise a_t is a sequences of $d \equiv -\mu_z$ and $b \equiv 1 - \mu_z$. Obviously, z_t described by Eq. (4.36) is a Boolean time series. No additional composition rules are necessary in this case.

ARIMA(0,1,0) Models. Boolean Time Series

From Eqs. (4.13) and (4.20) one finds:

$$z_t = z_{t-1} + a_t \tag{4.36}$$

where the white noise a_t consists of a series of −1, 0, and 1. If the usual addition is the composition law in the R.H.S. of Eq. (4.36) then the result is not a Boolean time series, as expected for z_t in the L.H.S. of Eq. (4.36). Additional composition rules are necessary in this case.

Table 4.5 shows the rules of the composition $w_t \equiv z_{t-1} * z_t \equiv z_t - z_{t-1}$, which is defined on two sets of Boolean values and applies in a set of three integer values. The composition $z_t \equiv z_{t-1} * a_t$ to be used for Eq. (4.36) is defined on one set of Boolean values and one set of three integer values and applies in a set of Boolean values.

Table 4.6 shows the rules of the composition $z_t \equiv z_{t-1} \circ a_t$ which comes from $z_t \equiv z_{t-1} + a_t$. They are called kernel rules and are related to the rules of Table 4.5. The kernel rules always apply. The other rules of the composition $z_t \equiv z_{t-1} \circ a_t$ are shown in Table 4.7. They cannot be derived from $z_t \equiv z_{t-1} + a_t$ and may be grouped into four sets. Choosing between these sets of rules should be done after tests.

Note that the set 1 of the rules $z_t \equiv z_{t-1} \circ a_t$ has an intuitive meaning: emptying an empty box keeps it empty (rule $0 \equiv 0 \circ (-1)$ and filling up a full box keeps it filled (rule $1 \equiv 1 \circ 1$).

Table 4.6 Kernel rules of the composition $z_t \equiv z_{t-1} \circ a_t \equiv z_{t-1} + a_t$

z_{t-1}	a_t	z_t
0	0	0
1	−1	0
0	1	1
1	0	1

Table 4.7 Additional sets of rules for the composition $z_t \equiv z_{t-1} \circ a_t$

Set of rules	z_{t-1}	a_t	z_t
1	0	−1	0
	1	1	1
2	0	−1	0
	1	1	0
3	0	−1	1
	1	1	1
4	0	−1	1
	1	1	0

ARIMA(0,2,0) Models. Boolean Time Series

From Eqs. (4.13) and (4.28) one finds:

$$z_t = 2z_{t-1} - z_{t-2} + a_t \tag{4.37}$$

Two different types of white noise are considered here. The first type white noise consists of a series of −1, 0, and 1, while the second type white noise consists of a series of −2, −1, 0, 1, and 2. For both cases, additional composition rules are necessary to obtain a Boolean time series, as expected for z_t in the L.H.S. of Eq. (4.37).

Table 4.8 shows the rules of the composition $w_t \equiv z_{t-2} * z_{t-1} * z_t \equiv z_{t-2} - 2z_{t-1} + z_t$, which is defined on three sets of Boolean values and applies in a set of five integer values. The composition $z_t \equiv z_{t-2} \circ z_{t-1} \circ a_t$ to be used for Eq. (4.37) is defined on two sets of Boolean values and one set of three integer values and applies in a set of Boolean values.

Column five in Table 4.9 shows the rules adopted for the composition law $z_t \equiv z_{t-2} \circ z_{t-1} \circ a_t$.. The upper part of Table 4.9 refers to the first type of white noise while the whole table refers to the second type of white noise. The kernel rules derived from the composition $w_t \equiv z_{t-2} - 2z_{t-1} + z_t$ are shown in bold. The other rules adopted in Table 4.9 constitute one particular set from the $2^{12} = 4096$ possible sets. This set has been adopted by using the intuitive reasoning described at the end of the preceding section.

Table 4.8 Rules for the composition $w_t \equiv z_{t-2} * z_{t-1} \times z_t \equiv z_{t-2} - 2z_{t-1} + z_t$

z_{t-2}	z_{t-1}	z_t	$w_t \equiv z_{t-2} - 2z_{t-1} + z_t$
0	0	0	0
1	0	0	1
0	1	0	−2
1	1	0	−1
0	0	1	1
1	0	1	2
0	1	1	−1
1	1	1	0

Table 4.9 Rules for the composition $z_t \equiv z_{t-2} * z_{t-1} * a_t$

z_{t-2}	z_{t-1}	a_t	$z_t \equiv -z_{t-2} + 2z_{t-1} + a_t$	$z_t \equiv z_{t-2} \circ z_{t-1} \circ a_t$
0	0	−1	−1	0
1	0	−1	−2	0
0	1	−1	**1**	**1**
1	1	−1	**0**	**0**
0	0	0	**0**	**0**
1	0	0	−1	0
0	1	0	2	1
1	1	0	**1**	**1**
0	0	1	**1**	**1**
1	0	1	**0**	**0**
0	1	1	3	1
1	1	1	2	1
0	0	−2	−2	0
1	0	−2	−3	0
0	1	−2	**0**	**0**
1	1	−2	−1	0
0	0	2	2	1
1	0	2	**1**	**1**
0	1	2	4	1
1	1	2	3	1

The upper part refers to the first type of white noise treated in section ARIMA(0,*d*,0) Models. White Noise for Integer-Value Time Series while the whole table refers to the second type of white noise. The rules derived from the composition $w_t \equiv z_{t-2} - 2z_{t-1} + z_t$ are shown in bold

4.5.2 Sunshine Stability Number. Sequential Characteristics

4.5.2.1 Sunshine Stability Number. Series of Daily Averaged Values

The yearly series of the daily averaged values of the sunshine stability number ζ has been analyzed. Several tests showed that ARIMA models without the adjustment constant θ_0 give better results than models with that constant included.

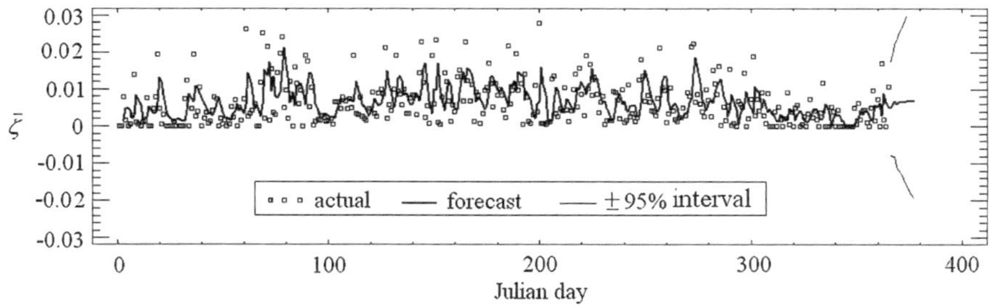

Fig. 4.14 ARIMA(2,2,1) model in the case of a time series of daily averaged values of the sunshine stability number $\bar{\zeta}(\Delta\tau, \Delta t)$. Data from Timisoara during January–December 2009 have been used. Forecast for the next 12 days are also shown together with the ±95 % confidence interval. From Badescu and Paulescu (2011), with permission from Springer

The series of daily averaged values of sunshine stability number during 2009 at Timisoara are best described by the ARIMA(2,2,1) model with $\varphi_1 = -0.602336$ (with a standard error of 0.0513597, a t-statistic of -11.7278, and a p-value of 0.000000), $\varphi_2 = -0.267903$ (with a standard error of 0.0510374, a t-statistic of −5.24915, and a p-value of 0.000000) and $\theta_1 = 0.99326$ (with a standard error of 0.000023, a t-statistic of 43595.5, and a p-value of 0.000000). The estimated white noise standard deviation is 0.006464 (for 360 degrees of freedom).

Figure 4.14 shows the time series of the daily averaged values of sunshine stability number $\bar{\zeta}(\Delta\tau, \Delta t)$ during 2009 as described by the ARIMA(2,2,1) model. The RMSE of the residuals is 0.006463. There is a good similarity between the sequential features of the observed time series (Fig. 4.6) and the synthetic time series (Fig. 4.14).

4.5.2.2 Sunshine Stability Number. Series of 15 s Lag Values

The time series of the sunshine stability number ζ have been analyzed for all days belonging to particular cloud shade classes. Further analysis is reported here for the cloud shade class 0.5–0.599. In this case differencing is not used, since the autocorrelation function of ζ is always appropriate. Differencing of various orders has been applied to the data. In all cases, the autocorrelation function is worse. Thus, ARIMA models without differencing ($d = 0$) are recommended when the generation of the sunshine stability number time series is considered. Also, models without adjusting constant have better autocorrelation function than those with adjusting constant. For all days belonging to the cloud shade class 0.5–0.599, the model ARIMA(0,0,0) without adjusting constant have the lowest values of RMSE. Consequently, this model is recommended to be used for generating time series of the sunshine stability number ζ. Section 4.5.1.3 shows the white noise associated with the ARIMA(0,0,0) model and describes the time series synthesis.

4.5.3 ARIMA Models Forecasting

4.5.3.1 ARIMA(0,d,0) Models. Forecasting Sunshine Number Time Series

The ARIMA(0,1,0) model have been used to simulate a sunshine time series in a day with a white noise standard deviation 0.1147, which corresponds to the day 13 April 2009. The kernel rules of the composition $z_t = z_{t-1} \circ a_t$ have been always used. The additional sets of rules have used one a time.

A qualitative estimation of model performance follows. Figure 4.15 shows the generated time series of sunshine number while Fig. 4.16 shows the associated diagrams of residual autocorrelation coefficients.

The sets 1 and 3 of additional rules yield a time series which has the closest visual similitude to the original time series (compare Figs. 4.15a and c, on one hand, and Fig. 4.9, on the other hand). The sets 2 and 4 (Figs. 4.15b and d) yield a larger variability of the sunshine number during the day than the other two sets (Figs. 4.15a and c). The autocorrelation diagrams show rather good performance for all additional sets of rules (Fig. 4.16).

A quantitative estimation of model performance is provided by Table 4.10. The ARIMA(0,1,0) model based on set 1 of additional rules in Table 4.7 gives the best results. All statistical indicators, except standard skewness are close to the corresponding values from the original time series. The second best is the set 2 of rules in Table 4.7.

One may conclude that an ARIMA(0,1,0) model based on the kernel rules in Table 4.6 and the set 1 of rules in Table 4.7 may be used to generate time series of sunshine number.

The ARIMA(0,2,0) model has been used to simulate a sunshine time series in a day which corresponds to the day 13 April 2009. A $d = 2$ differencing operation was applied to the sunshine number time series ($z_t \equiv \xi_t$) on that day. The resulting time series $w_t \equiv z_{t-2} - 2z_{t-1} + z_t$ consists of $N = 3167$ values and has the following statistical moments: $\mu_a = 0, \sigma_a = 0.1758, s_a = 11.6051$, and $k_a = 60.0539$. A white noise of the second type in section ARIMA(0,d,0) Models. White Noise for Integer-Value Time Series has been considered. Then, the values of the four statistical moments have been used as input to find the solutions of Eqs. (4.29)–(4.33) shown in Table 4.4. The results are $N_{-2} = 2$, $N_{-1} = 38$, $N_0 = 3090$, $N_1 = 32$, and $N_2 = 5$. These results were used as input for the white noise generation described by Eq. (4.34). The white noise a_t and the composition rules of Table 4.9 yielded the synthetic time series $\hat{z}_t \equiv \hat{\xi}_t$. Results are shown in Table 4.10.

There is good agreement between the statistical indicators of the synthetic and original time series. The ARIMA(0,2,0) model performs similarly well as the ARIMA(0,1,0) with the set 1 of additional composition rules in Table 4.7. This is to be expected. Indeed, the small values of N_{-2} and N_2 show that the second type of white noise adopted here consists mainly in a sequence of -1, 0, and 1. This is in fact a first type white noise, which has been used within the ARIMA(0,1,0) model.

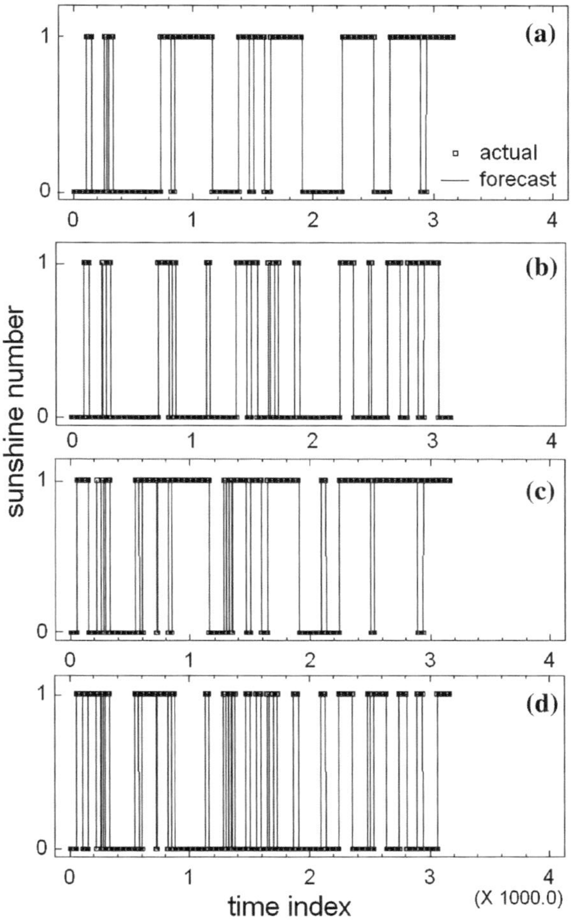

Fig. 4.15 Time series of sunshine number generated by using an ARIMA(0,1,0) model under several sets of assumptions for the composition $z_t \equiv z_{t-1} \circ a_t \equiv z_{t-1} + a_t$ in Tables 4.6 and 4.7. The standard deviation of the white noise for the original sunshine number time series during 13 April 2009 has been used. **a** Set 1; **b** set 2; **c** set 3; **d** set 4. From Badescu and Paulescu (2011), with permission from Springer

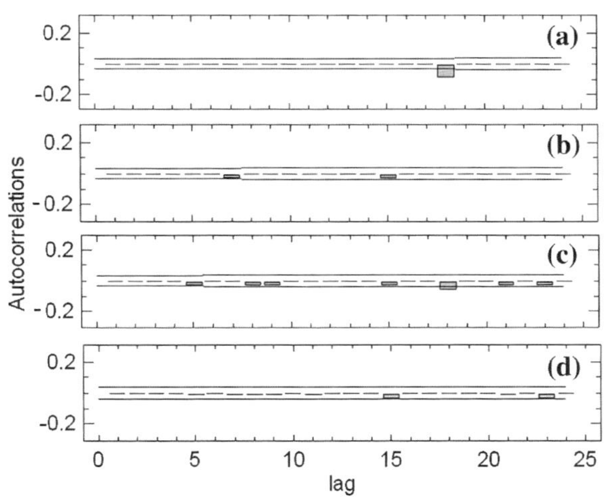

Fig. 4.16 Residual autocorrelation coefficients for the ARIMA(0,1,0) model applied under the four sets of assumptions of Fig. 4.15. The lags refer to intervals of 15 s. The *horizontal solid lines* show the 95 % confidence level. From Badescu and Paulescu (2011), with permission from Springer

Table 4.10 Statistical indicators for the original time series of sunshine number during 13 April 2009 and for the sunshine number time series generated by using an ARIMA(0,1,0) model under various additional sets of rules (see Table 4.7)

	Average	Standard deviation	Skewness	Kurtosis
Original series	0.506	0.500	−0.62	−22.99
ARIMA(0,1,0) Set 1	0.533	0.499	−3.09	−22.79
ARIMA(0,1,0) Set 2	0.293	0.455	20.88	−13.50
ARIMA(0,1,0) Set 3	0.663	0.472	−15.84	−17.52
ARIMA(0,1,0) Set 4	0.388	0.487	10.58	−20.56
ARIMA(0,2,0)	0.547	0.498	−4.40	−22.57

Results for the ARIMA(0,2,0) model with the rules in Table 4.9 are also shown

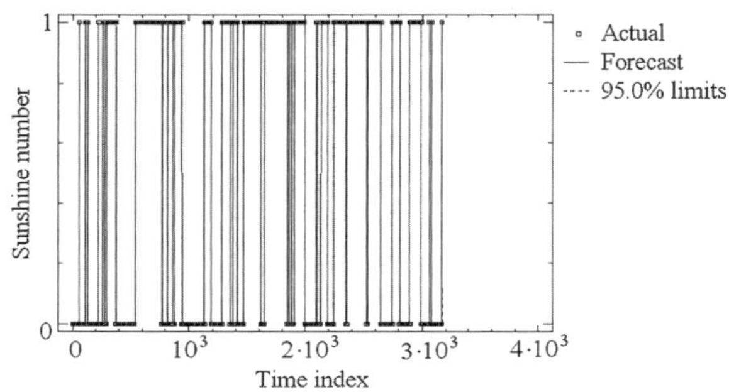

Fig. 4.17 Time series of sunshine number generated by using an ARIMA(0,2,0) model under the assumptions for the composition $z_t \equiv z_{t-2} \circ z_{t-1} \circ a_t$ in Table 4.9. From Badescu and Paulescu (2011), with permission from Springer

The synthetic time series generated by the ARIMA(0,2,0) model has a rather close visual similitude to the original time series and the series generated by the ARIMA(0,1,0) model with the first set of additional rules (compare Fig. 4.17, on one hand and Fig. 4.15a, on the other hand). One may conclude that the ARIMA(0,1,0) model with the first set of additional rules in Table 4.7 should be preferred to the ARIMA(0,2,0) model, taking into account its simplicity.

4.5.3.2 ARIMA(0,0,0) Models. Forecasting Time Series of Sunshine Stability Number

Figure 4.10 shows the sunshine stability number variation during days belonging to the cloud shade 0.4–0.499. An ARIMA(0,0,0) model has been used to generate time series of sunshine stability number during 4 days belonging to this class. These days have rather different regimes of radiative stability. Statistical properties for the observed and synthetic time series are reported in Table 4.11.

There is a good concordance between the statistical measures of the observed and simulated time series.

Figure 4.18 shows the sunshine stability number time series during 2 days with rather different radiative regimes, i.e., 23 April and 17 June 2009. They are ranked 1 and 17 (see Fig. 4.10 and Table 4.11). Visual observation shows a good similarity between the sequential characteristic of synthetic and observed time

Table 4.11 Statistical properties for sunshine stability number time series for 4 days belonging to the daily cloud shade class 0.4–0.499

Days	Rank[a]	Property	Observed	Simulated
2009-04-23	1	Average	0.0033	0.0030
(N = 3291)		Standard deviation	0.0577	0.0550
		Skewness	403	423
		Kurtosis	3450	3801
2009-04-13	8	Average	0.0066	0.0072
(N = 3171)		Standard deviation	0.0811	0.0850
		Skewness	279	271
		Kurtosis	1680	1549
2009-03-01	10	Average	0.0072	0.0072
(N = 2619)		Standard deviation	0.0848	0.0850
		Skewness	242	271
		Kurtosis	1390	1549
2009-06-17	17	Average	0.0112	0.0134
(N = 3719)		Standard deviation	0.1056	0.1151
		Skewness	230	210
		Kurtosis	1041	865

Information for observed and synthetic time series is shown. N denotes the number of recordings in a time series

[a] For days ranking see Sect. 4.4

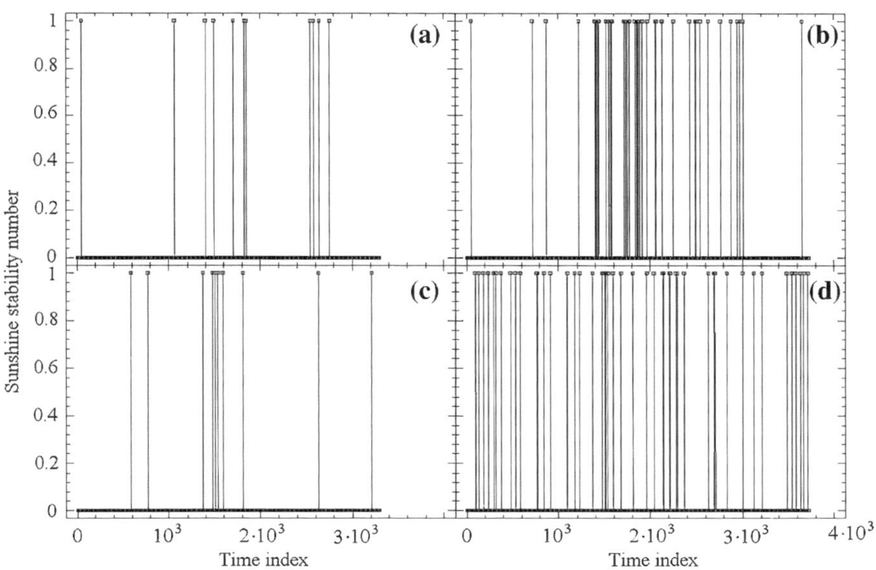

Fig. 4.18 Sunshine stability number time series during **a** 23 April and **b** 17 June 2009 (cloud shade class 0.4–0.499). Synthetic series are shown in **c** 23 April and **d** 17 June. The time index refers to intervals of 15 s

Table 4.12 The white noise standard deviation of the ARIMA(0,0,0) model without constant for the sunshine stability number

Days	White noise standard deviation	Daily average of sunshine stability number $\bar{\zeta}(\Delta\tau, \Delta t)$	Day ranking based on $\bar{\zeta}(\Delta\tau, \Delta t)$
2009-01-08	0.11807	0.01394	17
2009-01-19	0.13957	0.01948	22
2009-03-09	0.10846	0.01176	16
2009-03-10	0.15892	0.02526	26
2009-03-12	0.14620	0.02138	24
2009-03-19	0.15554	0.02419	25
2009-03-26	0.10746	0.01155	15
2009-03-30	0.13793	0.01903	21
2009-06-24	0.10233	0.01047	13
2009-06-25	0.13006	0.01692	18
2009-06-29	0.09128	0.00833	8
2009-07-01	0.09138	0.00835	9
2009-07-03	0.04926	0.00242	2
2009-07-08	0.13980	0.01954	23
2009-08-10	0.13786	0.01901	20
2009-08-11	0.10128	0.01026	12
2009-09-05	0.13405	0.01797	19
2009-09-08	0.09193	0.00845	10
2009-09-11	0.06278	0.00392	4
2009-09-19	0.10268	0.01054	14
2009-10-16	0.07343	0.00539	6
2009-10-21	0.08193	0.00671	7
2009-11-13	0.09631	0.00927	11
2009-11-17	0.07032	0.00494	5
2009-11-29	0.06137	0.00376	3
2009-12-21	0.03137	0.00098	1

Data from all the days in the daily cloud shade class 0.5–0.599 are shown

series in case of the day with rather stable radiative regime (i.e., 23 April) (compare Fig. 4.18a and c). The similarity is worse for the day with less stable radiative regime (compare Fig. 4.18b and d, respectively).

The ARIMA(0,0,0) model may be used for any particular day. Here we refer to the cloud shade class 0.5–0.599, which contains days of almost equal total durations of shade and bright sunshine, respectively. The only input parameter depending on the day is the white noise standard deviation. Also, that particular day is characterized by a daily averaged value $\bar{\zeta}(\Delta\tau, \Delta t)$ (where Δt is the daylight length). The days belonging to the cloud shade class 0.5–0.599 are stratified according to their value $\bar{\zeta}(\Delta\tau, \Delta t)$.

Table 4.12 shows the results. Figure 4.19 shows that the (daily) white noise standard deviation σ_a increases by increasing the daily averaged value $\bar{\zeta}(\Delta\tau, \Delta t)$ (and, of course, the day ranking number according to $\bar{\zeta}(\Delta\tau, \Delta t)$—see Table 4.12).

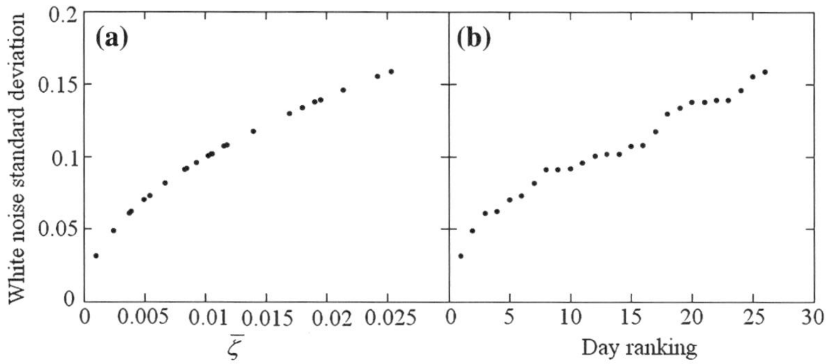

Fig. 4.19 Dependence of the white noise standard deviation σ_a for ARIMA(0,0,0) model without constant on **a** average daily sunshine stability number $\bar{\zeta}(\Delta\tau, \Delta t)$ and **b** day ranking according to $\bar{\zeta}(\Delta\tau, \Delta t)$. All days in the cloud shade class 0.5–0.599 during 2009 in Timisoara have been considered

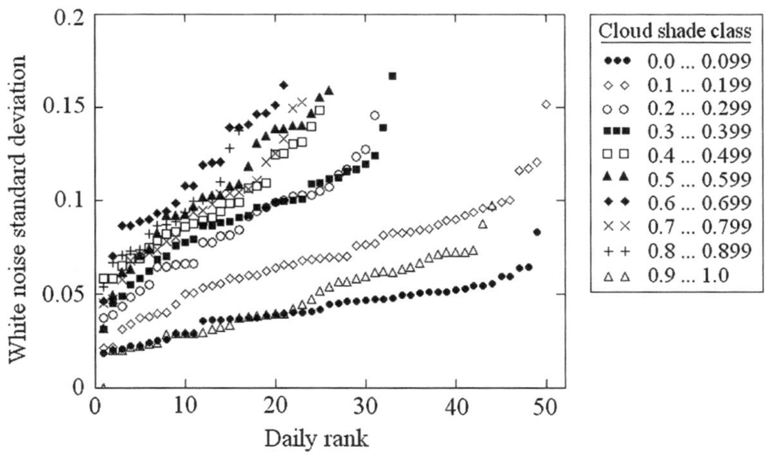

Fig. 4.20 Sunshine stability number variation during days belonging to the daily cloud shade class 0.4–0.499

Figures similar to Fig. 4.19 can be found for all the cloud shade classes in Badescu and Paulescu (2011).

The previous results apply for the days belonging to the daily cloud shade class 0.5–0.599. However, these results maintain for all the other cloud shade classes. Figure 4.20 shows that, whatever the cloud shade class is, the (daily) white noise standard deviation σ_a increases by increasing the day ranking number according to $\bar{\zeta}(\Delta\tau, \Delta t)$.

Figure 4.20 also shows that knowledge of the σ_a value for a given day does not provide enough information to find the cloud shade class to whom that day belongs. Mixing the days of all cloud shade classes was used to prepare Fig. 4.21.

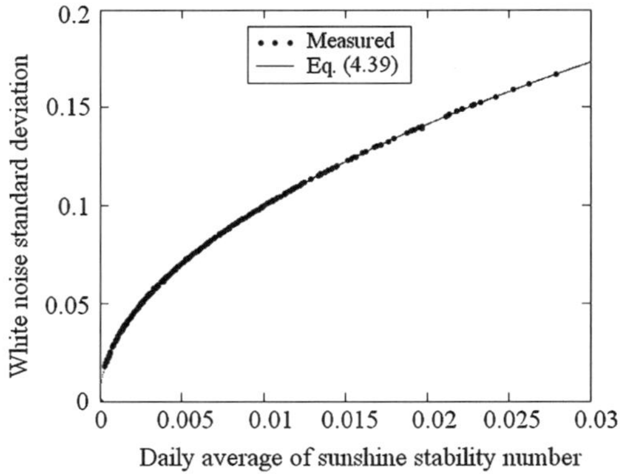

Fig. 4.21 White noise standard deviation σ_a for ARIMA(0,0,0) without constant versus daily averaged sunshine stability number $\bar{\zeta}(\Delta\tau, \Delta t)$. Measurements corresponding to all the 365 days in 2009 at Timisoara are represented. The *curve* described by Eq. (4.38) is also shown

There is an obvious dependence of the white noise standard deviation σ_a on the daily averaged sunshine stability number $\bar{\zeta}(\Delta\tau, \Delta t)$. This dependence is much clearer than in case of Fig. 4.19, where data for a single daily cloud shade class have been used.

Several curves were fitted to the measured data represented in Fig. 4.21. The best results are given by:

$$\sigma_a = 1.7235 \times 10^{-5} + 0.9996\sqrt{\bar{\zeta}(\Delta\tau, \Delta t)} \cong \sqrt{\bar{\zeta}(\Delta\tau, \Delta t)} \qquad (4.38)$$

There is a very good agreement between results derived from measurements and results predicted by Eq. (4.38) (see Fig. 4.21).

Knowing the daily averaged value of the sunshine stability number $\bar{\zeta}$ allows estimation of the standard deviation of the white noise σ_a from Eq. (4.38). This may be used as input into ARIMA(0,0,0) models to generate time series of the sunshine stability number as described in Section 4.5.1.3, item ARIMA (0,d,0) Models. Synthesis of Boolean Time Series.

4.5.4 Summary and Discussion

The procedure to obtain a proper ARIMA model is described in detail. It is based on finding appropriate transforms to allow obtaining statistical equilibrium for the data sets. Next, a criterion based on *t*-statistics and the parsimony principle has been systematically used to select the model with the smallest number of coefficients.

In case of the sunshine number time series, a first-order differencing is always necessary to remove (in part) the trends. ARIMA models without adjustment constant give better results than models with that constant included. The series of the daily averaged values $\bar{\xi}(\Delta\tau, \Delta t)$ of sunshine number are best described by an ARIMA (0,1,2) model.

The analysis recommends ARIMA(p,1,q) to be used in forecasting or synthesis of the sunshine number ξ time series. However, in general the forecasted time series contains noninteger numbers rather than a sequence of 0 and 1 as needed by a Boolean variable. The solution proposed and implemented in this chapter is based on ARIMA(0,d,0) models. Further testing of ARIMA(0,1,0) and ARIMA(0,2,0) models shows that the first model is to be preferred for practical reasons.

In case of the sunshine stability number ζ time series, ARIMA models without adjustment constant give better results than models with that constant included. The series of daily averaged values $\bar{\zeta}(\Delta\tau, \Delta t)$ of the sunshine stability number are best described by an ARIMA(2,2,1) model.

Differencing is not needed in case of the time series of the sunshine stability number ζ. Also, models without adjusting constant perform better than models with adjusting constant. The ARIMA(0,0,0) model is recommended to be used for generating time series of sunshine stability number. This model may be used for any particular day and the only parameter depending on the day is the white noise standard deviation σ_a. Results show that there is an obvious dependence of the white noise standard deviation σ_a on the daily averaged sunshine stability number $\bar{\zeta}(\Delta\tau, \Delta t)$. Equation (4.38) gives the best fit of this dependence. The performance of ARIMA(0,0,0) to generate time series of sunshine stability number is analyzed. The synthetic series has good statistic

References

Badescu V (1991) Studies concerning the empirical relationship of cloud shade to point cloudiness (Romania). Theor Appl Climatol 44:187–200

Badescu V, Paulescu M (2011) Autocorrelation properties of the sunshine number and sunshine stability number. Meteorol Atmos Phys 112:139–154

Boland J (2008) Time series modeling of solar radiation. In: Badescu V (ed) Modeling solar radiation at the Earth surface. Springer, Berlin, p 283

Box GEP, Jenkins GM (1970) Time series analysis. Forecasting and control. Holden-Day, San Francisco

Brown RG (1983) Introduction to random signal analysis and Kalman filtering. John Wiley and Sons, New York

Davison M, Shiner JS (2005) Extended entropies and disorder. Adv Complex Syst 8(1):125–158

Dubuc B, Quiniou JF, Roques-Carmes C, Tricot C, Zucker SW (1989) Evaluating the fractal dimension of profiles. Phys Rev A 39:1500–1512

Harrouni S (2008) Fractal classification of typical meteorological days from global solar irradiance: application to five sites of different climates. In: Badescu V (ed) Modelling solar radiation at the Earth surface. Springer, Berlin, p 29

Landsberg PT (1984) Can entropy and "order" increase together? Phys Lett A 102:171–173

Liu BY, Jordan RC (1960) The interrelationship and characteristic distribution of direct, diffuse and total solar radiation. Sol Energy 4:1–19

Papoulis A (1984) Probability, random variables, and stochastic processes, 2nd edn. McGraw-Hill, New York, pp 145–149

Paulescu M, Badescu V (2011) New approach to measure the stability of the solar radiative regime. Theor Appl Climatol 103:459–470

Shiner JS, Davison M, Landsberg PT (1999) Simple measure for complexity. Phys Rev E 59(2):1459–1464

Tomson T (2010) Fast dynamic processes of solar radiation. Sol Energy 84(2):318–323

Tovar-Pescador J (2008) Modelling the statistical properties of solar radiation and proposal of a technique based on Boltzmann statistics. In: Badescu V (ed) Modeling solar radiation at the Earth surface. Springer, Berlin, p 55

Yates R (2009) Mean of a white-noise process, digital signal labs. http://www.digitalsignallabs.com

Chapter 5
Modeling Solar Radiation at the Earth Surface

5.1 General Algorithm

The history of solar energy estimation algorithms is almost 100 years old. One of the first references is considered the work published by Ångström (1924), noting the correlation between global solar radiation and sunshine duration in the sky. Most actual methods for estimating the solar energy are still based on the equation established by Ångström. A typical algorithm to estimate the solar energy collected on oriented surfaces based on Ångström type-correlation is outlined in Fig. 5.1. The algorithm steps are:

1. On the top of the atmosphere, the solar energy flux is determined only by the Earth's rotation around its axis and revolution around the sun. Thus, the relations for calculating extraterrestrial solar radiation (ETR) are deducted only from astronomical considerations. The only parameters to be specified are the geographical coordinates and the time. The equations describing the periodic variation of ETR are the primary element of any model designed for calculating the solar energy collected at ground level under specified meteorological conditions.
2. The atmosphere modifies ETR, both in terms of spectral content and its spatial distribution by account of two physical phenomena: absorption and diffusion. The weight of these two phenomena is closely related to the path length of the radiation trough the atmosphere and the atmospheric composition. The path length can be calculated from astronomical considerations based on geographical coordinates and time, while the effects of atmospheric composition are random phenomena dependent of the wavelength of radiation. Thus, the inputs required by models are surface meteorological parameters like atmospheric pressure, water vapor column content, turbidity coefficient (as a measure of the aerosols content in the atmosphere), and ozone column content.

M. Paulescu et al., *Weather Modeling and Forecasting of PV Systems Operation*, Green Energy and Technology, DOI: 10.1007/978-1-4471-4649-0_5,

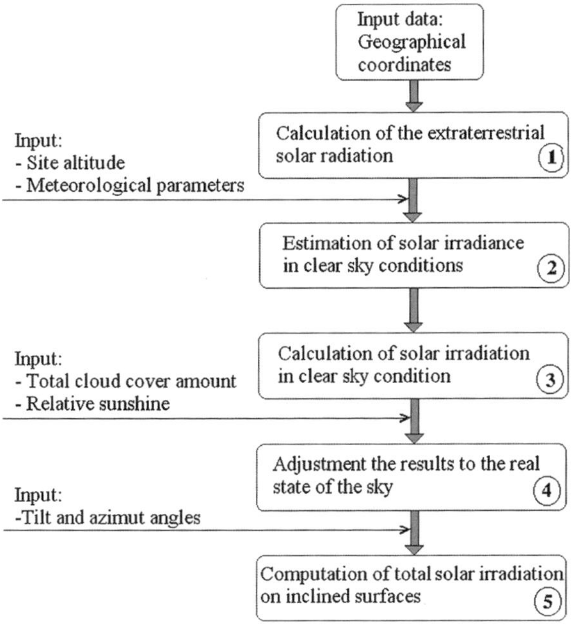

Fig. 5.1 Schematic of an algorithm for solar irradiation estimation on inclined surfaces

3. The estimation of solar irradiation in a given time interval is performed by integrating the clear sky model between the limits of the time interval considered. For calculating daily values of solar irradiation on horizontal surfaces, the integration can be done between sunrise and sunset times. An hourly sampling at least is required when estimating the collectable solar energy on oriented surfaces.
4. In general, solar radiation extinction due to the clouds is larger than that expected from any other atmospheric constituents, but it is always difficult to estimate because the distribution of clouds in the sky is random. The transmission of radiation by clouds depends on their nature, altitude, thickness, and extent. Usually, only the nature and altitude of the clouds are measured and known, therefore in algorithms clouds are considered homogeneous. Correction equations are deduced for the irradiation calculated in the previous step having as input parameters, such as the total cloud cover amount, the relative sunshine (see Chap. 3), and/or air temperature (see Chap. 8).
5. In this step, the solar irradiance calculated on a horizontal surface is converted by means of specific equations to solar irradiance on inclined surfaces. A new component appears, namely the radiation reflected from the ground. Usually, the ground reflected component is correlated with the global solar irradiation using the albedo and a factor of geometric configuration. In addition to the coordinates of the surface orientation (tilt and azimuth angle), no other input parameters are required. If the inclined surface is oriented toward a different direction than the south, particular attention should be paid to the calculation of the direct

component of solar irradiation: the integration limits must be the apparent moments of "sunrise" and "sunset" in relation to the collector surface.

For each stage of the algorithm sketched in Fig. 5.1, a key element in the selection of a model is the availability of input data. On the one hand, there are very few meteorological stations performing solar radiation measurements, on the other databases are not always accessible. Therefore, of frequent use are simplified models which require a minimum number of meteorological parameters as input. The algorithm in Fig. 5.1 presents a classical variant of calculation of solar energy collected on oriented surfaces which is only apparently simple. In reality, each step is a challenge to find a way toward a result characterized by acceptable accuracy, often an "art of compromise".

5.2 Variation of Extraterrestrial Radiation

There are two sources of variation in extraterrestrial solar radiation (ETR). The first is the Sun's output which has slight variations over short and long periods (Fröhlich 1991). For modeling purpose of solar radiation at the Earth surface, the power radiated by the sun is set to a constant value, the solar constant $G_{SC} = 1366.1\ \mathrm{W/m^2}$ (see Chap. 2). The second source of the ETR variation is the Earth's elliptical orbit, which determines a continuous variation of the Sun-Earth distance.

Assuming that in a day the Sun-Earth distance is constant, the density of solar flux energy incident on a surface normal to the Sun's rays may be computed by applying the Spencer's correction to the solar constant $G_{ext}(j) = G_{SC}\,\varepsilon$ (Spencer 1971):

$$\begin{aligned} G_{ext}(j) = G_{SC}(&1.00011 + 0.034221 \cos\theta_j \\ &+ 0.00128 \sin\theta_j - 0.000719 \cos 2\theta_j + 0.000077 \sin 2\theta_j) \end{aligned} \tag{5.1}$$

where the index $j = 1 \ldots 365$ stands for the Julian day and $\theta_j = 2\pi\,(j-1)/365$. The maximum deviation of $G_{ext}(j)$ from G_{SC} is of $\pm 3.4\ \%$.

The extraterrestrial solar irradiance on a horizontal surface is computed using the cosine law:

$$G_{0,ext} = G_{ext} \cos\theta_z \tag{5.2}$$

The zenith angle θ_z may be expressed in respect to geographical latitude ϕ, sun declination angle δ, and hour angle ω as:

$$\cos\theta_z = \sin\phi \sin\delta + \cos\phi \cos\delta \cos\omega \tag{5.3}$$

Before inferring Eq. (5.3), some explanations on the physical quantity involved are required. The declination δ is the angle between the rays of the Sun and the plane of the Earth's equator. Because the Earth's axial tilt is nearly constant $23°26'$, solar declination varies with the seasons and its period is one year. At the moment of each equinox, the Sun passes through the celestial equator and $\delta = 0°$. At the solstices, the angle between the rays of the Sun and the equatorial plane of the Earth reaches its maximum value of 23°26′. Therefore, the declination angle is $\delta = +23°26'$ at the northern summer solstice and $\delta = -23°26'$ at the northern winter solstice. The Sun's declination in a day is calculated by (Spencer 1971):

$$\begin{aligned}\delta(j) = {} & 0.006918 - 0.399912\cos\theta_j + 0.070257\sin\theta_j - 0.006759\cos 2\theta_j \\ & + 0.000907\sin 2\theta_j + 0.00148\sin 3\theta_j - 0.002697\cos 3\theta_j\end{aligned} \tag{5.4}$$

In Eq. (5.4) the declination angle is in radians.

The hour angle ω stands for the angular displacement of the Sun to East or West in respect to the local meridian. Mathematically, it is expressed as:

$$\omega = \frac{2\pi}{24}(t - 12) \tag{5.5}$$

where t (in hours) is the solar time which can be correlated with the local time t_l with the equation:

$$t = t_l - c + \frac{L_S - L}{15} + \mathrm{ET} \tag{5.6}$$

In Eq. (5.6) L_S is the local standard meridian, L is the local meridian, and c is the arbitrary correction of time by legal convention. The equation of time ET models the non-uniformity of the Earth's movements and can be expressed in hours with (Spencer 1971):

$$\begin{aligned}\mathrm{ET}_j = {} & 0.000075 + 0.001868\cos\theta_j - 0.032077\sin\theta_j - 0.14615\cos 2\theta_j \\ & - 0.04084\sin 2\theta_j\end{aligned} \tag{5.7}$$

where $j = 1\ldots365$.

The physical quantities involved in Eq. (5.3) are now elucidated and a short demonstration of Eq. (5.3) follows. Let us observe Fig. 5.2. We intend to calculate the zenith angle θ_z on the observer point P localized at latitude ϕ. Assuming that the Sun is in the meridian plane OES, the angle between OE and OS is just the declination angle δ. We choose a coordinate system Oxyz such that the Ox axis is the intersection of the equatorial and the observer meridian planes, Oz the polar axis, and Oy is perpendicularly on the Oxz plane. If R denotes the Earth's radius, the coordinates of the points P are:

$$x_P = R\cos\phi\,; \quad y_P = 0\,; \quad z_P = R\sin\phi \tag{5.8}$$

The coordinates of the point C (see Fig. 5.2) are:

$$x_C = R\cos\delta\cos\omega\,; \quad y_C = R\cos\delta\sin\omega\,; \quad z_C = R\sin\delta \tag{5.9}$$

Thus, the length of the segment PC is:

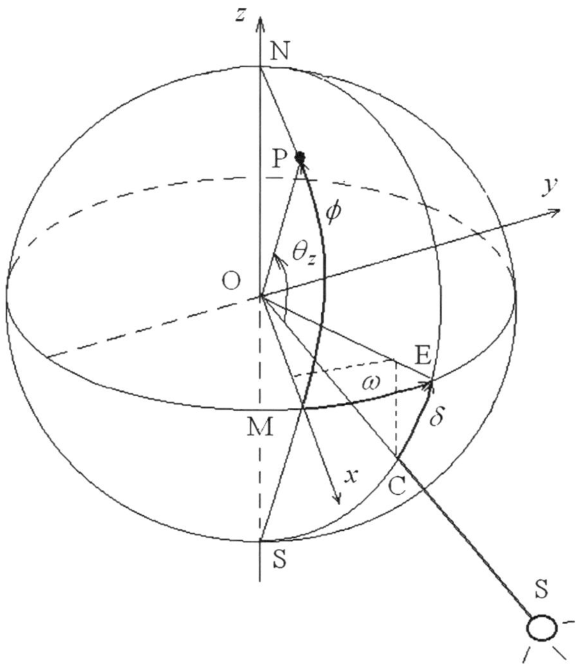

Fig. 5.2 Schematic to the calculation of the zenith angle

$$\begin{aligned}PC^2 &= [R\cos\delta\cos\omega - R\cos\phi]^2 + [R\cos\delta\sin\omega]^2 + [R\sin\delta - R\sin\phi]^2 \\ &= 2R^2(1 - \sin\phi\sin\delta - \cos\phi\cos\delta\cos\omega)\end{aligned} \quad (5.10)$$

By applying the generalized theorem of Pythagoras' in the triangle OPC one obtains:

$$PC^2 = 2R^2(1 - \cos\theta_z) \quad (5.11)$$

The difference of Eqs. (5.10) and (5.11) gives just the Eq. (5.3).

The hour angles corresponding to sunrise ($-\omega_0$) and sunset ($+\omega_0$) can be calculated from Eq. (5.3) imposing $\theta_z = \pi/2$

$$\omega_0 = \arccos(-\tan\phi\tan\delta) \quad (5.12)$$

The extraterrestrial solar irradiation on a horizontal surface in a time interval $\Delta t = t_2 - t_1$ corresponding to an hour angle interval $\Delta\omega = \omega_2 - \omega_1$, is calculated by integration of Eq. (5.2):

$$H_{0,\mathrm{ext}} = c \cdot G_{\mathrm{SC}}\,\varepsilon \int_{\omega_1}^{\omega_2} \cos\theta_z \mathrm{d}\omega \quad (5.13)$$

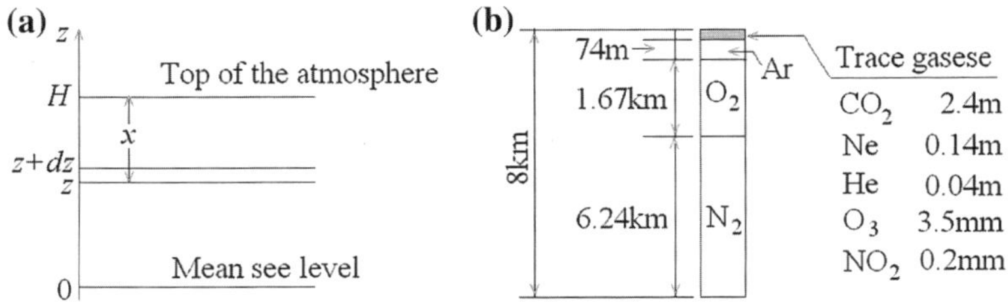

Fig. 5.3 **a** Schematic to the calculation of the attenuation of ETR passing a layer of thickness x from the top of the standard atmosphere to the altitude z. **b** The thickness of the various gas layers in the standard atmosphere

where the constant c adjusts the unit. If G_{SC} is in W/m^2 and $c = 12/\pi$ hour/rad then $H_{0,ext}$ is in Wh/m^2. If, for a given day, in Eq. (5.13) $\omega_1 = -\omega_0$ and $\omega_2 = \omega_0$, then $H_{0,ext}$ stands for the daily extraterrestrial solar irradiation.

5.3 Solar Radiation Through Earth's Atmosphere

Two processes modify ETR passing through the atmosphere toward the Earth's surface: scattering and absorption. The process of scattering occurs when atmospheric gas molecules and aerosols diffuse part of the incoming solar radiation in random directions without any alteration to the wavelength. At first instance, the amount of scattered radiation is dependent on two factors: wavelength of the incoming radiation and the size of the scattering molecules or particles. A significant proportion of scattered shortwave solar radiation is redirected back to space. The atmospheric constituents also have the ability to absorb part of the incoming solar radiation. While scattering is a continuous phenomenon in respect to wavelength, the atmospheric gas selectively absorbs solar radiation. Thus, the absorption effects are very complex in the whole solar spectrum and the process is very hard to be treated analytically.

In the following, first the effects of the cloudless atmosphere on the extraterrestrial radiation are studied and then cloud effects will be considered by specific equations.

5.3.1 Modeling the Effects of Cloudless Atmosphere on ETR

Having the purpose to compute the attenuation of the solar energy flux at wavelength λ by an atmospheric layer of thickness dz, located at the altitude z (see Fig. 5.3a), one can define the monochromatic extinction coefficient $t_\lambda(z)$ so that the variation of the solar radiation flux density between the planes z and z + dz is:

$$dG_\lambda = t_\lambda(z)G_\lambda dz \tag{5.14}$$

The solar radiation flux density at the altitude z_2 will be related to the one at the altitude z_1 by the equation:

$$G_\lambda(z_2) = G_\lambda(z_1)e^{\int_{z_1}^{z_2} t_\lambda(z)dz} \tag{5.15}$$

The extinction coefficient $t_\lambda(z)$ encapsulates both effects of absorption and scattering that occurs in atmosphere. Given that these effects are disconnected it can be written $t_\lambda(z) = t_a(\lambda, z) + t_d(\lambda, z)$, where $t_a(\lambda, z)$ models the absorption phenomena and $t_d(\lambda, z)$ models the scattering phenomena.

Since the gases that constitute the atmosphere are at low pressure, the computation of extinction coefficients may be simplified assuming that the various atmospheric particles are not interacting. Thus, for every particle species (indexed by the subscript i), each absorption or scattering process is characterized by a coefficient $a_i(\lambda)$ and $D_i(\lambda)$, respectively. The contribution to the extinction coefficient of a particle species i will be proportional with its concentration $n_i(z)$ at altitude z. With these assumptions it can be written:

$$t_a(\lambda, z) \approx \sum_i n_i(z)a_i(\lambda), \quad t_d(\lambda, z) \approx \sum_i n_i(z)D_i(\lambda) \tag{5.16a, b}$$

The real Earth atmosphere is not uniform; the particle density, the pressure and temperature varies with altitude. For modeling rationale, it is appropriate to replace the real atmosphere with the standard atmosphere. The term standard atmosphere stands for a homogenous gas layer with the same composition as the real atmosphere but with uniform pressure. At normal pressure the vertical height of the standard atmosphere is $H \approx 8$ km. The thickness of gas layers which compose the standard atmosphere is revealed in figure Fig. 5.3b, assuming they are separated. Given that in standard atmosphere the distribution of species i is homogenous with concentration $\bar{n}_i$ and that the coefficients $a_i(\lambda)$ and $D_i(\lambda)$ do not depend on the composition, for a vertical crossing of solar radiation through standard atmosphere (see Fig. 5.3a), it can be written:

$$\int_H^z \sum_i n_i(z)\, a_i(\lambda)dz = (z - H) \sum_i \bar{n}_i a_i(\lambda) \tag{5.17a}$$

$$\int_H^z \sum_i n_i(z)D_i(\lambda)dz = (z - H) \sum_i \bar{n}_i D_i(\lambda) \tag{5.17b}$$

By using the results (5.17) in Eq. (5.15) with $z_1 = H$ and $z_2 = z$ one obtains:

$$G_\lambda(z) = G_\lambda(H)e^{-(H-z)\left(\sum_i \bar{n}_i a_i(\lambda) + \sum_i \bar{n}_i D_i(\lambda)\right)} \tag{5.18a}$$

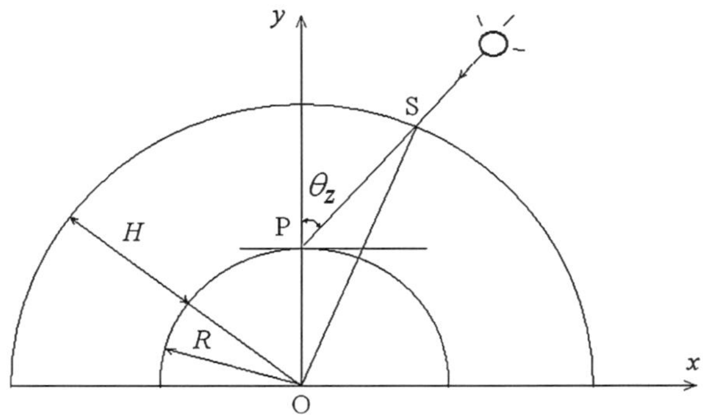

Fig. 5.4 Schematic to the calculation of the optical atmospheric air mass formula

Denoting $t_a(\lambda) + t_d(\lambda) = H\left(\sum_i \bar{n}_i a_i(\lambda) + \sum_i \bar{n}_i D_i(\lambda)\right)$, Eq. (5.18a) becomes:

$$G_\lambda(z) = G_{\text{ext}}(\lambda)e^{-\frac{d}{H}[t_a(\lambda)+t_d(\lambda)]} \tag{5.18b}$$

where $d = H - z$ is the path length of solar radiation through the atmosphere. Equation (5.18b) models the effect of the Earth atmosphere on extraterrestrial radiation when the atmosphere crossing is normal ($\theta_z = 0$). At a zenith angle different from zero, the path length of solar radiation through the atmosphere is different as will be calculated next.

5.3.2 Optical Air Mass

A measure of the path length of solar radiation through the atmosphere is the optical air mass (AM), defined as $m = d/H$. Optical AM normally indicates *relative AM*, the path length relative to that at the zenith and at sea level. So, the sea-level AM at the zenith is 1. Optical AM increases as the zenith angle increases, reaching a value of approximately 38 at the horizon. Optical AM can be less than one at an elevation greater than sea level; however, most expression for optical AM do not include the effects of elevation, so adjustment must usually be accomplished by other means.

A simple way to calculate the optical AM follows. Let us consider the standard atmosphere of vertical height *H*. In Fig. 5.4 the observer is located in the point *P* of the Earth surface, *S* is the point where ETR is incident on the top of the atmosphere, and SP is the path length of solar radiation through the atmosphere. At first, SP depends only on the zenith angle θ_z.

We consider a system of coordinates with the origin in the center of the Earth and a *y* axis oriented toward the zenith of the observer (see Fig. 5.4). The coordinates of the points *O* and *S* are $O(0,0)$ and $S(d\sin\theta_z,\ d\cos\theta_z + R)$ where *R* is the Earth radius and $d =$ SP. From the identity:

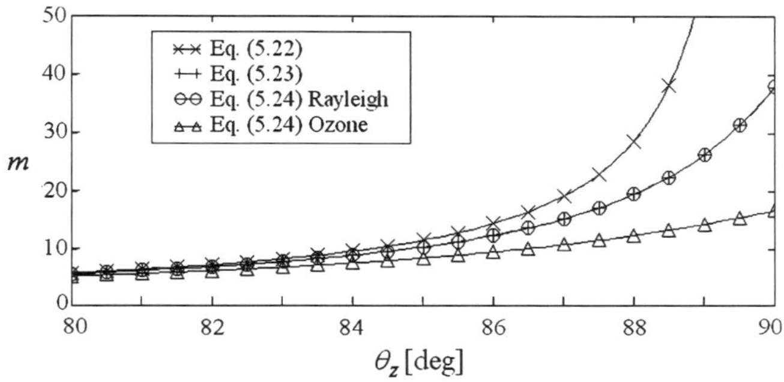

Fig. 5.5 Optical mass calculated with Eqs. (5.22), (5.23), and (5.24) with respect to the zenith angle. At zenith angles less than 80° all curves coincide

$$\mathrm{OS}^2 = (R+H)^2 = (x_S - x_O)^2 + (y_S - y_O)^2 \tag{5.19}$$

one obtains the equation $d^2 + 2dR\cos\theta_z - H^2 - 2RH = 0$ with the solution $d = (1/2)\left(-2R\cos\theta_z + \sqrt{4R^2\cos^2\theta_z + 4H^2 + 8RH}\right)$. Substituting the path length SP in the definition of optical AM, it becomes:

$$m = -\frac{R}{H}\cos\theta_z + \frac{R}{H}\cos\theta_z\sqrt{1 + \frac{2H}{R\cos^2\theta_z} + \frac{H^2}{R^2\cos^2\theta_z}} \tag{5.20}$$

At zenith angles lower than 85° one can expand the square root using the Taylor formula keeping only the first two terms:

$$\sqrt{1 + \frac{2H}{R\cos^2\theta_z} + \frac{H^2}{R^2\cos^2\theta_z}} \approx 1 + \frac{H}{R\cos^2\theta_z} + \frac{H^2}{2R^2\cos^2\theta_z} \tag{5.21}$$

By replacing the square root in (5.20) with (5.21), the equation for optical AM becomes:

$$m \cong \frac{1}{\cos\theta_z} + \frac{H}{2R\cos\theta_z} \cong \frac{1}{\cos\theta_z} = \frac{1}{\sin h} \tag{5.22}$$

In (5.22) the second approximation is feasible because $H \ll R$ and θ_z has been assumed lower than 85°. Equation (5.22) is the simplest formula for calculating the optical AM.

A more accurate equation should take into account the bending of light rays in real atmosphere due to the variation of the refractive index with air density. The following empirical expression reported in (Kasten and Young 1989) is frequently used in practice for calculating the optical AM:

Table 5.1 Coefficients for the optical masses given by Eq. (5.24) (Gueymard 1995)

Extinction process	a_{i1}	a_{i2}	a_{i3}	a_{i4}	$m_i(\theta_z = 90°)$
m_{O_3}	268.45	0.5	115.42	−3.2922	16.601
m_{NO_2}	602.30	0.5	117.960	−3.4536	17.331
$m_R \cong m_g$	0.45665	0.07	96.4836	−1.6970	38.136
$m_a \cong m_w$	0.031141	0.1	92.471	−1.3814	71.443

$$m = \frac{1}{\cos\theta_z + 0.50572(96.07995 - \theta_z)^{-1.6364}} \tag{5.23}$$

with θ_z in degrees. As shown in Fig. 5.5, Eq. (5.23) is correcting the simpler Eq. (5.22) near the sunrise and sunset times ($\theta_z > 80°$). An empirical adjustment with respect to altitude is present in the nominator of Eq. (5.23), which becomes $1 - 2.26 \cdot 10^{-4}z$. The altitude z is in meters.

Most simplified models use a single optical mass (usually the optical mass for air molecules or AM, as discussed above) to estimate the path for all the extinction processes in the atmosphere. Since each extinction process corresponds to a particular vertical molecule concentration profile, specific equations for the corresponding optical mass may be considered. Such a model for the optical mass is (Gueymard 1995):

$$m_i = \frac{1}{\cos\theta_z + a_{i1}\theta_z^{a_{i2}}(a_{i3} - \theta_z)^{a_{i4}}} \tag{5.24}$$

where m_i corresponds to m_{O_3} (ozone absorption), m_{NO_2} (nitrogen dioxide absorption), m_w (water vapor absorption), m_g (mixed gases absorption), m_R (Rayleigh scattering), m_a (aerosol extinction), and θ_z (degrees) is the zenith angle. The coefficients are listed in Table 5.1. Consideration of separate optical masses improves the model accuracy at large zenith angles, as they differ substantially above about 80° (see Fig. 5.5). Also, from Fig. 5.5 it can be seen that optical mass estimated by Kasten's equation (5.23) is superimposed with optical AM associated to Rayleigh scattering in Eq. (5.24).

The values of optical masses m_i for $\theta_z = 90°$ are also indicated in Table 5.1, showing a wide dispersion between 16.6 and 71.4. In particular, the optical AM thus calculated for $\theta_z = 90°$ is 38.1361, in good agreement with rigorously determined values, e.g. 38.0868 calculated with Eq. (5.23) (Kasten and Young 1989).

Some notations regarding to AM are used in solar energy field. The density of the extraterrestrial solar flux outside Earth's atmosphere is referred to as AM0, meaning *zero atmospheres*. Space solar cells are characterized under AM0 solar spectrum. The spectrum of solar radiation pass the atmosphere to sea level with the sun directly overhead is referred to as AM1. This means *one atmosphere*. AM1.5, *one point five atmospheres*, previous defined in Sect. 2.1, is the standardized solar spectrum for testing solar cells designed for terrestrial use.

5.3.3 Spectral Models for Atmospheric Transmittances

Equation (5.18b) has been inferred assuming a vertical crossing of the atmospheric layer by the solar radiation. At a different zenith angle, the ratio d/H in Eq. (5.18b) is identified with the optical AM. Thus, the quantities defined as:

$$\tau_a(\lambda) \equiv e^{-mt_a(\lambda)}; \quad \tau_d(\lambda) \equiv e^{-mt_d(\lambda)} \tag{5.25a, b}$$

are called *spectral atmospheric transmittance*s and are fundamental in modeling solar radiation transit through the atmosphere. As defined, $\tau_a(\lambda)$ and $\tau_d(\lambda)$ encapsulate on the whole the effects absorption and scattering processes in atmosphere. Taking into account Eqs. (5.17a, 5.17b), an atmospheric transmittance may be associated to every extinction process, absorption or scattering, as a result of solar radiation interacting with certain species of particles from the atmosphere. Few words on terrestrial atmospheric transmittance follow.

There are two main scattering processes of solar radiation: Rayleigh scattering and aerosol scattering. Rayleigh scattering occurs when solar radiation interacts with particles with dimensions much smaller than its wavelength. The particles may be individual atoms or molecules. For a given atmospheric constituent, the extinction coefficient of Rayleigh scattering $t_R(\lambda)$ is inverse proportional with the fourth power of wavelength (Leckner 1978):

$$t_R(\lambda) = c\lambda^{-4} \tag{5.26}$$

where the coefficient c is dependent on nature and the concentration of the respective atmospheric gas.

Aerosols are constituted by small particles in which dimensions ranging between 0.02 and 10 μm. In general aerosols are scattering solar radiation; only a small amount of solar radiation is being absorbed. In the simplest representation the extinction coefficient of aerosols can be expressed as (Ångström 1961):

$$t_a(\lambda) = \beta\lambda^{-\alpha} \tag{5.27}$$

where the exponent α is ranging between 0.5 and 2.5 depending on the particle sizes and the solar radiation wavelength. Roughly, its mean is estimated at 1.3. β is the Ångström turbidity coefficient (Ångström 1961), dependent on aerosol size and concentration.

Since the scattering processes are continuous phenomena with respect to wavelength they can be easily incorporated into the atmospheric transmittance models.

Modeling absorption is more complicated because atmospheric gases are selectively absorbing the solar radiation. Absorption spectra due to electronic transitions of atoms and molecules of oxygen, nitrogen, and ozone are extending into the visible and ultraviolet range. Most ultraviolet photons are absorbed by ozone; for wavelengths below 280 nm the absorption is complete. While in the visible solar spectrum the absorption is relatively low, it is strong in the infrared,

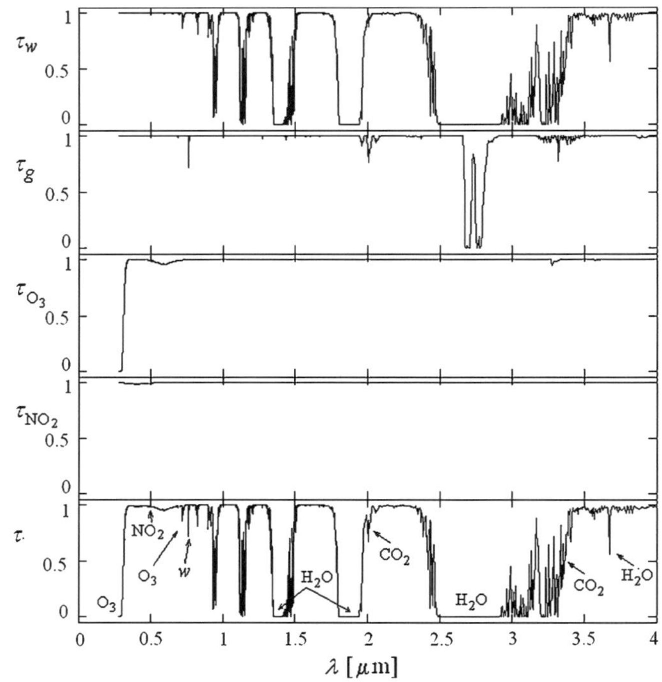

Fig. 5.6 Atmospheric transmittance as function of wavelength for an AM $m = 1$: water vapor (τ_w), mixed gases (τ_g), ozone (τ_{O_3}), NO_2 τ_{NO_2}, and all absorption processes $(\tau = \tau_w \tau_g \tau_{O_3} \tau_{NO_2})$. The calculation has been done using the absorption coefficients listed in the Appendix of Ref. Gueymard (1995) and the following set of surface parameters: normal atmospheric pressure, ozone column content of 0.350 cm·atm, NO_2 column content of 0.0002 cm·atm, water vapor column content of 2.3 g/cm^2

mainly due to vibrations and rotations of water molecules and carbon dioxide. Absorption effects are very complex in the entire solar spectrum and it is basically impossible to characterize absorption coefficients in a suitable analytical form. Moreover, each absorption band is related to meteorological parameters pressure and temperature and the band structure changes depending on the content of gases in the atmosphere. Figure 5.6 shows the atmospheric transmittance corresponding to absorption process, emphasizing various absorption bands and their weight. In a descending hierarchy, the stronger absorbers of atmospheric constituents are: water vapor, carbon dioxide, ozone, oxygen, NO_2.

Figure 5.7 shows the spectral distribution of the solar radiation flux density at sea level under clear sky conditions for two values of the optical AM $m = 1$ and 5. The weight of absorption and scattering processes are emphasized.

Spectral solar irradiance models that take into account the dependence of atmospheric transmittance on wavelength are employed in a variety of disciplines such as atmospheric science and photobiology but are less used in renewable energy science such as photovoltaic and solar-thermal conversion. However, as

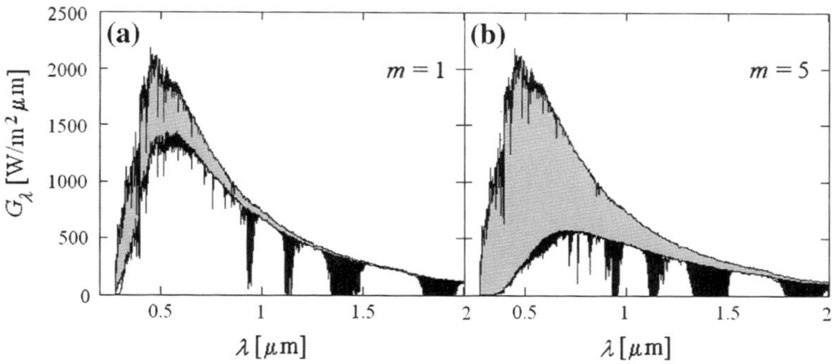

Fig. 5.7 Spectral distribution of the solar radiation flux density incident on a surface normal to the sun rays for an optical AM :**a** $m = 1$ and **b** $m = 5$. The upper curve corresponds to ETR while the lower curve corresponds to direct spectral solar irradiance at the ground. The scattered energy is in gray and the absorbed energy is in black. The calculation has been done using the spectral model SMARTS2 Gueymard (1995) and the following set of surface parameters: normal atmospheric pressure, ozone column content of 0.35 cm·atm, NO_2 column content of 0.0002 cm·atm, water vapor column content 2.3 g/cm^2, and Ångström turbidity coefficient $\beta = 0.089$

they constitute the starting point in deducing parametric models, two spectral solar irradiance models will be reviewed in the following: Leckner's spectral model (Leckner 1978) and Gueymard's Simple Model for the Atmospheric Radiative Transfer of Sunshine—SMARTS2 (Gueymard 1995).

5.3.3.1 Leckner's Spectral Solar Irradiance Model

Leckner's model can be considered a milestone in the history of spectral solar irradiance models developed for computerized engineering calculations. Many other models reported in literature (Bird and Riordan 1986; Gueymard 1993a) are based on Leckner's contribution.

The Leckner model considers five independent processes experienced by solar radiation passing through the atmosphere: Rayleigh scattering (τ_R), aerosols scattering (τ_a), water vapor (τ_w), ozone (τ_{O_3}), and mixed gases (τ_g) absorption. The corresponding transmittances are expressed as (Leckner 1978):

$$\tau_{O_3}(\lambda) = \exp(-ml_{O_3}K_{O_3}(\lambda)) \tag{5.28a}$$

$$\tau_w(\lambda) = \exp\left(-\frac{0.2385mwK_w(\lambda)}{[1 + 20.07mwK_w(\lambda)]^{0.45}}\right) \tag{5.28b}$$

$$\tau_g(\lambda) = \exp\left(-\frac{1.41mK_g(\lambda)}{[1 + 118.3mK_g(\lambda)]^{0.45}}\right) \tag{5.28c}$$

$$\tau_R = \exp\left(-0.008735\frac{mp}{p_0}(\lambda)^{-4.08}\right) \quad (5.28d)$$

$$\tau_a = \exp\left(-m\beta(\lambda)^{-1.3}\right) \quad (5.28e)$$

where m is the atmospheric AM that may be computed by means of Kasten and Young equation (5.23); l_{O3} is the ozone column content, β is the Ångström turbidity coefficient, p [hPa] is local atmospheric pressure, and w [g/cm^2] is the water vapor column content. The extinction coefficients $K_{O_3}(\lambda)$, $K_w(\lambda)$, and $K_g(\lambda)$ are tabulated in Leckner (1978).

In the Leckner model the direct solar irradiance at a wavelength λ is naturally expressed with the equation (in the sense of Eq. 5.18b):

$$G_b(\lambda) = G_{SC}\varepsilon\tau_{O_3}\tau_w\tau_g\tau_R\tau_a\cos\theta_z \quad (5.29)$$

The diffuse solar irradiance at wavelength λ incident on a horizontal surface is expressed as:

$$G_d(\lambda) = G_{SC}\varepsilon\gamma\tau_{O_3}\tau_w\tau_g(1-\tau_R\tau_a)\cos\theta_z \quad (5.30)$$

where γ is the downward fraction of scattered radiation. For a single-scattering Rayleigh atmosphere $\gamma = 0.5$.

The global solar irradiance is computed as the sum of the two components:

$$G = G_b + G_d \quad (5.31)$$

In addition to geographical coordinates and temporal information the computation of the spectral solar irradiance components requires a set of other four parameters: surface air pressure, the ozone column content, the Ångström turbidity coefficient, and the water vapor column content. A discussion on their availability follows. Surface air pressure is worldwide available, being measured by all meteorological stations. Ozone column content on daily lag can be retrieved from the NASA website Total Ozone Monitoring Spectrometer (TOMS 2012); so it is available everywhere. The aerosol optical depth is globally available in the Aerosol Robotic Network (AERONET 2012). Alternatively, the Ångström turbidity coefficient may be estimated. In a simple model, the yearly minimum (often occurring in the winter) mean and maximum (often occurring in the summer) values $\beta_{\min}$, β_m, and $\beta_{\max}$, respectively, can be computed with respect to the local latitude ϕ and altitude z (in km) with the formula proposed in Yang et al. (2001): $\beta_m = (0.025 + 0.1\cos\phi)\exp(-0.7z)$, $\beta_{\min} = \beta_m - \Delta\beta$, $\beta_{\min} = \beta_m + \Delta\beta$, and $\Delta\beta$ range between 0.02 and 0.06. The transition between extreme values can be modeled as a Gaussian function in respect to the Julian day $\beta = \beta_{\min} + (\beta_{\max} - \beta_{\min})\cdot\exp\left[-(j-c)^2/2\sigma\right]$ centered in the middle of the year $c = 182$ and with the dispersion $\sigma = 10^4$. Water vapor column content w can be estimated by using relationships between w and the surface air temperature T and relative humidity u, both currently measured by meteorological stations. Such a formula (Leckner 1978) is: $w = (0.39219 \cdot u/T)\exp(26.23 - 5416/T)$. Relative humidity is fraction of unity, T is in Kelvin and w is in g/cm^2.

5.3.3.2 SMARTS2 Model

SMARTS2 (Gueymard 1995) is a revised version of SMARTS1, a spectral model used to calculate direct beam and diffuse radiation (Gueymard 1993a).

The represented extraterrestrial spectrum is improved, both in accuracy and resolution, containing a total of 1881 wavelength points at 1 nm intervals between 0.28 and 1.7 μm and at 5 nm intervals between 1.705 and 4 μm This is to be compared to 122 wavelengths used in SPCTRAL2 and 70 wavelengths used in the Leckner's work. This highly increases the spectral resolution of the transmittance calculations. In comparison with Leckner's model, SMARTS2 include highly accurate absorption coefficients and introduce more accurate transmittance functions for all the atmospheric extinction processes also considering temperature and humidity effects. The NO_2 is added to the list of absorbers for the first time in such type of model.

The beam irradiance is calculated from spectral transmittance functions for the main extinction processes in the cloudless atmosphere: Rayleigh scattering, aerosol extinction, and absorption by ozone, uniformly mixed gases, water vapor, and NO_2. Temperature-dependent or pressure-dependent extinction coefficients have been developed for all these absorbing gases. Aerosol extinction is evaluated using a two-level Ångström approach.

The model is appropriate for studies in atmospheric research, being rather unpractical for engineering applications. The most important equations of the model are summarized in the following. For more details see Gueymard (1995).

The Beam Component

The beam irradiance received at ground level by a horizontal surface at wavelength λ is given by:

$$G_b(\lambda) = G_{SC}\varepsilon\tau_{O_3}\tau_{NO_2}\tau_w\tau_g\tau_R\tau_a \cos\theta_z \tag{5.32}$$

where the individual spectral transmittance for absorption processes are expressed by the Bouguer's law.

$$\tau_{O_3} = \exp(-m_{O_3} t_{O_3}) = \exp\left(-m_{O_3} l_{O_3} A_{\lambda,O_3}\right) \tag{5.33}$$

m_{O_3} is the optical mass of ozone (given by Eq. 5.24), l_{O_3} is the ozone column content (in atm·cm), and A_{λ,O_3} is the spectral absorption coefficient. The product $t_{O_3} = l_{O_3}A_{\lambda,O_3}$ is the ozone optical thickness.

Like for ozone, the NO_2 transmittance is expressed as:

$$\tau_{NO_2} = \exp\left(-m_{NO_2} l_{NO_2} A_{\lambda,NO_2}\right) \tag{5.34}$$

m_{NO_2} is the optical mass of NO_2 (Eq. 5.24), l_{NO_2} is the NO_2 column content (in atm·cm), and A_{λ,NO_2} is the spectral absorption coefficient.

The transmittance for water vapor absorption is given by the equation:

$$\tau_w = \exp\left[-\left((m_w l_w)^{1.05} f_w^n B_w A_{\lambda,w}\right)^c\right] \tag{5.35}$$

where m_w is the water vapor optical mass (Eq. 5.24) and l_w is the water vapor column content. The exponents n and c in Eq. (5.35) are given by the following equations, respectively: $n = 0.88631 + 0.025274\lambda - 3.5945\exp(-4.5445\lambda)$ and $c = 0.5381 + 0.003262\lambda + 1.5244\exp(-4.2892\lambda)$. f_w is a pressure scaling factor that compensates for inhomogeneities in the water vapor path length: $f_w = k_w[0.394 - 0.26946\lambda + (0.46478 + 0.23757\lambda)p/p_0]$ where p is the actual atmospheric air pressure and p_0 is the normal air pressure, $k_w = 1$ if $\lambda \leq 0.67$ μm, or else $k_w = (0.98449 + 0.023889\lambda) l_w^{-0.02454+0.037533\lambda}$. The correction factor B_w is introduced to improve the parameterization away from the absorption band center in varying humidity condition (Gueymard 1995).

The mixed gas transmittance is defined as:

$$\tau_g = \exp\left[-\left(m_g l_g A_{\lambda,g}\right)^a\right] \tag{5.36}$$

where m_g is the gas optical mass (Eq. 5.24), $A_{,g}$ is the spectral absorption coefficient and l_g is the altitude-dependent path length. O_2 and CO_2 are the main constituents of the so-called mixed gas. In accord with their absorption spectra, l_g for O_2 is used below $\lambda = 1$ μm and the value for CO_2 is used above. The exponent a is 0.5641 for $\lambda < 1$ or 0.707 otherwise.

All the absorption coefficients A_{λ,O_3}, A_{λ,NO_2}, $A_{\lambda,w}$, and $A_{\lambda,g}$ are listed in the Appendix of the Ref. Gueymard (1995) which is available online (see references of this chapter). The effective path lengths l_{O_3}, l_{NO_2}, l_w, and l_g for ten reference atmospheres are listed in Table 3.1 of the same Ref. Gueymard (1995).

The transmittance of Rayleigh scattering is calculated with the equation:

$$\begin{aligned}\tau_R &= \exp(-m_R t_R)\\ &= \exp\left(-\frac{m_R(p/p_0)}{117.2594\lambda^4 - 1.3215\lambda^2 + 3.2073\cdot 10^{-4} - 7.6842\cdot 10^{-5}\lambda^{-2}}\right)\end{aligned} \tag{5.37}$$

where m_R is the optical mass in the Rayleigh extinction, p is the site-level pressure, and p_0 is the standard air pressure. If the local pressure correction factor (p/p_0) is not known, it can be estimated from site altitude and latitude according to the procedure provided by Gueymard (1993a) (see Sect. 5.3.4.3 for details)

Table 5.2 Wavelength exponents from Eq. (5.39) related to relative humidity RH in percents. The correlation coefficient r^2 is also displayed

Aerosol model	Equation	r^2
Rural	$\alpha_1 = \frac{0.933 - 0.03123RH + 0.00034915RH^2 - 1.30102 \cdot 10^{-6}RH^3}{1 - 0.033457701RH + 0.0003738RH^2 - 1.391372 \cdot 10^{-6}RH^3}$	0.999
	$\alpha_2 = \frac{1.446 - 0.0143891RH}{1 - 0.0097703RH - 1.3599 \cdot 10^{-6}RH^2}$	0.963
Urban	$\alpha_1 = \frac{0.8208 - 8.02098 \cdot 10^{-5}RH^2}{1 - 0.0001041RH^2 - 9.42836 \cdot 10^{-10}RH^4}$	0.999
	$\alpha_2 = \frac{1.167 - 0.0384989RH + 0.0004266RH^2 - 1.583813 \cdot 10^{-6}RH^3}{1 - 0.0329536RH + 0.00036455RH^2 - 1.3503619 \cdot 10^{-6}RH^3}$	0.984
Maritime	$\alpha_1 = \frac{0.468 - 0.0162806RH + 0.0001883RH^2 - 7.2254931 \cdot 10^{-7}RH^3}{1 - 0.03442618RH + 0.000393485RH^2 - 1.486223 \cdot 10^{-5}RH^3}$	0.999
	$\alpha_2 = \frac{0.626 - 0.2033687RH^{0.5} + 0.0220084RH - 0.000793152RH^{1.5}}{1 - 0.3225736RH^{0.5} + 0.03461857RH - 0.000793152RH^{1.5}}$	0.999
Troposphere	$\alpha_1 = \frac{1.0095 - 9.2398809 \cdot 10^{-5}RH^2}{1 - 9.2502683 \cdot 10^{-5}RH^2 + 4.7956244 \cdot 10^{-10}RH^4}$	0.993
	$\alpha_2 = \frac{2.389 - 0.0797479RH + 0.0008873RH^2 - 3.2843024 \cdot 10^{-6}RH^3}{1 - 0.0333393RH + 0.00037041RH^2 - 1.3681572 \cdot 10^{-6}RH^3}$	0.999

The aerosol transmittance is considered from a two band model below and above $\lambda = 0.5$ μm (Bird, 1984):

$$\tau_a = \exp(-m_a t_a) \tag{5.38}$$

where the aerosol optical thickness is:

$$t_a = \begin{cases} \beta 2^{\alpha_2 - \alpha_1} \lambda^{-\alpha_1} & \text{if} \quad \lambda \leq 0.5 \mu m \\ \beta \lambda^{-\alpha_2} & \text{otherwise} \end{cases} \tag{5.39}$$

m_a is the aerosol optical mass (Eq. 5.24) and β is the Ångström's turbidity coefficient. For aerosol model of Shettle and Fenn (1979), the wavelength exponents α_1 and α_2 may be computed with respect to relative humidity with the equations listed on Table 5.2. These equations are the result of the fit of the discrete values α_1 and α_2 provided by Shettle and Fenn (1979).

The Diffuse Component

In SMARTS2 model, the diffuse solar irradiance is considered the sum of three components due to: Rayleigh scattering $G_{d,R}(\lambda)$, aerosol scattering $G_{d,a}(\lambda)$, and ground/sky backscattering $G_{d,b}(\lambda)$:

$$G_d(\lambda) = G_{d,R}(\lambda) + G_{d,a}(\lambda) + G_{d,b}(\lambda) \tag{5.40}$$

The Rayleigh scattered component is calculated as:

$$G_{d,R}(\lambda) = G_{SC}\varepsilon\,\gamma_R\left(1 - \tau_R^{0.9}\right)\Gamma_{O_3}\tau_{NO_2}\tau_w\tau_g\tau_{aa}\cos\theta_z \tag{5.41}$$

where $\gamma_R = 0.5\,\gamma_2$ is the downward fraction of scattered radiation. The factor 0.5 is the downward fraction for a single-scattering Rayleigh atmosphere. The correction factor γ_2 for the multiple scattering effects of air molecules is expressed (Skartveit and Olseth 1988): $\gamma_2 = 1$ if $t_R < t_{Rm}$ and $\gamma_2 = \exp\left[-\left(\left(t_{\lambda,R} - t_{Rm}\right)/\sigma_R\right)^{0.72+\sin h}\right]$ otherwise, with $t_{Rm} = 0.17[1 - \exp(-8\sin h)]$ and $\sigma_R = 3.65 - 2.3\exp(-4\sin h)$. All the transmittance functions in Eq. (5.41) have been defined previously, except τ_{aa} and Γ_{O_3}. The transmittance of aerosol absorption process τ_{aa} is defined as:

$$\tau_{aa} = \exp[-m_a(t_a - t_{as})] = \exp[-m_a t_a(1 - \varpi)] \tag{5.42}$$

$t_{as} = \varpi\, t_a$ is the optical depth for aerosol scattering and ϖ is the single scattering albedo, which is a fundamental optical characteristics of aerosols, showing the weight of scattering process in the aerosol extinction through the atmosphere. The single scattering albedo parameter is usually not directly available, but information on its regional and temporal distribution may be retrieved from satellite-based measurements (TOMS 2012); there are a lot of papers dealing with this, for example (Hu et al. 2007). An effective ozone diffuse transmittance for downward scattering Γ_{O_3} appears in Eq. (5.41) instead of the direct ozone transmittance τ_{O_3} that has been used in most simplified models since Leckner's work. The reason for this substitution is that the diffuse solar irradiance estimated by simple models decreased too much with wavelength in the UVB, where ozone absorption by far dominates all other extinction processes (Gueymard 1995). The equation for Γ_{O_3} is:

$$\Gamma_{O_3} = \begin{cases} \exp\left(-\gamma_1 t_{O_3}^{0.95} - \gamma_2 t_{O_3}\right), & \text{if}_{O_3} \le 2 \\ \exp[-\gamma_3 - \gamma_4(t_{O_3} - 2)] & \text{otherwise} \end{cases} \tag{5.43}$$

where the coefficients γ_i are $\gamma_1 = (-11.012 + 12.392 m_{O_3})/(1 + 0.23644 m_{O_3})$, $\gamma_2 = 3.2656\left[1 - \exp\left(-0.46464 m_{O_3}^{1.25}\right)\right] - 0.965936\,\gamma_1, .\gamma_3 = 1.93187\gamma_1 + 2\gamma_2$, and $\gamma_4 = \exp\left(0.31045 + 0.001684 m_{O_3} - 0.28549 m_{O_3}^{-4}\right)$.

The aerosol-scattered solar irradiance is computed as:

$$G_{d,a}(\lambda) = G_{SC}\varepsilon\,\gamma_a(1 - \tau_{as})\Gamma_{O3}\tau_{NO_2}\tau_w\tau_g\tau_R\tau_{aa}\cos\theta_z \tag{5.44}$$

where all factors are known except the fraction of scattered flux γ_a. In SMARTS2 γ_a is computed like γ_R as the product of single-scattering fraction γ_1 dependent on aerosol asymmetry factor and a multiple-scattering correction factor γ_2 depending on wavelength and zenith angle.

For engineering applications, close to the subject of this book, a simplified equation may be used, such as the equation of Robinson (1962):

$$\gamma_a = 1 - \exp(-0.6931 - 1.8326 \cos \theta_z) \tag{5.45}$$

The backscattered component is calculated assuming an infinite series of repeated reflections between the ground and the atmosphere. It is modeled as:

$$G_{d,b} = \frac{\rho_s(\rho_b G_b + \rho_d G_d)}{1 - \rho_d \rho_s} \tag{5.46}$$

where ρ_b is the zonal ground spectral reflectance, ρ_d is the counterpart for diffuse radiation, and ρ_s is the overall reflectance of the sky. These quantities are evaluated in Gueymard (1995).

Model Performance

First of all, SMARTS2 performance is assessed in the original work of Gueymard against both radiative transfer models and measured data. There are many studies that test the SMARTS model. For example: The paper Alados et al. (2002) compares estimates of direct, diffuse, and global photosynthetically active solar radiation (PAR) calculated using the spectral models SPCTRAL2 and SMARTS2 with measurements made at two Spanish location with different climates: Granada, an inland location and Almeria, a coastal Mediterranean location. The results show that both spectral codes SPCTRAL2 and SMARTS2 provide proper estimates of the different components of the PAR density flux. Tadros et al. (2005), testing three spectral models against data measured at two station from Egypt, found that SMARTS2 model is most suitable to compute solar irradiance in most spectral bands.

The high quality of this model has been proven, SMARTS2 being the basis of most studies (e.g. Power 2001; Tasumi et al. 2008; Paulescu et al. 2012).

Considering the results from literature on the quality of SMARTS2 model, it can be successfully implemented for nowcasting of solar irradiance under clear sky condition. Success depends on the availability of forecasted parameters at input. The number of these parameters determines the number of SMARTS's components that can be implemented in a computational procedure. Overall accuracy is reliant on the quality of the forecasted data. Given that in general the input parameters do not vary significantly over an hour, their availability with hourly sampling would guarantee an accurate prediction of solar irradiance under clear sky.

5.3.4 Parametric Models for Solar Irradiance

Spectral solar irradiance models provide high accuracy in estimating solar irradiance. However, a spectral model like SMARTS2 seems to be too difficult to implement in engineering applications, like nowcasting of PV plants power output. In such applications, parametric models for solar irradiance could be the optimal solution. Many of these models are derived from spectral codes by averaging the spectral atmospheric transmittances. The name "parametric" for these models signifies the use of meteorological parameters at input. The use of surface meteorological data at input adjusts the model outputs to the climate specific of the application area. In practice, a major consideration in choosing a model, spectral, or parametric, is the availability of the meteorological surface data needed for input.

In the parametric models, the global solar irradiance G is expressed as the sum of the beam G_b and diffuse G_d components:

$$G = G_b + G_d = G_{SC}\,\varepsilon\bar{\tau}_b\cos\theta_z + G_{SC}\,\varepsilon\bar{\tau}_d\cos\theta_z \qquad (5.47)$$

The key terms in Eq. (5.47) are $\bar{\tau}_b$ and $\bar{\tau}_d$, the average beam and diffuse atmospheric transmittances. They are computed as in a spectral solar irradiance model from which a parametric model is inferred, but for each atmospheric attenuator the specific spectral transmittance $\tau_i(\lambda)$ is replaced by its energy-weighted average:

$$\bar{\tau}_i = \frac{\int_{\lambda_{\min}}^{\lambda_{\max}} \tau_i(\lambda)G_{\text{ext}}(\lambda)\text{d}\lambda}{\int_{\lambda_{\min}}^{\lambda_{\max}} G_{\text{ext}}(\lambda)\text{d}\lambda} \qquad (5.48)$$

The subscript i indicate the atmospheric process, i.e. $i = O_3$ (ozone absorption), NO_2 (dioxide nitrogen absorption), w (water vapor absorption), g (mixed gases trace absorption), R (Rayleigh scattering), a (aerosol extinction). The wavelengths $\lambda_{\min}$ and $\lambda_{\max}$ set the limits of the band in which the spectral transmittances are averaged. For broadband solar irradiance models, usually $\lambda_{\min} = 0.28\,\mu\text{m}$ and $\lambda_{\max} = 4\mu\text{m}$. It is important to note that if the integration in Eq. (5.48) is carried out in a narrow band like UV ($\lambda_{\min} = 0.28\mu\text{m}$ and $\lambda_{\max} = 0.4\mu m$) then in Eq. (5.47) G_{SC} will represent the *effective solar constant* in the band, i.e. $\int_{\lambda_{\min}}^{\lambda_{\max}} G_{\text{ext}}(\lambda)\text{d}\lambda$.

Equation (5.48) stands for the common approach of deriving parametrical models from solar spectral codes. Four parametric models are reviewed in the following: (1) Hybrid (Yang et al. 2001), (2) PS (Paulescu and Schlett 2003), (3) PSIM (Gueymard 1993b), and (4) ESRA (Rigollier et al. 2000). While the first two models originate from the averaging the spectral transmittance of Leckner's model with Eq. (5.48), the other two are constructed in a different manner.

5.3.4.1 Hybrid Model

The original equations for the beam and diffuse solar irradiance of (Yang et al. 2001), slightly updated in Yang et al. (2006), are:

$$G_b = G_{SC}\varepsilon\left(\bar{\tau}_{O3}\bar{\tau}_w\bar{\tau}_g\bar{\tau}_R\bar{\tau}_a - 0.013\right)\cos\theta_z \qquad (5.49a)$$

Table 5.3 The variable x and the coefficients a, b, c, and d from Eq. (5.51) for the specific atmospheric transmittances τ

τ	x	a	b	c	d
$\bar{\tau}_{O_3}$	ml_{O_3}	0.0184	−0.0004	0.022	−0.66
$\bar{\tau}_w$	ml_w	−0.002	$1.67 \cdot 10^{-5}$	0.094	−0.693
$\bar{\tau}_g$	m	-5.410^{-5}	$-3.8 \cdot 10^{-6}$	0.0099	−0.62
$\bar{\tau}_R$	m_c	0.709	0.0013	−0.5856	0.058
$\bar{\tau}_a$	$m\beta$	1.053	−0.083	0.3345	−0.668

$$G_d = \frac{1}{2} G_{SC}\varepsilon\left[\bar{\tau}_{O3}\bar{\tau}_w\bar{\tau}_g(1 - \bar{\tau}_R\bar{\tau}_a) + 0.013\right]\cos\theta_z \tag{5.49b}$$

where the averaged atmospheric transmittances are expressed as:

$$\bar{\tau}_{O_3} = \exp\left[-0.0365 \cdot (\mathrm{ml}_{O_3})^{0.7136}\right] \tag{5.50a}$$

$$\bar{\tau}_w = \min[1,\ 0.909 - 0.036\ln(\mathrm{ml}_w)] \tag{5.50b}$$

$$\bar{\tau}_g = \exp\left[-0.0117 m^{0.3139}\right] \tag{5.50c}$$

$$\bar{\tau}_R = \exp\left[-0.008735 m_c\left(0.547 + 0.014 m_c - 3.8 \cdot 10^{-4} m_c^{c2} + 4.6 \cdot 10^{-6} m_c^3\right)^{-4.08}\right] \tag{5.50d}$$

$$\bar{\tau}_a = \exp\left[-m\beta\left(0.6777 + 0.1464 m\beta - 0.00626(m\beta)^2\right)^{-1.3}\right] \tag{5.50e}$$

In Eq. (5.50a) m is the optical AM that can be calculated with the Kasten and Yang formula (5.23) and m_c is the pressure corrected optical AM $m_c = mp/p_0$ where $p_0 = 1.013 \cdot 10^5$ Pais the normal pressure and p[Pa] is the surface atmospheric pressure.

Gueymard (2003a, b) evaluated 21 models and concluded that the PS model is one of the best broadband models whose accuracy is comparable to spectral radiative transfer models for calculating beam irradiance under clear sky. Paulescu and Schlett (2003, 2004) and Madkour et al. (2006) also ascertain the high performance of this model.

5.3.4.2 PS model

The PS model (Paulescu and Schlett 2003) is also a simplification by means of Eq. (5.48) of the Leckner's spectral. All the averaged transmittances are calculated with equation:

$$\bar{\tau} = \exp\left[-x\left(a + bx + cx^{-d}\right)\right] \tag{5.51}$$

where the coefficients for the atmospheric attenuators are listed in Table 5.3.

Beam and diffuse solar irradiance is expressed as:

$$G_b = G_{SC}\varepsilon\bar{\tau}_{O_3}\bar{\tau}_w\bar{\tau}_g\bar{\tau}_R\bar{\tau}_a\cos\theta_z \tag{5.52a}$$

$$G_d = \gamma G_{SC}\varepsilon\bar{\tau}_{O_3}\bar{\tau}_w\bar{\tau}_g(1-\bar{\tau}_R\bar{\tau}_a)\cos\theta_z \tag{5.52b}$$

At a first instance, for the downward fraction of the scattered radiation, a Rayleigh single-scattering value of $\gamma = 0.5$ can be used. Otherwise, γ may be treated as an empirical parameter and estimated from historical recorded data. In Paulescu and Schlett (2003), for the particular location of Timisoara (latitude 45°46′N, longitude 21°25′E, altitude 86 m above mean sea level), using data collected during 3 years, a value $\gamma = 0.432$ has been fitted.

Test results reported in Paulescu and Schlett (2003) demonstrate that the PS model performs with acceptable accuracy when global solar irradiance is estimated (RMSE < 10 %). Badescu et al. (2012) tested 54 solar irradiance models and concluded that the PS model is one of the best broadband models. Due to its good performance, the PS model has been adopted as a tool for calculating solar irradiance in various applications, e.g. Kim and Hogue (2008).

5.3.4.3 Parametric Solar Irradiance Model: PSIM

PSIM (Gueymard 1993b) has been derivated from CPCR2 (Gueymard 1989) a two-band physical solar irradiance model. In this model, direct normal G_n and global G solar irradiances are expressed as:

$$G_n = G_{SC}\varepsilon F_{bg}(l_w)F_b(p,\beta)\sum_{i=0}^{5} b_i \sin^i h \tag{5.53a}$$

$$G = G_{SC}\varepsilon F_{bg}(l_w)F_g(p,\beta)\sum_{i=0}^{4} g_i \sin^i h \tag{5.53b}$$

G is calculated by (5.53b) for a zonal albedo of 0.2. For a different albedo an adjustment factor is required $F_{bg}(l_w)$is a function of water vapor column content:

$$F_{bg}(l_w) = \exp\left[0.155\left(1 - l_w^{0.24}\right)\right] \tag{5.54}$$

and $F_b(p,\beta)$, $F_g(p,\beta)$ are functions of the station's pressure and Ångström turbidity coefficient β, that becomes 1.0 for a sea-level pressure:

$$F_b(p,\beta) = 1 + (0.1594 - 0.266\beta)(1 - p/p_0) \tag{5.55}$$

$$F_g(p,\beta) = 1 + (0.0752 - 0.107\beta)(1 - p/p_0) \tag{5.56}$$

The numerical coefficients b_i and g_i in Eq. (5.53) are expressed as a function of a modified turbidity coefficient $T = \ln(1 + 10\beta)$, such that:

Table 5.4 Coefficients c_{ki} used in Eq. (5.57a, b).(Gueymard 1993b)

β	i	0	1	2	3	4	5
<0.175	c_{0i}	−1.62364	2.94298	−8.12160	12.5571	−9.8044	3.00487
	c_{1i}	−6.87270	5.23504	−15.8000	25.4400	−20.3172	6.31760
	c_{2i}	–	−18.23861	69.2345	−123.3933	103.9906	−33.38910
	c_{3i}	–	11.16520	−45.1637	83.1014	−71.3091	23.15470
≥0.175	c_{0i}	–	4.41547	−18.4519	31.2506	−25.1876	7.64179
	c_{1i}	–	−5.09266	38.3584	−74.5384	64.3575	−20.41687
	c_{2i}	–	1.47187	−22.7449	48.3550	−43.6586	14.20502
	c_{3i}	–	0	4.3189	−9.8657	9.2315	−3.06053

$$b_0 = \begin{cases} \exp(c_{00} + c_{10}T) & \text{if} \quad \beta < 0.175 \\ 0 \quad \text{otherwise} \end{cases} \tag{5.57a}$$

$$b_i = \sum_{k=0}^{3} c_{ki} T^k, \quad i = 1,\ 2,\ 3,\ 4, 5 \tag{5.57b}$$

$$g_i = \begin{cases} \sum_{k=0}^{3} d_{ki} T^k\,, & i = 1,\ 2,\ 3,\ 4 \\ 0.006 & i = 0 \end{cases} \tag{5.58}$$

The numerical values of the coefficients c_{ki} and d_{ki} are listed in Tables 5.4 and 5.5, respectively.

If the station's pressure is not known, it can be computed from altitude z (in km) and latitude ϕ (in degree) using the following equation:

$$p = P(\phi) p_0 \exp\left(0.00177 - 0.11963z - 0.00136z^2\right) \tag{5.59}$$

where the factor $P(\phi)$ is:

$$P(\phi) = \sum_{i=0}^{2} c_{1i}\phi^i + z \sum_{i=0}^{2} c_{2i}\phi^i \tag{5.60}$$

with the coefficients $c_{10} = 0.993$, $c_{11} = 2.0783 \cdot 10^{-4,}$ $c_{12} = -1.1589 \cdot 10^{-6,}$ $c_{20} = 8.855 \cdot 10^{-3}$, $c_{21} = -1.5236 \cdot 10^{-4}$, and $c_{22} = -9.2907 \cdot 10^{-7}$. The Eq. (5.59) has been fitted for the northern hemisphere (0–90°N). It may be assumed that this equation also holds for the southern hemisphere, in which case the absolute value of ϕ should be used.

Table 5.5 Coefficients d_{ki} from Eq. (5.58) (Gueymard 1993b)

i	1	2	3	4
d_{0i}	0.38702	1.35369	−1.59816	0.66864
d_{1i}	−0.38625	1.53300	−1.90377	0.80172
d_{2i}	0.09234	−1.07736	1.63113	−0.75795
d_{3i}	0	0.2378	−0.38770	0.18895

5.3.4.4 ESRA: The Clear-Sky Irradiance Model of the European Solar Radiation Atlas

The ESRA model (Rigollier et al. 2000) has been developed with the main objective to estimate the solar radiation at ground level from satellite images with the Heliosat method. The Linke turbidity factor (Linke 1922) is a key point in this model. Before describing the model equations, a few words on the Linke turbidity factor follows.

The Linke turbidity factor (T_L, for an AM equal to 2) is a convenient approximation to represent the atmospheric absorption and scattering of solar radiation under clear sky. It describes the optical depth of the atmosphere due to both absorption by water vapor and absorption and scattering by the aerosols relative to a dry and clean atmosphere (Kasten 1996). In other words, T_L stands for number of normal atmospheres, clean and dry, necessary to reduce the extraterrestrial solar radiation to the value of the direct component at ground level. The increase of the T_L leads to the larger attenuation of the direct solar radiation by the clear-sky atmosphere. Therefore, T_L is an appropriate measure for the atmospheric turbidity. T_L can be obtained directly from observations performed during very clear-sky periods, but this kind of experimental data is rarely available, thus in general T_L is an estimated parameter.

In the ESRA model, the direct solar irradiance on a horizontal surface is given by:

$$G_b = G_{SC}\,\varepsilon \exp(-0.8662 m T_L K_R) \cos\theta_z \tag{5.61}$$

The factor $\exp(-0.8662 m T_L)$ represents the beam atmospheric transmittance, $\bar{\tau}_b$ in Eq. (5.47). K_R is the Rayleigh optical thickness for which the authors recommend the Kasten (1996) parameterization:

$$K_R(m) = \begin{cases} \dfrac{1}{6.6296 + 1.75513m - 0.1202m^2 + 0.0065m^3 - 0.00013m^4} & \text{if } m \le 20 \\ \dfrac{1}{10.4 + 0.718m} & \text{otherwise} \end{cases}$$

The diffuse solar irradiance is expressed as:

$$G_d = G_{SC}\,\varepsilon\,\bar{\tau}_{dz}(T_L) F_d(\theta_z, T_L) \tag{5.62}$$

where $\bar{\tau}_{dz}(T_L)$ is the diffuse transmittance at zenith ($\theta_z = 0$):

$$\bar{\tau}_{dz}(T_L) = -0.015843 + 0.030543 T_L + 3.797 \cdot 10^{-4} T_L^2 \tag{5.63}$$

Characteristic $\bar{\tau}_{dz}(T_L)$ ranges from 0.05 for very clear sky ($T_L = 2$) to 0.22 for very turbid atmosphere ($T_L = 7$). The factor $F_d(\theta_z, T_L)$ corrects $\bar{\tau}_{dz}(T_L)$ with respect to the actual zenith angle:

$$F_d(\theta_z,\ T_L) = A_0 + A_1 \cos\theta_z + A_2(\cos\theta_z)^2 \tag{5.64}$$

The unitless coefficients A_0, A_1, and A_2, only depend on the Linke turbidity factor and they are given by:

$$A_0 = \begin{cases} 0.26463 - 0.061581 T_L + 3.1408 \cdot 10^{-3} T_L^2 & \text{if} \quad A_0 \bar{\tau}_{dz} > 2 \cdot 10^{-3} \\ 2 \cdot 10^{-3} / \bar{\tau}_{dz} & \text{otherwise} \end{cases} \tag{5.65a}$$

$$A_1 = 2.0402 + 0.018945 T_L - 0.011161 T_L^2 \tag{5.65b}$$

$$A_2 = -1.3025 + 0.039231 T_L + 8.5079 \cdot 10^{-3} T_L^2 \tag{5.65c}$$

The global irradiance $G = G_b + G_c$ decreases as the turbidity increases and as the solar elevation decreases. In this model the global solar irradiance is not equal to zero at sunset or sunrise because of the diffuse component which is still noticeable even if the sun is below the horizon.

The four parametric models presented, Hybrid, PS, PSIM, and ESRA, are compared in Figs. 5.8 and 5.9. The same set of input parameters has been used for all models. A horizontal surface located at $\phi = 45°$N latitude and mean sea level altitude $z = 0$ has been considered. The following set of meteorological parameters has been assumed: normal atmospheric pressure $p = p_0$, ozone column content $l_{O_3} = 0.35$ cm · atm, water vapor column content $l_w = 2.3\ \text{g}/cm^2$, and Ångström turbidity coefficient $\beta = 0.079$. For running the ESRA model, the Linke turbidity factor has been determined by using the conversion function of Ineichen (2008) that relate it to the water vapor column content and the aerosol optical thickness.

Figure 5.8 shows the beam G_b, diffuse G_d, and global G solar irradiance as function of hour angle calculated with the four models in the days January 1 ($j = 1$) and July 1 ($j = 182$). Visual inspection shows that no major difference occurs when G_b is estimated in winter (Fig. 5.8a) but some differences in summer are noticed at small hour angles (Fig. 5.8b). The curves generated by the Hybrid and ESRA models are overlapped. These models also generate the highest values for G_b, while PS generate the smallest values. The PSIM model predicts G_b between these two extremes. At large hour angles all the models predict the same values of G_b. In the case of diffuse radiation no significant difference in the models estimations are noticed in the winter (Fig. 5.8c). Different, in the summer at small hour angles a significant scattering of the values G_d estimated by the four models occur (Fig. 5.8d). The highest values of G_d are estimated by the PSIM model,

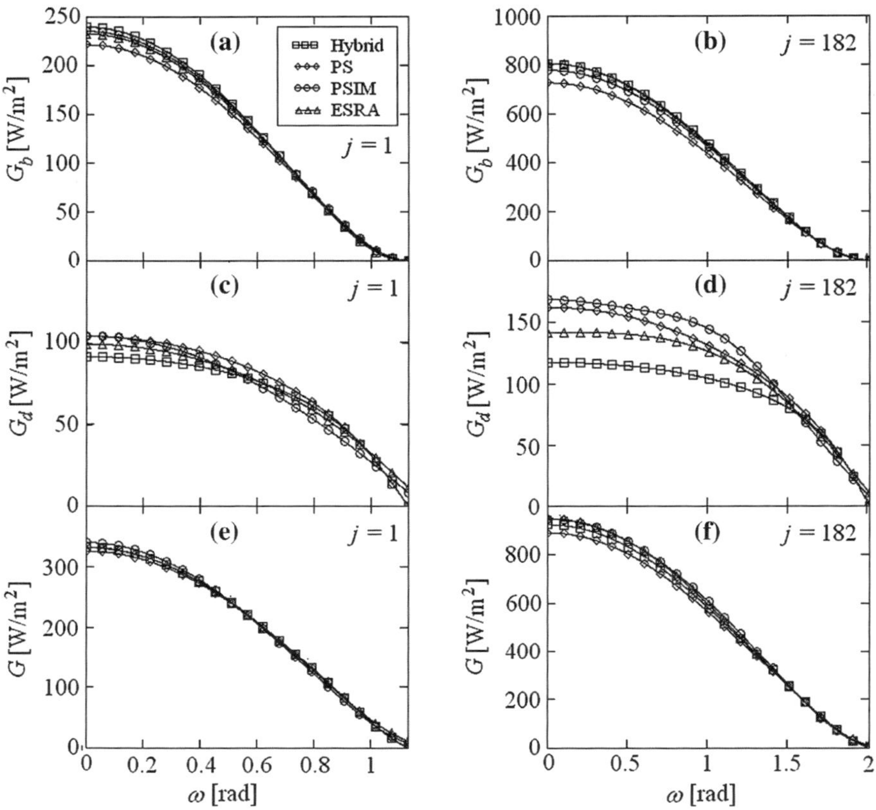

Fig. 5.8 Solar irradiance estimated by the four models Hybrid (Yang et al. 2001), PS (Paulescu and Schlett 2003), PSIM (Gueymard 1993b), and ESRA (Rigollier et al. 2000) as function of hour angles in two different days January 1 (**a**, **c**, **e**) and July 1 (**b**, **d**, **f**). Beam (**a**, **b**), diffuse (**c**, **d**), and global (**e**, **f**) components are displayed

closely followed by PS. The Hybrid model estimates the lowest values of G_d while the ESRA model returns something like the average between PSIM and Hybrid. When the global solar irradiance is estimated, the differences between the four models fade (Fig. 5.8e–f). To make a clear picture of the order of magnitude of differences between the predictions of the four models, Table 5.6 collects estimates at hourly angles 0, $\omega_0/2$, and $3\omega_0/4$.

Figure 5.9 displays the variation of the beam G_b, diffuse G_d, and global G solar irradiance in midday as function of Julian day in the first half of a year. Visual inspection shows that even if a large spreading of diffuse solar irradiance estimated by the four models is noted (Fig. 5.9b), the models estimate almost the same value for beam and global solar irradiance (Fig. 5.9a, c). From Fig. 5.9c PS model estimates the lowest values of solar irradiance while Hybrid and ESRA estimate highest values. Evenly spaced from the two extremes is the curve estimated by PSIM.

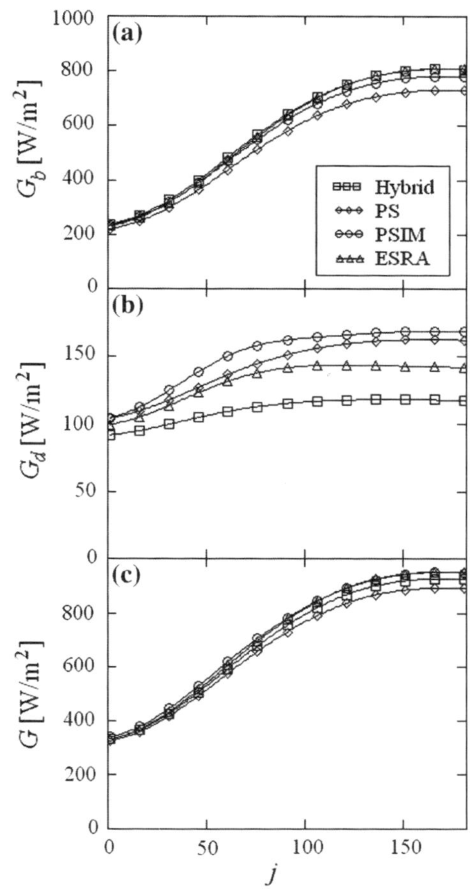

Fig. 5.9 Beam G_b, diffuse G_d, and global G solar irradiance in the midday as function of Julian day in the first half of the year

To conclude, this section introduced the class of parametric models mostly used in estimating solar irradiation. Like the spectral models, these models require meteorological parameters as input. The main advantage results from the simpler form of equations, which reduces the calculus to simple algebra. This allows a fast implementation in computer applications designated for various applications, including the forecast of the output power of PV systems. The comments regarding the successful implementation made for the case of spectral models at the end of Sect. 5.3.3.2 apply here as well.

5.3.5 Empirical Models for Solar Irradiance

Many clear-sky global solar irradiance models are constructed empirically, mainly by fitting the data recorded from an evanescent area which entails a close connection to their parental site. These models require only the geographical location

Table 5.6 Beam, diffuse, and global solar irradiance components estimated by the models Hybrid (Yang et al. 2001), PS (Paulescu and Schlett 2003), PSIM (Gueymard 1993b), and ESRA (Rigollier et al. 2000) in two different days at three different hour angles: zero, half ($\omega_0/2$), and three quarters of sunset hour angle ($3\,\omega_0/4$)

Model	Irradiance [W/m^2]	January 1			July 1		
		0	$\omega_0/2$	$3\,\omega_0/4$	0	$\omega_0/2$	$3\,\omega_0/4$
Hybrid	G_b	239.5	143.6	50.8	805.5	476.3	173.0
	G_d	91.8	78.8	55.3	117.4	104.4	79.6
	G	331.3	222.4	106.2	923.1	580.8	252.6
PS	G_b	221.4	135.4	50.5	727.5	432.1	161.5
	G_d	104.2	85.4	56.9	161.9	131.0	88.0
	G	325.6	220.3	107.5	889.4	563.1	249.6
PSIM	G_b	235.1	142.3	53.0	778.3	462.1	170.5
	G_d	104.5	78.2	47.0	168.2	144.3	84.2
	G	339.6	220.5	100.1	946.5	606.5	254.8
ESRA	G_b	232.5	140.2	51.9	86.3	468.1	168.2
	G_d	99.5	79.8	52.7	141.9	126.0	83.2
	G	332.0	220.1	104.6	948.3	594.1	251.4

and the time without any measured weather parameter. This is a considerable advantage when the user has limited access to measurements. Being very simple, empirical models are still widely preferred in various applications. Many papers deal with checking empirical equations, e.g. Badescu (1998) or Paulescu and Schlett (2004).

In the following, six solar irradiation models fitted with data collected in various parts of the world are listed. The first two models have two entries, solar zenith angle and altitude while the next four models have only one entry, namely the solar zenith angle.

1. *H*—model (Hottel 1976), with good accuracy and simple use, estimates the transmittance of direct solar radiation through clear sky. The model equation is:

$$G_b^H = G_{SC}\varepsilon\left[a_1 + a_2 \exp\left(-\frac{a_3}{\cos\theta_z}\right)\right]\cos\theta_z \tag{5.66}$$

The numerical coefficients are functions only of the altitude z (in km):

$$a_1 = 0.4327 - 0.00821(6 - z)^2 \tag{5.67a}$$

$$a_2 = 0.5055 + 0.00595(6.5 - z)^2 \tag{5.67b}$$

$$a_3 = 0.2711 + 0.01858(2.5 - z)^2 \tag{5.67c}$$

Hottel's model is applicable up to 2.5 km altitude.

2. *S*—model (Samimi 1994) is also a model for direct solar irradiance which takes into account the site altitude:

$$G_b^S = G_{SC}\varepsilon\left[(1-0.14z)\exp\left(-\frac{0.357}{(\cos\theta_z)^{0.678}}\right)+0.14zf(\theta_z)\right]\cos\theta_z \qquad (5.68)$$

where $f(\theta_z) = 1 - \exp\left[-\frac{36}{\pi}\left(\frac{\pi}{2}-\theta_z\right)\right]$, with the zenith angle θ_z in radians. This model uses a simple step function to compute the diffuse solar irradiance $G_d^S = 0.1G_b^S$. Then, the global solar irradiance is computed as:

$$G^S = 1.1\,G_b^S \qquad (5.69)$$

The model has been tested against long-term data measured in Teheran (Iran) proving a very good accuracy when clear-sky global solar irradiance is computed.

3. *B*—model (Bugler 1977) allows calculating the diffuse component of solar irradiance through a very simple function of zenith angle (in degrees):

$$G_d^B\left[\mathrm{W}/m^2\right] = 16\sqrt{90-\theta_z} - 0.4(90-\theta_z) \qquad (5.70)$$

Coupled with Hottel's model it provides satisfactory results when calculating the global solar irradiance in clear sky conditions.

4. *A*—model (Adnot et al. 1979) computes global solar irradiance in clear-sky conditions with the equation:

$$G^A\left[\mathrm{W}/m^2\right] = 951.39(\cos\theta_z)^{1.15} \qquad (5.71)$$

5. *PS*—model (Paulescu and Schlett 2004), is fitted using data recorded in clear sky conditions at Timişoara (Romania) and computes the global solar irradiance in W/m^2:

$$G^{PS} = G_{SC}\,\varepsilon[1-0.4645\cdot\exp(-0.69\cos\theta_z)]\exp\left(-\frac{0.05211}{\cos\theta_z}\right)\cos\theta_z \qquad (5.72)$$

6. *DPP*—model (Daneshyar 1978; Paltridge and Proctor 1976) calculates the direct solar irradiance with the equation:

$$G_b^{\mathrm{DPP}}\left[\mathrm{W}/m^2\right] = 950.2[1-\exp(-0.075(90-\theta_z))]\cos\theta_z \qquad (5.73)$$

where θ_z is in degrees. In this model the diffuse solar irradiance is expressed as:

$$G_d^{\mathrm{DPP}}\left[\mathrm{W}/m^2\right] = 14.29 + 21.04\left(\frac{\pi}{2}-\theta_z\right) \qquad (5.74)$$

Beam solar irradiance $G_b(\theta_z)$ estimated by the models *H*, *S*, and DPP in summer day ($j = 182$) as function of solar zenith angle is plotted in Fig. 5.10a. The same shape of curves for all three models can be noticed. Models *H* and *S* depends on altitude. Two curves are displayed for each of the two models, one at

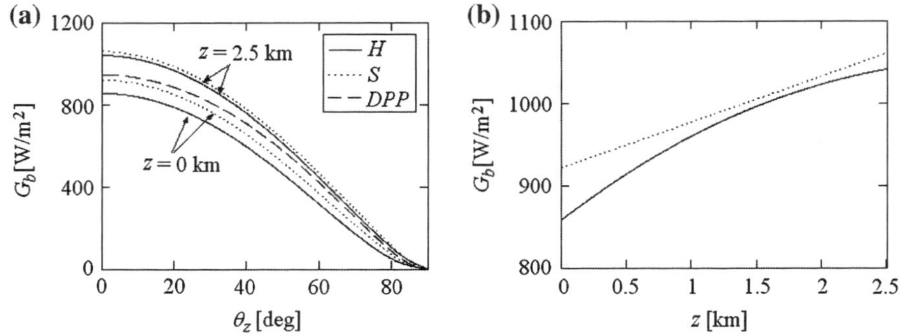

Fig. 5.10 Beam solar irradiance calculated with the models *H* (Hottel 1976), *S* (Samimi 1994), and *DPP* (Daneshyar 1978; Paltridge and Proctor 1976) with respect to **a** zenith angle and **b** site altitude

sea level and the other at 2.5 km altitude. The effect of altitude coverage into the model can be observed: as the altitude increase, the plot of $G_b(\theta_z)$ is raised to higher irradiance values. At small zenith angles, the difference between $G_b(\theta_z)$ at sea level and $G_b(\theta_z)$ at 2.5 km altitude amounts to about 200 W/m^2. The DPP model, being independent of altitude, generates a single curve $G_b(\theta_z)$ placed roughly midway between the curves generated by the model *H* at 0 and 2.5 km altitudes, overlapping the *H* model at 1 km altitude. By contrast, the beam irradiance predicted by the *DPP* model is close to the one estimated by the *S* model at sea level.

The dependence on altitude ($\theta_z = 0$) of the models *H* and *S* is presented in Fig. 5.10b, where for the lower elevations one can notice a significant difference between the estimated values of the two models.

Figure 5.11 shows the diffuse solar irradiance $G_d(\theta_z)$ estimated with the models *B*, *S* and DPP (Daneshyar 1978; Paltridge and Proctor 1976) with respect to the zenith angle. It can be seen that the curves are dispersed in a relative large domain of $G_d(\theta_z)$ values. For example, in almost all range of the zenith angle the diffuse irradiance estimated by the model *B* is greater than twice the estimation according to the DPP model.

Figure 5.12 shows the global solar irradiance $G(\theta_z)$ in a summer day ($j = 182$) with respect to the zenith angle, estimated by the models *HB* ($G_b(\theta_z)$ and $G_d(\theta_z)$ are calculated with *H* and *B* models, respectively), *S*, *A*, *PS*, and DPP. The altitude depending models have been run with $z = 0$. A visual inspection shows that the estimates for all models are close, regardless of the zenith angle.

To conclude, it follows from the analysis above that altitude coverage in an empirical model for estimating the solar irradiance under clear-sky conditions expands the area of application. While all the considered models give good estimations of global solar irradiance, significant differences are found in the estimation of direct and diffuse components. Therefore, the use of an empirical model in an application must be made after thoroughly testing against measured data.

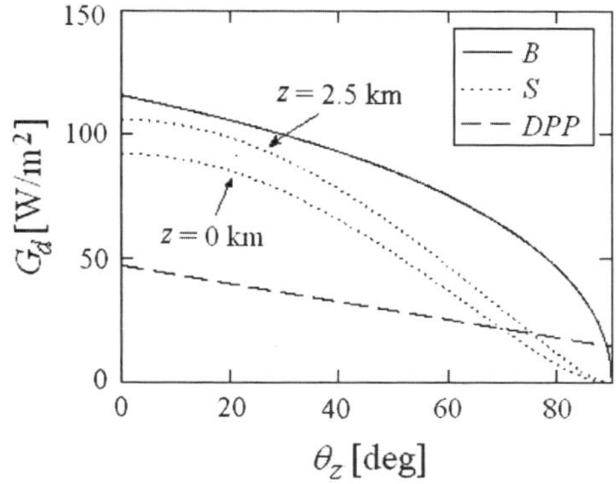

Fig. 5.11 Diffuse solar irradiance calculated with the models *B* (Bugler 1977), *S* (Samimi 1994), and *DPP* (Daneshyar 1978; Paltridge and Proctor 1976) with respect to the zenith angle

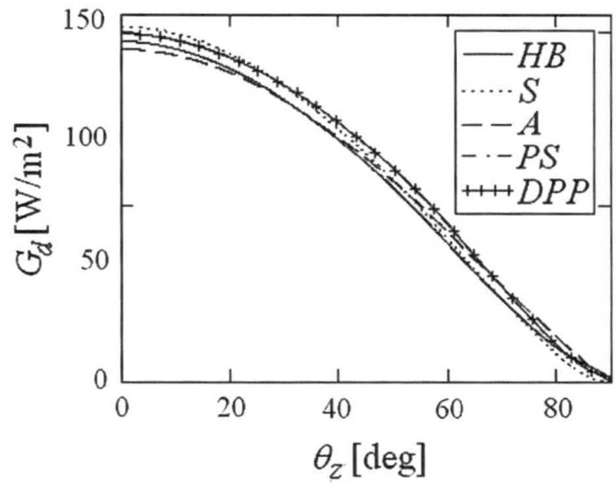

Fig. 5.12 Global solar irradiance calculated with the models *HB* (Hottel 1976; Bugler 1977), *S* (Samimi 1994), *A* (Adnot et al. 1979), *PS* (Paulescu and Schlett 2004), and *DPP* (Daneshyar 1978; Paltridge and Proctor 1976) with respect to the zenith angle

5.4 Computation of the Clear-Sky Solar Irradiation

The solar irradiation in a time interval $\Delta t = t_2 - t_1$ is usually calculated by integrating a solar irradiance model between the limits t_1 and t_2. For the calculation of daily solar irradiation, the integration is done between the times of sunrise and sunset; the collectable solar energy in a longer period of time is obtained by summing global irradiation daily values.

Thus, for a day of the year j, solar irradiation is calculated by the relationship:

$$H(j) = C \int_{-\omega_{0j}}^{\omega_{0j}} G(j, \omega) \mathrm{d}\omega \tag{5.75}$$

where the sunrise and sunset hour angles are given by Eq. (5.12). In (5.75), the constant C serves to adjust the units. For $C = 12/\pi$ hours/radian and G expressed in W/m^2, H results in Wh/m^2.

Equation (5.75) is general in a sense that it can be applied to all components of solar irradiance beam, diffuse, or global.

5.5 Cloud Amount Influence On Solar Radiation

The extinction of solar radiation due to the clouds is more significant than that due to any other atmospheric constituent, but it is always difficult to be modeled because of the random distribution of clouds on the sky. Moreover, the optical transmittance of clouds is in relation with their type, altitude, depth, and extension.

There are many measures related to the state of the sky, as previously discussed in Chap. 3. The most usual indicator is the total cloud cover amount (sometimes called cloudiness degree) C which represents the fractions of the celestial vault covered by clouds expressed in tens or octas. The total cloud cover amount definition assumes that the clouds are identical, homogenous, and have the same thickness. A major drawback of this type of indicator for the state of the sky is that C does not take into account the relative position of the sun and the clouds; for $C = 0.5$, half of the sky is covered with clouds, but C does not indicate whether the sun is shining or not in the sky. This problem has been solved in Badescu (2002) by introducing the sunshine number $\xi(t)$. Sunshine number properties are comprehensively revised in Chap. 3 of this book. Taking into account the sunshine number definition (Eq. 3.23), in a given time interval $\Delta t = t_2 - t_1$ the beam solar irradiation can be calculated with the equation:

$$H_b = \int_{t_1}^{t_2} G_b(t)\, \xi(t)\, \mathrm{d}t = \int_{s(\Delta t)} G_b(t)\mathrm{d}t \tag{5.76}$$

where $s(\Delta t)$ is the number of time units with the sun shining during Δt. One denotes by H_{b0} the beam solar irradiation in the hypothesis of clear-sky during the entire interval Δt. Then, a mean value of the beam solar irradiance during Δt can be defined:

$$H_{b0} = \int_{\Delta t} G_b \mathrm{d}t = \bar{G}_b\, \Delta t \tag{5.77}$$

Assuming that in the time Δt the beam solar irradiance $G_b(t)$ does not exhibit large variations (i.e. Δt is a proper short interval), Eq. (5.76) can be rewritten as:

$$H_b \cong H_{b0} \frac{s(\Delta t)}{\Delta t} = H_{b0}\, \sigma \tag{5.78}$$

where $\sigma = s(\Delta t)/\Delta t$ stands for the relative sunshine duration.

The same procedure may be applied for computing the diffuse solar irradiation. Apart from the sunshine number definition (see Eq. 3.23), here we introduce the cloud transmittance number $\varsigma(t)$ defined as a time-dependent two states variable, as follows:

$$\varsigma(t) = \begin{cases} \tau_C & \text{if the sun is covered by clouds at time } t \\ 1 & \text{otherwise} \end{cases} \tag{5.79}$$

In Eq. (5.79) τ_C is the clouds layer transmittance. Thus, the diffuse solar irradiation may be calculated with the equation:

$$H_d \cong H_{d0}\sigma + (H_{d0} + H_{b0})\bar{\tau}_C(1 - \sigma) \tag{5.80}$$

where H_{d0} is the diffuse solar irradiation in the time interval Δt in clear-sky conditions. When the sun is covered by clouds, the second term in Eq. (5.80) takes into consideration the contribution to H_d of both scattered by atmosphere and beam components of solar radiation.

Global solar irradiation is computed by summing up the results (5.78) and (5.80):

$$H = H_b + H_d \cong H_0[\tau_C + \sigma(1 - \tau_C)] \tag{5.81}$$

$H_0 = H_{b0} + H_{d0}$ is the clear-sky global solar irradiation in the time interval Δt.

Expression (5.81) is the famous Ångström equation (Ångström 1924) which linearly relate the ratio H/H_0 with the relative sunshine σ. In the original Ångström' equation:

$$\bar{H}/\bar{H}_0 = [\tau_C + \vec{\sigma}(1 - \tau_C)] \tag{5.82}$$

$\bar{H}$ stands for monthly mean of daily global solar irradiation and $\bar{H}_0$ stands for the monthly mean of daily global irradiation in clear-sky condition while $\bar{\sigma}$ represents the monthly mean of daily relative sunshine duration. The parameter τ_C is closely connected to location and varies according to the season. τ_C is of order 0.25 in the continental temperate climate and of order 0.3 in the tropics (Perrin de Brichambaut et al. 1988).

The Ångström equation has been modified by Prescott (1940) who replaced the daily clear-sky global irradiation H_0 with the daily extraterrestrial global solar irradiation, H_{ext}, establishing the classical form of the correlation:

$$\bar{H}/\bar{H}_{\text{ext}} = a + b\bar{\sigma} \tag{5.83}$$

where $\bar{H}$ and $\bar{H}_{\text{ext}}$ are the monthly average of daily solar irradiation at the ground and extraterrestrial level, respectively, and $\bar{\sigma}$ is the monthly mean of daily relative sunshine. a and b are empirical constants obtaining by fitting Eq. (5.83) to measured data. Usually, a takes values in the interval $0.2 \div 0.3$ and the sum $a + b$ lies between 0.65 and 0.8 (Perrin de Brichambaut et al. 1988). In Ref. Akinoglu and Ecevit (1990), based on the published values for 100 locations from all over the world, a global relation between the parameters a and b is reported:

$$a = 0.783 - 1.509b + 0.892b^2 \tag{5.84}$$

The linearity of Ångström–Prescott equation (5.83) leads to significant errors when the sky is mostly cloudy or overcast.

5.5.1 Relative Sunshine-Based Correlations

Over the years many papers have been published on this matter and in order to improve the accuracy of solar irradiation estimation Eq. (5.83) has been modified and related to other meteorological data. Some of proposed Ångström–Prescott (hereinafter denoted Å-P) are summarized below.

1. *P*—model (Page 1961) has given the coefficients of Å-P equation as follows:

$$\bar{H}/\bar{H}_{\text{ext}} = 0.23 + 0.48\bar{\sigma} \tag{5.85}$$

Eq. (5.85) is one of the most used worldwide.

2. *O*—model (Ögealman et al. 1984) added a nonlinear term to the Å-P equation, which transforms into a quadric expression:

$$\bar{H}/\bar{H}_{\text{ext}} = a + b\bar{\sigma} + c\bar{\sigma}^2 \tag{5.86}$$

Akinoglu and Ecevit (1990) reported conclusions of testing Eq. (5.86) with data collected over Turkey and found the best results occurring for the following parameters: $a = 0.195$, $b = 0.676$, and $c = -0.142$.

3. *S*—model (Samuel 1991) have suggested following polynomial correlation equations of order 3:

$$\bar{H}/\bar{H}_{\text{ext}} = -0.14 + 2.52\bar{\sigma} - 3.71\bar{\sigma}^2 + 2.24\bar{\sigma}^3 \tag{5.87}$$

The coefficients have been fitted with data collected at meteorological station in Sri Lanka.

4. *J*—model (Jin et al. 2005) based on the radiation data and the geographical information including latitude and altitude at 69 stations in China, have proposed nine equations like type Å-P, one being listed below:

$$\bar{H}/\bar{H}_{\text{ext}} = 0.1094 - 0.0014\phi + 0.0212z + (0.5176 + 0.0012\phi + 0.015z)\,\bar{\sigma}. \tag{5.88}$$

Equation (5.88) differs from the majority of Å-P equations. In this equation the coefficients a, b from Eq. (5.83) are functions of latitude ϕ (in degrees) and altitude z (in kilometers).

5. *A*—model (Almorox et al. 2005) reported the monthly-specific Å-P equations for estimating global solar irradiation from relative sunshine for Toledo, Spain. The monthly coefficients a, b are given in Table 5.7.

Table 5.7 Monthly coefficients in Eq. (5.83)

Month	Jan	Feb	Mar	Apr	May	Jun	Jul	Aug	Sep	Oct	Nov	Dec
a	0.285	0.272	0.291	0.266	0.286	0.311	0.329	0.313	0.271	0.259	0.279	0.282
b	0.444	0.465	0.491	0.495	0.475	0.439	0.406	0.410	0.479	0.465	0.431	0.428

6. *PS*—model Paulescu and Schlett (2003), in order to improve the Å-P equation for a wide range of cloudy and overcast situations, propose a power equation:

$$H/H_0 = a + b\sigma^c \tag{5.89}$$

In this equation H stands for daily global solar irradiation, H_0 stands for the daily clear-sky global solar irradiation, and σ is the daily relative sunshine. The coefficients $a = 0.2881$, $b = 0.7429$, and $c = 0.6168$ (Paulescu and Schlett 2003) have been fitted with daily data collected in Romania.

7. *Y*-model Yang et al. (2001). The hybrid model contains separate linear equations for each component of solar irradiation, beam, and diffuse. Thus, the daily global solar irradiation is computed as:

$$H = (a + b\,\sigma)H_b + (c + d\sigma)H_d \tag{5.90}$$

where the daily beam and diffuse solar irradiation is computed by the integration of Eqs. (5.49a) and (5.49b) between sunrise and sunset. The coefficients have been fitted using daily data from 16 stations spread over Japan territory: $a = 0.391$, $b = 0.518$, $c = 0.308$, and $d = 0.320$ if $\sigma > 0$ or $a = 0.222$ and $b = 0.199$ if $\sigma = 0$. The discontinuity at $\sigma = 0$ has been introduced to improve the correlation in overcast situation.

8. *I*—model (Iqbal 1979) used data measured in three locations from Canada to obtain the following Å-P correlation relating the ratio of monthly average of daily diffuse solar irradiation $\bar{H}_d$ and the monthly average of daily extraterrestrial solar irradiation $\bar{H}_{ext}$ to monthly average of daily relative sunshine $\bar{\sigma}$:

$$\bar{H}_d/\bar{H}_{ext} = 0.763 - 0.478\bar{\sigma} - 0.655\,\bar{\sigma}^2 \tag{5.91}$$

9. *L*—model (Lewis 1983), also for diffuse irradiation, has been obtained by fitting data collected from three stations in Zimbabwe:

$$\bar{H}_d/\bar{H}_{ext} = 0.754 - 0.654\bar{\sigma} \tag{5.92}$$

Few words on the above Å-P equations follow. Figure. 5.13a shows the graphs of Å-P correlation *P*, *O*, *S*, and *J*. The *J* model has been run for $\phi = 45°$ and $z = 0$. In the usual interval for$\bar{\sigma}$, 0.3…0.7, all four models exhibit the same almost linear behavior. The highest values of global irradiation are estimated by model *O* while the lowest values are estimated by model *J*. Figure 5.13b illustrates the dependence of the *J* model output on the geographical parameters. The model has been run for $\bar{\sigma} = 0.5$. As both parameters latitude and altitude increase, the estimation

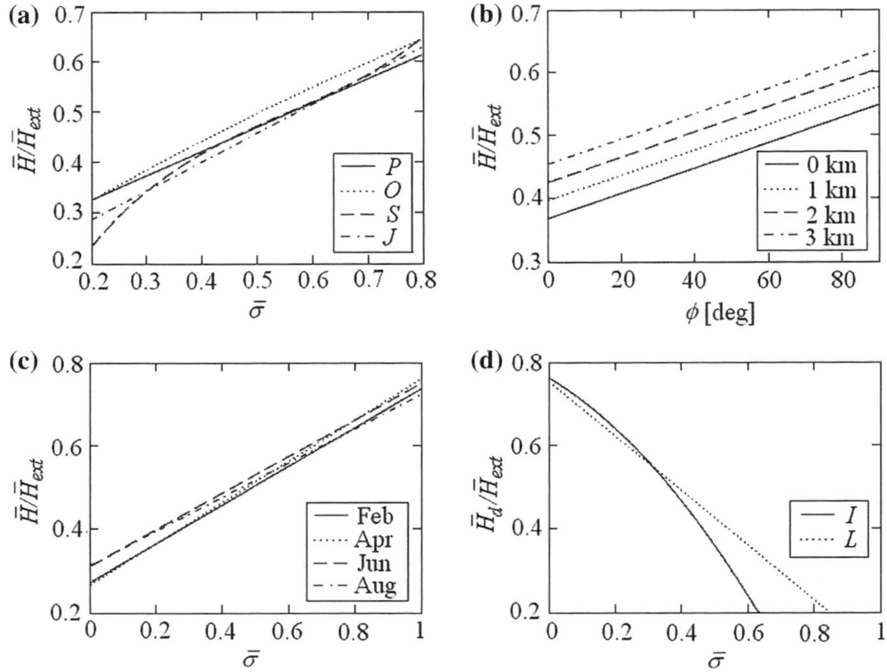

Fig. 5.13 Ångström–Prescott equations for global solar irradiation given by the models. **a** P (Page 1961), O (Ögealman et al. 1984), S (Samuel 1991), and J (Jin et al. 2005). **b** Dependence of the J model on latitude ϕ and altitude z. **c** Ångström–Prescott equations given by the model A for different moths. **d** Ångström–Prescott equations for diffuse solar irradiation given by the models I (Iqbal 1979) and L (Lewis 1983)

$\bar{H}/\bar{H}_{\text{ext}}$ is linearly shifted to higher values. Figure 5.13c shows the effect of generating monthly-specific parameters for Å-P equation by means of A model. A comparison between Å-P equations for diffuse irradiation given by the I and L models is presented in Fig. 5.13d. A noticeable difference occurs at higher values of $\bar{\sigma}$.

A historical perspective on Å-P correlation is presented by Martinez–Lozano et al. (1984). A review of 62 Å-P correlations based on relative sunshine is reported in Ref. Ahmad and Tiwari (2010).

5.5.2 Cloud Cover Amount Based Correlations

In many Ångström–Prescott like-type equations the total cloud cover amount C is used instead of the relative sunshine (Haurvitz 1945; Lumb 1964; Kasten and Czeplak 1980; Brinsfield et al. 1984). For example, Kasten and Czeplak (1980) correlate the ratio of daily global to daily clear-sky irradiation with total cloud cover amount by means of a power function:

$$H/H_0 = 1 - 0.72C^{3.2} \tag{5.93}$$

5.5.3 Air Temperature-Based Correlations

Although the air temperature is the most measured surface meteorological parameter around the world, it is not commonly used in solar irradiation models. However, the results reported in literature show that various indicators for the state of the sky may be based on air temperature. The temperature based Ångström–Prescott can be separated into two classes: one consisting of models which take into account the air temperature among other parameters when computing solar irradiation, and another consisting of models based only on air temperature as input. Fewer models are comprised in the second group than in the first. The subject of estimating and forecasting solar radiation is extensively treated in Chap. 8. Here, only a brief illustration of the two classes follows.

The models in the first group usually include daily extreme air temperature besides daily mean cloudiness (Supit and Van Kappel 1998; El-Metwally 2003) or sunshine duration (Chandel et al. 2005). Embedding the air temperature into the models is meant to increase the prediction quality, having the practical experience that prediction accuracy decays with increasing cloudiness. A representative model for this group is the Supit and Van Kappel equation:

$$H = H_{\text{ext}}\left(a\sqrt{t_{\max} - t_{\min}} + b\sqrt{1 - C}\right) + c \tag{5.94}$$

where t_{ma} and t_{min} are the daily maximum and minimum air temperatures and a, b, c are empirical parameters. Equation (5.94) was tested with a very large set of data recorded at 90 stations in Europe (Supit and Van Kappel 1998). The conclusion reached by the authors is that the Ångström equation, using relative sunshine, provides better estimates than their own equation. A similar conclusion was shared by El Metwally after three correlation tests at 7 stations in Egypt (El Metwally 2004). However, both works emphasizes that solar irradiation can be calculated with acceptable accuracy via air temperature and cloudiness, if no records of relative sunshine are at hand.

Let us illustrate the role of the air temperature in this type of models. With a dataset consisting of daily samples: maximum air temperature $t_{\max}$, minimum air temperature $t_{\min}$, mean value of total cloud amount C, solar irradiation H_j all recorded of Timisoara station (45.76°; 21.25°; 85 m), Romania, in 1998–2000, we calculate the best approximations for daily H/H_{ext} (Paulescu et al. 2011):

$$H(C)/H_{\text{ext}} = 0.5744 - 0.3866 \cdot C^{3.9122} \tag{5.95a}$$

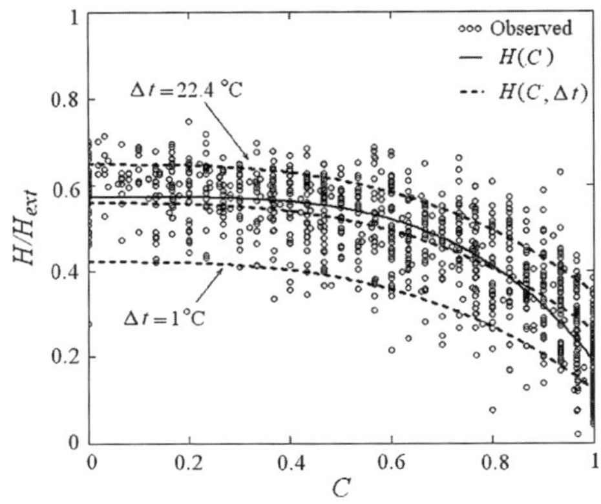

Fig. 5.14 H/H_{ext} computed with Eqs. (5.95a) (*line*) and (5.95b) (*dotted line*) in respect to cloud cover amount C. The *dotted lines* correspond to $\Delta t = 1$ °C (*down*), $\Delta t = 11.2$ °C (*middle*) and $\Delta t = 22.4$ °C (*up*). The 1,095 observed points used in the fitting processes are displayed in background. (From Paulescu et al. 2011, with permission from Wiley)

$$H(C, \Delta t)/H_{\mathrm{ext}} = 0.4233 - 0.3C^3 + 0.0232(\ln \Delta t)^2 \tag{5.95b}$$

The difference between Eqs. (5.95a) and (5.95b) consists of the presence of daily temperature amplitude $\Delta t = t_{\max} - t_{\min}$ in the second one. The effect of taking into account air temperature amplitude in the model is displayed in Fig. 5.14 where the Eqs. (5.95a, 5.95b) graph was superimposed on the observed points. We may notice that the curve $H(C)/H_{\mathrm{ext}}$ generated with Eq. (5.95a) is shifted by Eq. (5.95b) in a band $H(C, \Delta t)/H_{\mathrm{ext}}$ able to cover the almost entire observed dataset. The boundary curves correspond to the extremes Δt in the considered database ($\Delta t = 1$ °C down and $\Delta t = 22.4$ °C up) while the middle one, very close to the model which does not consider air temperature as input, corresponds to mean Δt. The result points out that for a given C, Δt acts as a refinement taking into account weather condition and it alone may justify the research in this field.

The models in the second group use only air temperature as input (Bristow and Campbell 1984; Donatelli and Bellocchi 2001; Paulescu et al. 2011). These models calculate daily solar clearness index mainly by using the daily air temperature average and amplitude as inputs. For example, Bristow–Campbell model exploits an exponential type relationship between the solar irradiation and the difference between a daily maximum and minimum temperature Δt (Bristow and Campbell 1984):

$$H/H_{\mathrm{ext}} = a[1 - \exp(-b(\Delta t)^c)] \tag{5.96}$$

More details on air temperature-based models are given in Chap. 8. Models of both groups, i.e. with and without other meteorological parameters besides air temperature at input, are reviewed. A section is dedicated to assessing models accuracy, a critical issue in using such models to forecast solar irradiation.

5.6 Solar Irradiance on Tilted Surfaces

The solar energy collected by a PV module depends not only on the density of the incident flux but also on the incidence angle. This dependence is a corollary of the Lambert's cosine law, which states that the measure of radiant energy from a surface that exhibits Lambertian reflectance is directly proportional to the cosine of the angle formed by the line of sight and the surface normal. It follows therefore, that solar irradiance incident on a surface varies similarly in respect to the cosine of the incidence angle. In the case of a horizontal surface, discussed so far in this chapter, the incidence angle is the same with the zenithal angle and the cosine law is described by Eq. (5.2). When the sunrays are perpendicular to the PV module surface, the flux density will always be at its maximum; for different incident angles the power on a horizontal PV module is less than that of the incident solar flux. This is the reason for which a large number of power plants are equipped with PV modules facing south and tilted to an angle close to the local latitude. The tilted angle is calculated to maximize the yearly amount of collected solar energy.

More than that, sometimes the PV modules are mounted on solar trackers. The main objective of all solar trackers is to minimize the angle of incidence, thus maximizing the collected solar energy.

Thus, in order to forecast PV power output an important task is the calculation of the energy collected by PV modules with an arbitrary spatial orientation. A key point in this task is the calculation of the incidence angle.

The inferring of the expression for the incidence angle θ for an arbitrarily oriented surface is treated in many papers, e.g. Braun and Mitchell (1983). Here we reproduce only the results.

$$\cos\theta = \cos\theta_z \cos\beta + \sin\theta_z \sin\beta \cos(\mu_s - \mu) \tag{5.97}$$

where β is the tilted angle of the surface with respect to the horizontal plane, θ_z is the zenithal angle, and μ and μ_s are the surface and sun azimuth angles, respectively. All of these angles are graphically illustrated in Fig. 2.2.

Important note: In the following, the angles that characterize the surface spatial orientation and the sun position have the meaning as in Fig. 2.2.

5.6.1 Estimation of Total Solar Irradiance

Total solar irradiance $G_t(\beta, \mu)$ on a surface with slope β and azimuth μ is calculated as a sum of three terms:

$$G_t(\beta, \mu) = G_b(\beta, \mu) + G_d(\beta, \mu) + G_r(\beta, \mu) \tag{5.98}$$

$G_b(\beta,\mu), G_d(\beta,\mu)$ and $G_r(\beta,\mu)$ are the beam, diffuse, and reflected components on the inclined surface, respectively. A description of the algorithm for converting horizontal solar irradiance components to the irradiance components on a tilted surface follows. $G_b(0), G_d(0)$ and $G(0)$ are the beam, diffuse, and global solar irradiation on the horizontal plane, respectively.

From simple geometric considerations, the beam term $G_b(\beta,\mu)$ can be expressed as function of the incidence angle θ of solar radiation on the tilted plane and the zenithal angle θ_z of the sun:

$$G_b(\beta,\mu) = G_b(0)\frac{\cos\theta}{\cos\theta_z} \tag{5.99}$$

The solar irradiance reflected by ground and received by the tilted surface $G_r(\beta,\mu)$ is expressed as:

$$G_r(\beta,\mu) = \rho\frac{1-\cos\beta}{2}G(0) \tag{5.100}$$

where ρ is the ground albedo. For ρ there are various models, ranging from the most simple isotropic assumption $\rho = 0.2$ to seasonal accounting for latitude and month of the year (Gueymard 1993b) or anisotropic effects (Gueymard 1987).

Even if different models can be selected for the calculation of the local albedo, the models governed by Eq. (5.98) differ in nature by the way in which the diffuse component is calculated. Five models of diffuse irradiance on tilted surfaces from horizontal irradiance are briefly described next.

1. The isotropic sky model (Liu and Jordan 1960) is the simplest model assuming that all diffuse radiation is uniformly distributed over the sky vault. The conversion factor $F_{\mathrm{LJ}} = G_d(\beta,\mu)/G_d(0)$ depends only on the surface tilt angle:

$$F_{\mathrm{LJ}} = \frac{1+\cos\beta}{2} \tag{5.101}$$

F_{LJ} given by Eq. (5.101) is plotted in Fig. 5.15.

2. In the Hay model (Hay 1979), diffuse solar radiation is calculated as a sum of the isotropic and circumsolar components. The circumsolar component is assumed to come from the sun's position. An anisotropy index:

$$F_a = G_b/G_{\mathrm{ext}} \tag{5.102}$$

is used to quantify the fraction of the diffuse radiation treated as circumsolar. As defined, the anisotropy index stands for the atmospheric transmittance associated to the beam component of solar radiation. In this model the conversion factor of diffuse solar irradiance is expressed as:

$$F_H = \frac{1+\cos\beta}{2}(1-F_a) + \frac{\cos\theta}{\cos\theta_z}F_a \tag{5.103}$$

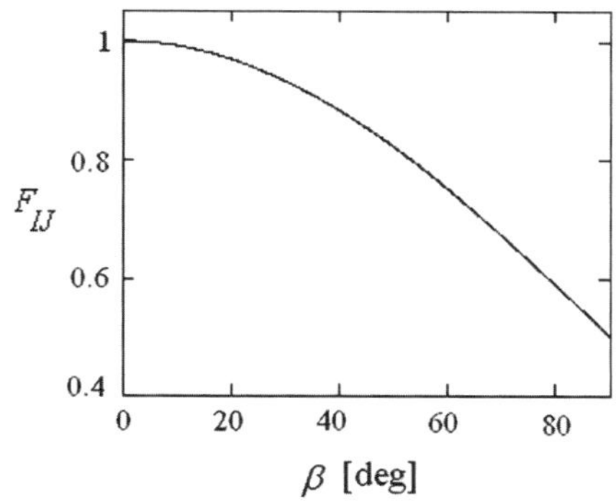

Fig. 5.15 Conversion factor F_{LJ} as function of surface tilt angle β given by Eq. (5.101)

In clear sky ($F_a > 0.5$) the model assumes that almost all energy comes from the sun's position while in overcast sky ($F_a \to 0$) the model assumes that the solar radiation is uniformly distributed on the sky. In all other cases the diffuse radiation is calculated by a weighted average of these two extreme states. Hay's model reduces to the Liu and Jordan model in the case of overcast sky. Hay's model (Eq. 5.103) introduces a much more complex conversion factor compared to the isotropic model of Liu and Jordan (Eq. 5.101). By the means of the zenithal angle F_H it is dependent of latitude and declination while by the anisotropy index F_a it is dependent on the state of the sky. Also, the incidence angle is included in this equation. To illustrate the way F_H is varying, Fig. 5.16 displays it as function of β in two cases. In both cases, it was assumed that the sun and surface azimuth are the same, $\mu_S = \mu$, which reduces Eq. (5.97) to $\theta = \theta_z - \beta$. The curves in Fig. 5.16a have been calculated for $F_a = 0.4$ and with the zenithal angle as parameter. Hay's model estimates that a tilted surface will receive more diffuse radiation as the zenithal angle increase. The curves in Fig. 5.16b have been calculated for $\theta_z = 20°$ (corresponding to the summer solstice at 45° latitude) and have parameter as the anisotropy index F_a. It can be seen that there is a threshold around $\beta_p = 70°$: for angles $\beta < \beta_p$ the estimated diffuse irradiance on the tilted surface increase with increasing atmospheric transparency, while for $\beta > \beta_p$ it slightly decreases.

3. In addition to isotropic diffuse and circumsolar radiation, the Reindl model (Reindl et al. 1990) accounts for horizon brightening and employs the same definition of the anisotropy index F_a as described in Eq. (5.102). The conversion factor reads:

$$F_R = \frac{1+\cos\beta}{2}(1-F_a)\left(1+\left(1-\frac{G_d(0)}{G(0)}\right)^{1/2}\sin^3\frac{\beta}{2}\right)+\frac{\cos\theta}{\cos\theta_z}F_a \qquad (5.104)$$

Due to the additional term in Eq. (5.104) representing horizon brightening, the Reindl model provides slightly higher diffuse irradiances than the Hay model.

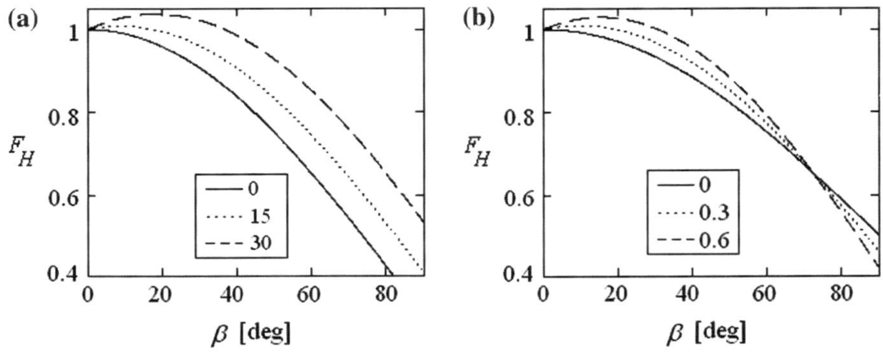

Fig. 5.16 Conversion factor of diffuse solar irradiance F_H in Hay's model (Eq. 5.103) as function of surface tilt angle. The curves parameter is: **a** zenithal angle θ_z [deg] and **b** anisotropy index F_a

4. The Muneer model (Muneer 1990) uses a conversion factor as:

$$F_M = F_T(1 - F) + F_T \frac{\cos\theta}{\cos\theta_z} \tag{5.105}$$

where the anisotropy indexes are $F_T = \frac{1+\cos\beta}{2} + N_1N_2$ and $F = \frac{G_b(0)\ \cos\theta_z}{G_{ext}}$ with $N_2 = \sin\beta - \beta\cos\beta - \pi\sin^2\frac{\beta}{2}$ and $N_1 = 0.00263 - 0.712F - 0.6883F^2$.

5. The Perez et al. model (1987) assumes that on an isotropic background there are superimposed circumsolar and brightening effects:

$$F_P = (1 - F_1)\frac{1 + \cos\beta}{2} + F_1\frac{\cos\theta}{\cos\theta_z} + F_2\sin\beta \tag{5.106}$$

where

$$\Delta = \frac{mG_d(0)}{G_{\text{ext}}},\ \varepsilon = \frac{G_d(0) + G_n(0)}{G_d(0)} \tag{5.107a, b}$$

and

$$F_1(\varepsilon) = f_{11} + f_{12}\Delta + f_{13}\theta_z, F_2(\varepsilon) = f_{21} + f_{22}\Delta + f_{23}\theta_z \tag{5.108a, b}$$

In the above relation, m stands for the atmospheric AM, θ_z (in radians) is the zenith angle, and $G_n(0)$ the normal component of horizontal solar irradiance. The quantities Δ (Eq. 5.107a, b) and ε (Eq. 5.107a, b) are parameters measuring sky brightness and atmospheric clearness, respectively. The coefficients $f_{i,j}$ in Eq. (5.108a, b) are listed in Table 5.8.

Figure 5.17 shows the variations of generic circumsolar (Eq. 5.108a, b) and horizon brightening (Eq. 5.108a, b) parameters as function of zenithal angle. Three sets of curves are plotted corresponding to ε classes 1, 4, and 7. This is a picture for $\Delta = 0.2$, other values of Δ will significantly change this.

Table 5.8 Coefficients f_{ij}, $i, j = 1,2,3$ in Eq. (5.108a, b) (Perez et al. 1987)

ε class	ε	f_{11}	f_{12}	f_{13}	f_{21}	f_{22}	f_{23}
1	0 ÷ 1.056	0.041	0.621	−0.105	−0.040	0.074	−0.031
2	1.056 ÷ 1.253	0.054	0.966	−0.166	−0.016	0.144	−0.045
3	1.253 ÷ 1.586	0.227	0.866	−0.250	0.069	−0.002	−0.062
4	1.586 ÷ 2.134	0.486	0.670	−0.373	0.148	−0.137	−0.056
5	2.134 ÷ 3.230	0.819	0.106	−0.465	0.268	−0.497	−0.029
6	3.230 ÷ 5.980	1.020	−0.260	−0.514	0.306	−0.804	0.046
7	5.980 ÷ 10.080	1.009	−0.708	−0.433	0.287	−1.286	0.166
8	>10.080	0.936	−1.121	−0.352	0.226	−2.449	0.383

Similar to the Hay model, in the Perez model the conversion factor from the horizontal to tilted diffuse solar irradiance F_P (Eq. 5.106) is influenced in a complex manner by the state of the sky and atmospheric parameters. Figure 5.18 illustrates the variation of F_P as function of surface tilt angle. At the first inspection, the curves profile is the same as in the Hay model. The difference between the graphs in Fig. 5.18a–c and e–f is given by the sky brightness, $\Delta = 0.2$ and $\Delta = 0.3$, respectively, while the difference between graphs in Figs. 5.18a, d and b, e and c, f is the zenithal angle $\theta_z = 0°$, .$\theta_z = 15°$ and $\theta_z = 30°$, respectively. The increase of sky brightness and zenith angle (toward sunrise or sunset times) determines in general an increase of the diffuse radiation received on the tilted plane. A clearer atmosphere reduces the diffuse component. These observations show that the Perez model is compatible with the common perception on solar radiation scattering in atmosphere; the merit of Perez model is the way in which it combines these influences to achieve a position among the most accurate models to compute solar radiation on tilted surfaces.

6. Artificial Intelligence techniques (see Chap. 7 for details) are also used for modeling solar irradiation on tilted surfaces. For example, Gazela and Tambouratzis (2002) reported a model for hourly average solar radiation on tilted surface via Artificial Neural Networks. Gomez and Casanovas (2003) proposed a model for solar irradiance on arbitrarily-oriented inclined surfaces based on fuzzy logic procedures. The model considers the circumsolar and horizon zones with the geometry of the Perez et al. (1987) model. The model likewise considers different sky categories for characterizing the different sky atmosphere transparency classes, but these classes are non-disjunctive. By using fuzzy logic, the clustering procedure is optimized and a reduced number of sky classes can be determined. The results of model assessment reported by authors (see Sect. 7.2.2) show that the fuzzy model offers performance similar to that of Perez's model.

There are many studies reported in literature from the beginning of PV systems to the present day dedicated to probing equations for calculating solar irradiance/irradiation on tilted surfaces (e.g. Klucher 1979; Santamouris et al. 1990; Behr 1997; Olmo et al. 1999; Kamali et al. 2006; Loutzenhiser et al. 2007; Noorian et al. 2008; Paulescu et. al 2010; Ibrahim et al. 2011; Chandel and Aggarwal 2011).

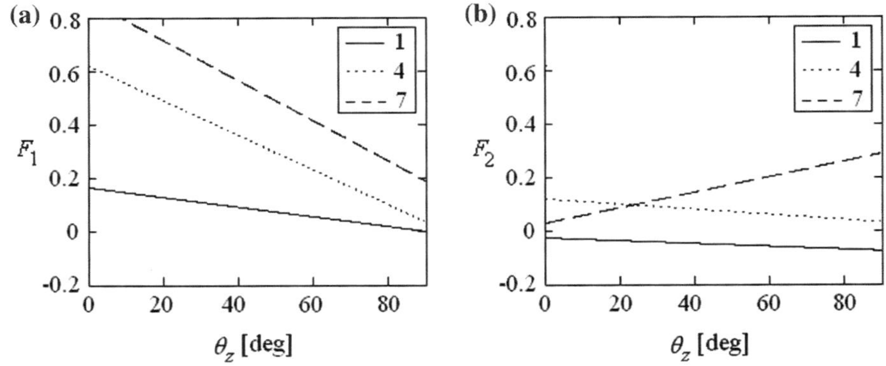

Fig. 5.17 Variation of the Perez model parameters: **a** circumsolar (Eq. 5.108a, b) and **b** horizon brightening (Eq. 5.108a, b) with respect to zenith angle. The curve parameter is the class of atmospheric clearness (Table 5.8)

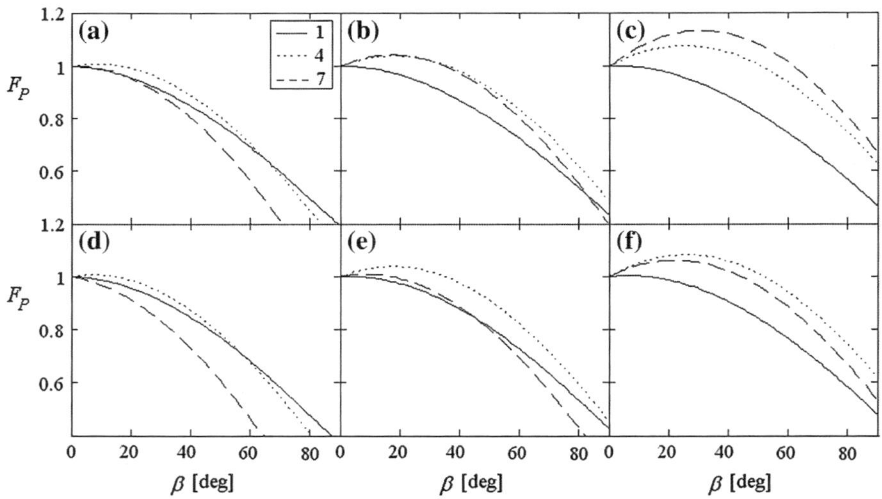

Fig. 5.18 Conversion factor of diffuse solar irradiance F_P in Perez's model (Eq. 5.106) as function of surface tilt angle for sky brightness: (**a**, **b**, **c**) $\Delta = 0.2$ and (**d**, **e**, **f**) $\Delta = 0.3$ and zenith angle (**a**, **b**) $\theta_z = 0$, (**c**, **d**) $\theta_z = 15$,and $\theta_z = 30°$. The curve parameter is the class of atmospheric clearness (Table 5.8)

In general, the conclusions are different and, thus, it is very difficult to recommend a model as universally applicable with high accuracy. Most discussed models are empirical and therefore perform with different accuracy in different places. On other hand, the models are tested in various places against data measured in different conditions and characterized by different uncertainty. Data quality may influence the results. However, on the whole, in numerous locations of the world, the model Perez et al. (1987) was frequently found as the best performing.

5.6.2 Solar Irradiance on Surfaces Tracking the Sun

In many applications, PV modules are mounted in sun tracking systems. The motivation consists in increasing of the amount of collected solar energy by minimizing the angle of incidence of beam radiation on the PV module surface and thus maximizing the incident beam radiation. To compute the solar irradiance available on these moving surfaces, the incidence angle should be calculated.

The tracking systems are classified in two classes by their motion. Single axis trackers have one degree of freedom that acts as an axis of rotation. The axis of rotation of single axis trackers is typically aligned along a true North meridian. Dual axis trackers have two degrees of freedom that act as axes of rotation. These axes are typically normal one to another. The axis that is fixed with respect to the ground can be considered a primary axis.

5.6.2.1 Single Axis Trackers

Equations for the incidence angle θ can be found in many sources, like the very clear presentation in chapter one from Duffie and Beckman (1991).

There are several common implementations of single axis trackers. These include horizontal single axis trackers, vertical single axis trackers, tilted single axis trackers, and polar aligned single axis trackers.

Equation (5.97) is general and gives the angle of incidence for both fixed and tracking surfaces in terms of surface slope and azimuth.

For a plane rotated around a horizontal east–west axis with a single daily adjustment so that the beam radiation is normal to the surface at noon each day, the incidence angle is:

$$\cos\theta = \sin^2\delta + \cos^2\delta\cos\omega \tag{5.109}$$

The slope β of this surface $\beta = |\phi - \delta|$ is fixed in each day. The surface azimuth angle μ for a day depends on latitude ϕ and declination δ: $\mu = 0°$ if $\phi - \delta > 0$ or else $\mu = 180°$.

For a plane rotated around a horizontal east–west axis with continuous adjustment the angle of incidence is:

$$\cos\theta = \left(1 - \cos^2\delta\sin^2\omega\right)^{1/2} \tag{5.110}$$

The slope of this surface is given by $\tan\beta = \tan\theta_z|\cos\mu_s|$ and the surface azimuth μ angle will switch between 0° and 180° if the solar azimuth angle μ_s passes through ±90°: $\mu = 0°$ if $|\mu_s| < 90°$ or else $\mu = 180°$.

For a plane with a fixed slope β rotated around a vertical axis, the angle of incidence is minimized when the surface and solar azimuth angles are equal $\mu = \mu_s$. From Eq. (5.97) the angle of incidence is:

$$\cos\theta = \cos(\theta_z - \beta) \quad (5.111)$$

For a plane rotated around the north–south axis parallel to the earth's axis with continuous adjustment

$$\cos\theta = \cos(\delta) \quad (5.112)$$

The surface slope varies continuously and is $\tan\beta = \tan\phi/\cos\mu$; the surface azimuth angle is:

$$\mu = \arctan\left(\frac{\sin\theta_z \sin\mu_s}{\cos\gamma \sin\phi}\right) + 180\, c_1 c_2 \quad (5.113)$$

where $\cos\gamma = \cos\theta_z \cos\phi + \sin\theta_z \sin\phi$ and:

$$c_1 = \begin{cases} 0 & \text{if} \quad \arctan\left(\dfrac{\sin\theta_z \sin\mu_s}{\cos\gamma \sin\phi}\right) + \gamma_s = 0 \\ 1 & \text{otherwise} \end{cases}, \qquad c_2 = \begin{cases} 1 & \text{if} \quad \mu_s \geq 0 \\ -1 & \text{otherwise} \end{cases} \quad (5.114\text{a, b})$$

5.6.2.2 Two Axis Trackers

There are several common implementations of dual axis trackers. They are classified by the orientation of their primary axes with respect to the ground. Two common implementations are tip-tilt dual axis trackers and azimuth-altitude dual axis trackers.

For a two axis tracking plane, the angle of incidence is minimized:

$$\cos\theta = 1 \quad (5.115)$$

and $\beta = \theta_z$ and $\mu = \mu_s$.

At present, sun tracking mechanisms are little used in the photovoltaic industry but it is expected that in future they become more common. There are several PV power plants in the world that exploit sun trackers, such as the Nellis Solar Power Plant, located within Nellis Air Force Base military base in Clark County, Nevada, north–west of Las Vegas. The system was inaugurated in 2007 and has an installed capacity of 14 MWp.

5.6.3 Comparison of Energy Collected on Surfaces with Different Orientations

The question that remains is what we gain if we collect the solar radiation on a surface that follows the path of the sun compared to a fixed surface. The answer is given in Fig. 5.19 where R is the ratio of solar energy collected in a year on four different surfaces to the energy collected on the same surface placed horizontally, depending on latitude. The four different surfaces are: two axis tracker, horizontal single axis tracker, polar axis tracker, and one fixed, facing south and optimally inclined. The graphs are raised with measured data reported in Ref. Lorenzo (2003) for 22 sites located in the northern hemisphere between 0 and 60° latitude. At first sight, with increasing latitude, R increases. Power equations that best fit the experimental data are also represented in Fig. 5.19. These are: (a) Two axis: $R(\phi) = \phi^{1.1}$; (b) Horizontal axis: $R(\phi) = -0.3 + 0.915\phi^{0.11}$; (c) $R(\phi) = -0.92 + 1.757\phi^{0.076}$; (d) Fixed, optimally tilted: $R(\phi) = -0.8 + 1.907\phi^{0.031}$. Obviously, the best results are obtained for two axis tracker (Fig. 5.19a), with R ranging between 1.22 at 6.2°N latitude and 1.72 at 59.9°N latitude. The inclination of the collector surface even at optimum angle does not bring always a benefit (Fig. 5.19d). One can find sub-unitary values for lower than 20° latitude, but such values can also be met at higher latitudes. The ratio R is determined mainly by the weight of the direct component in the global radiation on horizontal plane. An explanation in this regard follows considering the clearness index.

Figure 5.20 displays the ratio R as function of clearness index and the best fit linear equations: (a) Two axis: $R(k_t) = 0.75k_t + 1.06$; (b) Horizontal axis: $R(k_t) = 0.8k_t + 0.9$; (c) $R(k_t) = 0.82k_t + 0.92$; (d) Fixed, optimally tilted: $R(k_t) = 1.59k_t + 0.19$. Solar energy collected in a year with a two axis tracker increases less with increasing k_t, but even in cloudier environment reaches a high value. The energy collected by a sun tracker either in polar or horizontal configuration is almost equal given the same value of k_t. In this situation, the rate of R increase with k_t to a certain extent greater than for the two axis tracker. Optimally tilted fixed surfaces have the greatest rate of R increase, but also the lowest values in cloudier environments. This behavior is determined by the value of the incidence angle θ of the beam normal solar flux. For two axis tracker, θ is always zero (Eq. 5.115) and, therefore, the direct beam solar flux G_n is always perpendicular to the collector surface. In the case of a tilted surface, the incidence angle (Eq. 5.97) varies continuously and the direct beam solar flux is diminished with cosθ. Thus, in a cloudier environment, a two axis sun tracker always receives G_n integrally while a fixed tilted surface receives a percent of G_n which is strongly dependent on location, season, and day moment.

Finally, it is important to add that large PV systems have several rows of modules mounted above the ground. If the separation between rows is increased, fewer shadows are cast by some rows on the others and more energy is produced. But it also affects the cost, as greater separations lead to more land occupation, longer cables, and more expensive civil works. There is a widely held view that

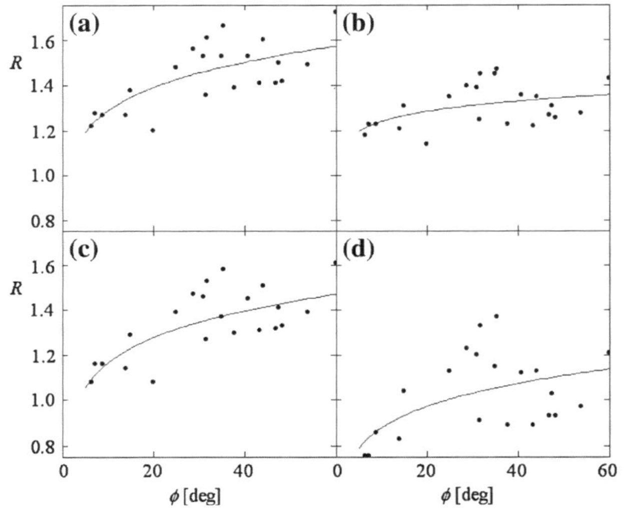

Fig. 5.19 Yearly solar irradiation availability (ratio to global horizontal yearly radiation) on different surfaces as function of latitude. **a** Two axis tracker. **b** Horizontal axis. **c** Polar axis. **d** Fixed, optimal oriented. Data collected in 22 northern sites around the world (*points*) and the fitted equations (*lines*) are shown. Source of data Lorenzo (2003)

tracking generally requires considerable more land than static arrangements (Lorenzo 2003). However, this is not necessarily the case with horizontal one-axis tracking. In this topic, the outstanding paper of Narvarte and Lorentzo (2008) presents the results of simulating the energy yield of flat modules for different tracking strategies as a function of land occupation.

5.7 Summary and Conclusion

The algorithm presented in Fig. 5.1 can be put into practice with the models described in this section and, in principle, allows calculation of solar irradiation on any surface in any arbitrary time. Algorithm can be implemented quickly using simplified models or software packages can be developed using more complex models. Basically, the choice of models depends first on the physical situation (i.e. solar irradiance, irradiation, on horizontal, tilted, or tracking surface) but mostly on the readiness of the required input data.

In principle, the models covered by this chapter are models for estimating solar radiation. However, they can be used directly or indirectly in the forecasting of solar radiation. For example, very simple empirical models (Sect. 5.3.5) may predict direct clear-sky global solar irradiance. More complex models (Sects. 5.3 and 5.4) may be used for nowcasting solar irradiance, given that the forecasted values of weather parameters are available. This is still an open research area. The

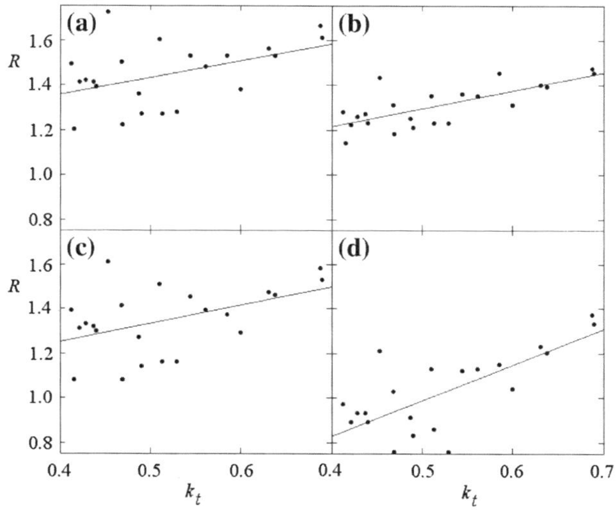

Fig. 5.20 Yearly solar irradiation availability (ratio to global horizontal yearly radiation) on different surfaces as function of the clearness index. **a** Two axis tracker. **b** Horizontal axis. **c** Polar axis. **d** Fixed, optimal oriented. Data collected in 22 northern sites around the world (*points*) and the fitted equations (*lines*) are shown. Source of data Lorenzo (2003)

equations in Sect. 5.6 for translating solar irradiance from horizontal plane to various oriented surface of interest in solar power systems can also be applied to forecasted values.

References

Adnot J, Bourges B, Campana D, Gicquel R (1979) Utilisation des courbes de frequence cumulees pour le calcul des installation solaires. In: Analise Statistique des Processus Meteorologiques Appliquee a l'Energie Solaire, Lestienne R. Paris, pp 9–40

AERONET (2012) Aerosol robotic network. Aerosol optical depth. Available at http://aeronet.gsfc.nasa.gov/new_web/aerosols.html

Ahmad MJ, Tiwari GN (2010) Solar radiation models—review. Int J Energy Environ 1(3):513–532

Akinoglu BG, Ecevit A (1990) Construction of a quadric model using modified Ångström coefficients to estimate global solar radiation. Sol Energy 45(2):85–92

Alados I, Foyo-Moreno I, Olmob FJ, Alados-Arboledas L (2002) Improved estimation of diffuse photosynthetically active radiation using two spectral models. Agr Forest Meteorol 111:1–12

Almorox J, Benito M, Hontoria C (2005) Estimation of monthly Angstrom–Prescott equation coefficients from measured daily data in Toledo, Spain. Renewable Energy 30:931–936

Ångström A (1924) Solar and terrestrial radiation. Q J Roy Meteorol Soc 50:121

Ångström A (1961) Techniques of determining the turbidity of the atmosphere. Tellus 13:214–223

Badescu V (1998) Verification of some very simple clear and cloudy sky models to evaluate global solar irradiance. Sol Energy 61:251–264

Badescu V (2002) A new kind of cloudy sky model to compute instantaneous values of diffuse and global solar irradiation. Theor Appl Climatol 72:127–136

Badescu V, Gueymard CA, Cheval S, Oprea C, Baciu M, Dumitrescu A, Iacobescu F, Milos I, Rada C (2012) Computing global and diffuse solar hourly irradiation on clear sky. Review and testing of 54 models. Renew Sust Energy Rev 16:1636–1656

Behr HD (1997) Solar radiation on tilted south oriented surfaces: validation of transfer-models. Sol Energy 61:399–413

Bird RE (1984) A simple solar spectral model for direct-normal and diffuse horizontal irradiance. Sol Energy 32:461–471

Bird RE, Riordan C (1986) Simple solar spectral model for direct and diffuse irradiance on horizontal and tilted planes at the Earth's surface for cloudless atmospheres. J Climate Appl Meteorol 25:87–97

Braun JE, Mitchell JC (1983) Solar geometry for fixed and tracking surfaces. Sol Energy 31:439–444

Brinsfield R, Yaramangolu M, Wheaton F (1984) Ground level solar radiation prediction including cloud cover effects. Sol Energy 33:493–499

Bristow KL, Campbell GS (1984) On the relationship between incoming solar radiation and daily maximum and minimum temperature. Agr Forest Meteorol 31:159–166

Bugler JW (1977) The determination of hourly insolation on an inclinated plane using a diffuse irradiance model based on hourly measured global horizontal insolation. Sol Energy 19:477–491

Chandel SS, Aggarwal RK, Pandey AN (2005) New correlation to estimate global solar radiation on horizontal surfaces using sunshine hour and temperature data for Indian sites. J Sol Energy Eng 127(3):417–420

Chandel SS, Aggarwal RK (2011) Estimation of hourly solar radiation on horizontal and inclined surfaces in western Himalayas. Smart Grid Renew Energy 2:45–55

Daneshyar M (1978) Solar radiation statistics for Iran. Sol Energy 21:345–349

Donatelli M, Bellocchi G (2001) Estimate of daily global solar radiation: new developments in the software Rad-Est3.00. In: Proceedings of 2th International Symposium Modelling Cropping Systems, Florence, pp. 213–214

El-Metwally M (2003) Simple new methods to estimate global solar radiation based on meteorological data in Egypt. Atmos Res 69:217–239

Fröhlich C (1991) History of solar radiometry and the world radiation reference. Metrologia 28:111–115

Gazela M, Tambouratzis T (2002) Estimation of hourly average solar radiation on tilted surface via ANNs. Int J Neural Syst 12(1):1–13

Gomez V, Casanovas A (2003) Fuzzy modeling of solar irradiance on inclined surfaces. Sol Energy 75:307–315

Gueymard C (1987) An anisotropic solar irradiance model for tilted surfaces and its comparison with selected engineering algorithms. Sol Energy 38:367–386

Gueymard C (1989) A two-band model for the calculation of clear sky solar irradiance, illuminance, and photosynthetically active radiation at the Earth surface. Sol Energy 43:253–265

Gueymard CA (1993a) Development and performance assessment of a clear sky spectral radiation model. In: Proceedings of 22nd ASES Conference on Solar 1993, Washington, DC, American Solar Energy Society, pp. 433–438

Gueymard CA (1993b) Mathematically integrable parameterization of clear-sky beam and global irradiances and its use in daily irradiation applications. Sol Energy 50:385–397

Gueymard CA (1995) SMARTS2-A simple model of the atmospheric radiative transfer of sunshine: algorithms and performance assessment. In: Florida Solar Energy Center Rep. FSEC-PF-270-95. Available at http://instesre.org/GCCE/SMARTS2.pdf

Gueymard CA (2003a) Direct solar transmittance and irradiance predictions with broadband models. Part I Detailed theoretical performance assessment. Sol Energy 74:355–379

Gueymard CA (2003b) Direct solar transmittance and irradiance predictions with broadband models. Part II Validation with high quality measurements. Sol Energy 74:381–395

Hay JE (1979) Calculation of monthly mean solar radiation for horizontal and inclined surfaces. Sol Energy 23:301–330
Haurvitz B (1945) Insolation in relation cloudiness and cloud density. J Meteorol 2:154–156
Hottel HC (1976) A simple model for estimating the transmittance of direct solar radiation through clear atmosphere. Sol Energy 18:129–139
Hu R-M, Martin RV, Fairlie TD (2007) Global retrieval of columnar aerosol single scattering albedo from space-based observations. J Geophys Res 112:D02204
Ibrahim A, El-Sebaii A, Ramadan MRI, El-Broullesy SM (2011) Estimation of solar irradiance on inclined surfaces facing south in Tanta. Egypt Int J Renew Energy Res 1(1):18–25
Iqbal M (1979) Correlation of average diffuses and beam radiation with hours of bright sunshine. Sol Energy 23:169–173
Ineichen P (2008) Conversion function between the Linke turbidity and the atmospheric water vapor and aerosol content. Sol Energy 82:1095–1097
Jin Z, Yezheng W, Gang Y (2005) General formula for estimation of monthly average daily global solar radiation in China. Energy Convers Manage 46:257–268
Kamali Gh, Moradi I, Khalili A (2006) Estimation solar radiation on tilted surfaces with various orientations: a study case in Karaj (Iran). Theor Appl Climatol 84:235–241
Kasten F, Czeplak G (1980) Solar and terrestrial radiation dependent on the amount and type of cloud. Sol Energy 24:177–189
Kasten F, Young AT (1989) Revised optical air mass tables and approximation formula. Appl Opt 28:4735–4738
Kasten F (1996) The Linke turbidity factor based on improved values of the integral Rayleigh optical thickness. Sol Energy 56:239–244
Kim JY, Hogue TS (2008) Evaluation of a MODIS based potential evapotranspiration product at the point scale. J Hydrometeorol 9(3):444–460
Klucher TM (1979) Evaluation of models to predict insolation on tilted surface. Sol Energy 23:111–114
Leckner B (1978) The spectral distribution of solar radiation at the earth's surface—elements of a model. Sol Energy 20:143–150
Lewis G (1983) Diffuse irradiation over Zimbabwe. Sol Energy 31:125–128
Linke F (1922) Transmissions-Koeffizient und Trübungsfaktor. Beitraege zur Physik der Freien Atmosphaere 10:91–103
Liu BYH, Jordan RC (1960) The interrelationship and characteristic distribution of direct, diffuse and total solar radiation. Sol Energy 4:1–12
Lorenzo E (2003) Energy collected and delivered by PV modules. In: Luque A, Hegedus S (eds) Handbook of photovoltaics science and engineering. Wiley, Chichester, pp 905–970
Lumb FE (1964) The influence of cloud on hourly amounts of total solar radiation at the sea surface. Q J Roy Meteorol Soc 90:383–393
Loutzenhiser PG, Manz H, Felsmann C, Strachan PA, Frank T, Maxwell GM (2007) Empirical validation of models to compute solar irradiance on inclined surfaces for building energy simulation. Sol Energy 81:254–267
Madkour MA, El-Metwally M, Hamed AB (2006) Comparative study on different models for estimation of direct normal irradiance (DNI) over Egypt atmosphere. Renewable Energy 31:361–382
Martinez-Lozano JA, Tena F, Onrubia Je, Delarubia J (1984) The historical evolution of the Ångström formula and its modifications: review and biography. Agr Forest Meteorol 33:109–128
Muneer T (1990) Solar radiation model for Europe. Build Serv Eng Res Technol 11(4):153–163
Narvarte L, Lorentzo E (2008) Tracking and ground cover ratio. Prog Photovoltaics Res Appl 16:703–714
Noorian AM, Moradi I, Kamali GA (2008) Evaluation of 12 models to estimate hourly diffuse irradiation on inclined surfaces. Renewable Energy 33:1406–1412
Ögealman H, Ecevit A, Tasdemiroglu E (1984) A new method for estimating solar radiation from bright sunshine data. Sol Energy 33(6):619–625

Olmo FJ, Vida J, Foyo I, Castro-Diez Y, Alados-Arboledas L (1999) Prediction of global irradiance on inclined surfaces from horizontal global irradiance. Energy 24:689–704
Page JK (1961) The estimation of monthly mean values of daily total short wave radiation on vertical and inclined surface from sunshine records for latitudes 40°N–40S. Proc UN Conf New Sour Energy 4:378–390
Paltridge GW, Proctor D (1976) Monthly mean solar radiation statistics for Australia. Sol Energy 18:235–243
Paulescu M, Schlett Z (2003) A simplified but accurate spectral solar irradiance model. Theor Appl Climatol 75:203–212
Paulescu M, Schlett Z (2004) Performance assessment of global solar irradiation models under Romanian climate. Renewable Energy 29:767–777
Paulescu M, Dughir C, Tulcan-Paulescu E, Lascu M, Gravila P, Jurca T (2010) Solar radiation modeling and measurements in Timisoara, Romania: data and model quality. Environ Eng Manage J 8:1089–1095
Paulescu M, Tulcan-Paulescu E, Stefu N (2011) A temperature based model for global solar irradiance and its application to estimate daily irradiation values. Int J Energy Res 35:520–529
Paulescu E, Stefu N, Gravila P, Boata R-St, Pop N, Paulescu M (2012) Procedure of embedding biological action functions into the atmospheric transmittance. Theor Appl Climatol. doi:10.1007/s00704-011-0581-y
Perez R, Seals R, Ineichen P, Stewart P, Menicucci D (1987) A new simplified version of the Perez diffuse irradiance model for tilted surfaces. Sol Energy 39:221–223
Perrin de Brichambaut C, Bourges B, Renaudin MG, Villien C (1988) Ressources energetique solaires. Evaluation et caracteristiques statistiques. Application a la conception de systemes. In: Cahiers Scientifiques et techniques 2, COFEDES, Paris
Power HC (2001) Estimating atmospheric turbidity from climate data. Atmos Environ 35:125–134
Prescott JA (1940) Evaporation from water surface in relation to solar radiation. Trans Roy Soc South Aust 64:114–118
Reindl DT, Beckman WA, Duffie JA (1990) Evaluation of hourly tilted surface radiation models. Sol Energy 45:9–17
Rigollier C, Bauer O, Wald L (2000) On the clear—sky models of the ESRA—European radiation Atlas, with respect to the Heliosat method. Sol Energy 68:33–48
Robinson GD (1962) Absorption of solar radiation by atmospheric aerosol as revealed by measurements from the ground. Theor Appl Climatol 12(1):19–40
Samimi J (1994) Estimation of height-dependent solar irradiation and application to the solar climate in Iran. Sol Energy 52(5):401–409
Samuel TDMA (1991) Estimation of global radiation for Sri Lanka. Sol Energy 47:333–337
Santamouris M, Tselepidaki I, Dris N (1990) Evaluations of models to predict solar radiation on tilted surfaces for the Mediterranean region. Solar Wind Technol 7:585–589
Shettle EP, Fenn RW (1979) Models for the aerosols of the lower atmosphere and the effects of humidity variations on their optical properties. Rep. AFGL-TR-79-0214, Air Force Geophysics Lab. Hanscom, MA. Available at http://www.dtic.mil/cgi-bin/GetTRDoc?AD=ADA085951
Skartveit A, Olseth JA (1988) Some simple formulas for multiple Rayleigh scattered irradiance. Sol Energy 41:19–20
Spencer W (1971) Fourier series representation of the position of the sun. Search 2:172
Supit I, Van Kappel RR (1998) A simple method to estimate global radiation. Sol Energy 63:147–160
Tadros MTY, El-Metwally M, Hamed AB (2005) A comparative study on SPCTRAL2, SPCTR-1881 and SMARTS2 models using direct normal solar irradiance in different bands for Cairo and Aswan, Egypt. J Atmos Sol Terr Phy 67:1343–1356
Tasumi M, Allen RG, Trezza R (2008) At-surface reflectance and albedo from satellite for operational calculation of land surface energy balance. J Hydrol Eng 13:51–63

TOMS (2012) Total Ozone Mapping Spectrometer. Availabe at http://toms.gsfc.nasa.gov/ozone/ozone_v8.html

Yang K, Huang GW, Tamai N (2001) A hybrid model for estimating global solar irradiance. Sol Energy 70:13–22

Yang K, Koike T, Ye B (2006) Improving estimation of hourly, daily, and monthly solar radiation by importing global data sets. Agr Forest Meteorol 137:43–55

Chapter 6
Time Series Forecasting

A discrete time series is a sequence of time ordered data values, measured in general at fixed time intervals. Time series analysis consists of techniques for drawing inferences from such series. First, it is necessary to set up a hypothetical probability model to represent the data. In first approximation, a time series model presumes that past pattern will appear in the future. After a suitable family of models has been chosen, it follows the task of estimating parameters and checking the accuracy of the data fit. Once an adequate model has been developed, it may be used in a variety of ways, depending on the application target. Time series forecasting means the use of a model to predict future values based on past values. An introduction to time series forecasting may be read in Brockwell and Davis (2002), for example. Also, a large collection of papers on this topic are acknowledged by Jan D. De Gooijer and Rob J. Hyndman in the anniversary review "25 Years of Time Series Forecasting" (De Gooijer and Hyndman 2006).

Various time series are encountered in many domains, such as engineering (Palit and Popovic 2005), science (Sprott 2003), sociology (Gottman 1982), and economics (Chen and Peace 2010). The most popular forecasting methods used in the solar energy field are: ARIMA, Markov chains, Bayesian inference, and artificial intelligence.

Artificial neural networks and fuzzy logic, both belonging to the generic group of artificial intelligence methods, are discussed in the next chapter.

In Bayesian inference, a posterior probability is derived from a prior probability and a "likelihood function" is derived from a model describing the data. Bayes' rule is used to update the probabilistic estimation as additional data is gathered. In the philosophy of decision theory, Bayesian probability provides a rational method for updating beliefs. An outstanding introduction to the Bayesian inference is given in Box and Tiao (1992). There are some papers on solar radiation modeling using the Bayesian inference method (e.g. Paoli et al. 2010; Iizumi et al. 2012).

A Markov process is a stochastic process where the description of the present state fully contains the information for the future evolution. In addition, the present state encapsulates the history of the process. Both determined by the present state

M. Paulescu et al., *Weather Modeling and Forecasting of PV Systems Operation*, Green Energy and Technology, DOI: 10.1007/978-1-4471-4649-0_6,

of the system, future and past are independent, the Markov process is said to be "memoryless". Future states are reached through a probabilistic process instead of a deterministic one. Among many books on Markov processes, Gillespie (1991) is one that especially meets demands from physicists and engineers. In prediction of solar radiation some authors have developed models based on Markov processes, in particular the Markov chain (e.g. Poggi et al. 2000).

The acronym ARIMA refers to an *auto-regressive integrated moving average* model, a combination of autoregressive (AR) and moving average models (MA). These models are fitted to time series data either to better understand the data or to predict future points in the series. ARIMA models can be applied to nonstationary data. In this case, an initial differencing step should be applied to remove the nonstationarity. This step corresponds to the *integrated* part of the model. ARIMA models form an important part of the Box-Jenkins approach to time series modeling (Box and Jenkins 1970).

An introduction to ARIMA models has been done previously in this book, in Sect. 4.5, where elementary and statistical properties of the sunshine number are presented. Additional reference on ARIMA practice can be found in, e.g., Pankratz (1983); Brockwell and Davis (2002).

The objective of this chapter is to illustrate the ARIMA modeling approach applied to forecasting solar radiation. First, the problem of nowcasting solar irradiance on very short time intervals (15 s–30 min) is addressed, focusing on both stages: constructing and operating the ARIMA model. The influence on the prediction accuracy due to the time horizon, stability of the radiative regime, and season is discussed. Second, the problem of one day ahead forecasting of the daily global solar irradiation is addressed.

The choice of the two forecasting domains is not accidental. Both timescales are relevant in controlling a PV plant operation. Nowcasting of solar radiation is required for compensating the fluctuation of output power due to the variability in short time intervals of the state of the sky (passing clouds). Forecasting the output power on the next day is required for proper scheduling grid operation and even for preparing bid offers for the energy market. The subsequent chapters are devoted to the same subject, nowcasting solar irradiance on very short time intervals and forecasting solar irradiation one day ahead, using different approaches: fuzzy logic (Chap. 7) and via air temperature-based models for estimating solar irradiation (Chap. 8). Thus, the reader has the opportunity to compare different approaches from various points of view.

6.1 ARIMA Modeling of Solar Irradiance

This section is focused on the practice of clearness index nowcasting on very short time intervals using ARIMA modeling. The solar cell reacts almost instantaneously to light changes. The ramp of the PV plant output follows the changes of solar irradiance in seconds and there are situations when the solar radiative regime fluctuates in a timescale of seconds. A discussion on this topic is included in Sect. 1.3. Thus, the example reported here is conducted on a 15 s basis. ARIMA models for forecasting solar irradiance from 30 s to 30 min ahead are also discussed.

6.1.1 Database

Data measured continuously in Timisoara for 10 days between 12 and 21 June 2010 are used to develop the ARIMA model. The data consist of global and diffuse solar irradiance values, measured with a 15 s lag during day light (see Sect. 9.4.1). From these data, the time series of beam solar irradiance values was constructed with Eq. (2.2).

Only data measured for a sun elevation angle greater than 5° have been kept in the daily series. A brief explanation follows. The pyranometers's accuracy depends on the sun elevation angle, being smaller in the vicinity of sunrise and sunset. The standard deviation of measurement errors decreases smoothly with the elevation angle, initially steep and then more slowly. Modeling this accuracy variation upon the elevation angle is difficult. The usual Box-Jenkins ARIMA models have weaknesses when applied to heteroscedastic (i.e. with different variabilities) data. Removing these problems requires the use of heteroscedastic innovation error ARIMA models, which in turn needs the knowledge of the function giving the error variance with dependence to the sun elevation angle. Such a function can be fitted, e.g., via Kalman filtering of the database (Kalman 1960). An application of Kalman filtering in the solar radiation field can be found in Gallego and Camacho (2012). This additional tool is not used here. To diminish the heteroscedasticity of the database we adopt the usual procedure, i.e., keeping only those measurements associated to sun elevation angles larger than a given threshold (in this case, 5°). First, a subdatabase has been obtained for each day by removing solar irradiance values associated to sun elevation angle less than 5° (this also includes the night-time interval). Next, the database for the series of 10 consecutive days is obtained by concatenation of the subdatabases prepared for each day in part.

6.1.1.1 Daily Average of Solar Radiative Regime

Before discussing ARIMA models, it is opportune to take a quick look at the average characteristics of the solar regime in the considered 10 days. This will further help us understand the results and draw conclusions.

Instantaneous values of sunshine number and sunshine stability number have been calculated with Eqs. (2.5) and (4.5), respectively. From these values, daily mean values of sunshine number $\bar{\xi}$ and sunshine stability number $\bar{\zeta}$, quantifying the solar radiative regime and its stability, respectively, were worked out. These values are displayed in Table 6.1. It can be seen that the days 12, 15, and 17 were

Table 6.1 Daily mean values of sunshine number $\bar{\xi}$ and sunshine stability number $\bar{\zeta}$ in Timisoara, during the period 12–21 June 2010

Days	12	13	14	15	16	17	18	19	20	21
$\bar{\xi}$	0.8538	0.5750	0.7665	0.9121	0.4623	0.8637	0.4745	0.4577	0.4693	0.3035
$\bar{\zeta}$	0.0005	0.0052	0.0049	0.0002	0.0069	0.0023	0.0075	0.0127	0.0141	0.0173

mostly sunny, while 13, 14, 16, 18, 19, and 20 were partly sunny. Day 21 was different from all others, being mostly cloudy. The solar radiative regime was the most stable on day 15 and most unstable on day 21.

A more detailed picture of the solar regime stability in these days is given in Fig. 6.1, where the instantaneous values of sunshine number in each day are shown. Visual inspection reveals that on 15 June, for a very short period at sunrise the sun was covered by clouds after which it was shining all the day. On the morning of 20 June the radiative regime was rather stable, followed by a very high level of instability at noon with fast passing clouds (the black band). The instability persisted at afternoon but at a lower level than at noon. The radiative regime on the entire day of 21 June was characterized by instability. Periods of fast alternating sunshine and clouds were noticed.

6.1.1.2 Clearness Index

Instantaneous values of beam and global clearness index have been computed with Eq. (4.1). The resulted time series of clearness index, beam and global, are plotted in Fig. 6.2. Thus, it is expected that ARIMA models with nonzero differencing order will better describe the data series.

Each graph consists of 34,509 points. It can be seen that the time series of instantaneous clearness index exhibits a daily periodic pattern. At first sight, this is unexpected since the clearness index is defined to quantify the effects on solar radiation passing the atmosphere. However, this pattern can be easily explained starting just from the definition of clearness index Eq. (4.1) which can be expanded as:

$$k_t = \frac{G}{G_{\text{ext}}} = \exp(-mt_e). \tag{6.1}$$

In Eq. (6.1), $G = G_{\text{SC}}\varepsilon \exp(-mt_e) \sin h$ and $G_{\text{ext}} = G_{\text{SC}}\varepsilon \sin h$ are global solar irradiance at the ground and extraterrestrial levels, respectively, where $G_{\text{SC}}\ \varepsilon$ is the solar constant corrected with the Earth's orbit eccentricity, h is the sun elevation angle, m is the atmospheric air mass, and t_e a generic extinction coefficient modeling all the processes experienced by the solar radiation through the atmosphere. Assuming a clear day and a stationary atmosphere (the optical parameters do not change significantly over the day), the clearness index defined in Eq. (6.1) depends on time only by means of the atmospheric air mass. Expressing this relation in terms of the hour angle and taking into account the simplest equation for atmospheric air mass $m \approx 1/\sin h$ (Eq. 5.22) it can be written as:

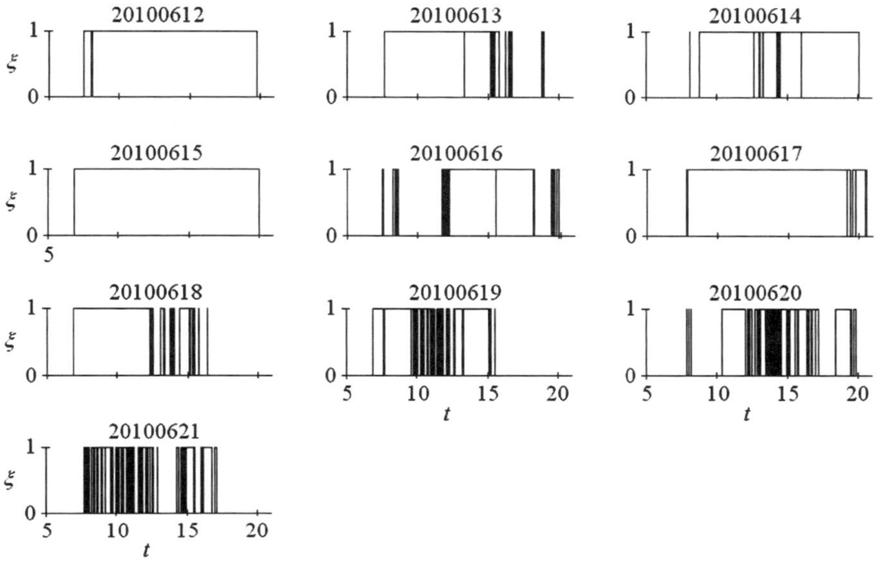

Fig. 6.1 Sunshine number variations in Timisoara from 12 to 21 June 2012. The date is indicated at the *top* of each graph as *yyyymmdd*

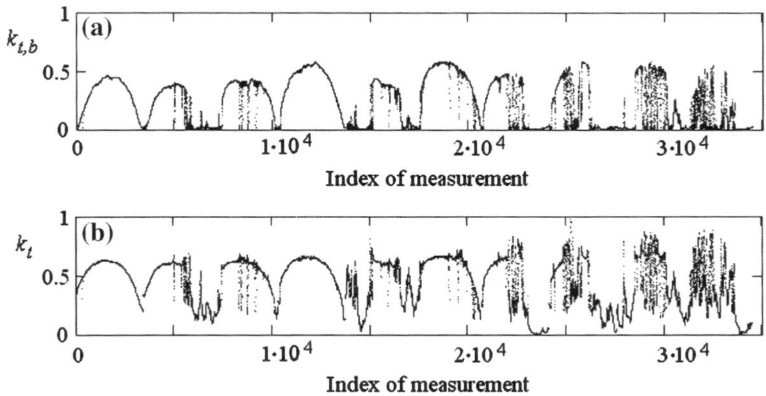

Fig. 6.2 Measured clearness index during 12–21 June 2010 at Timisoara **a** beam $k_{t,b}$ and **b** global k_t. Concatenated daily series for elevation angle greater than 5° are displayed

$$k_t(\omega) \cong \exp\left(-\frac{t_e}{\sin h(\omega)}\right) \tag{6.2}$$

Expanding the exponential in Taylor series around noon ($\omega = 0$) and keeping only the first three terms, Eq. (6.2) becomes a polynomial equation of order two with respect to the hour angle:

$$k(t) \cong a + b\omega + c\omega^2 \tag{6.3}$$

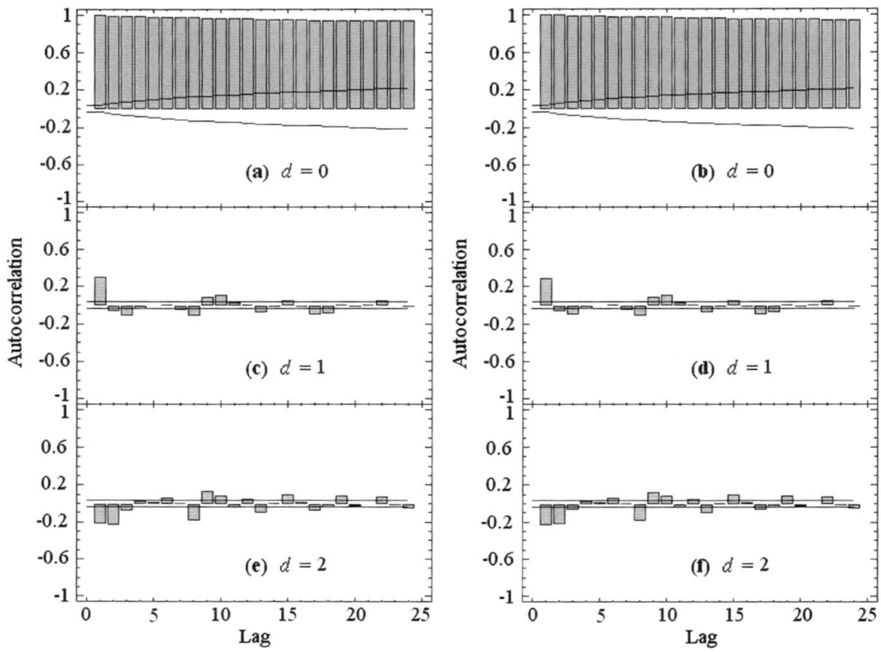

Fig. 6.3 Autocorrelation coefficients for the time series consisting of $k_{t,b}$ values measured on 13 June 2010 (**a**, **c**, **e**) at Timisoara. Frames (**b**, **d**, **e**) display the autocorrelation coefficients for the $k_{t,b}$ series measured on 13 June 2010 preceded by the last 240 records at 15 s lag from the previous day (i.e., 12 June). Differencing order d is indicated on the graphs

where a, b, and c are constant coefficients of Taylor series. Equation (6.3) authenticates the causal dependence on time of the instantaneous clearness index and its profile in Fig. (6.2). This is more visible in the graphs corresponding to the sunny days 12 and 15 June, where the curves are roughly parabolic. In other days, the curves depicted in Fig. 6.2 are shaped by passing clouds which reduce the atmospheric transmittance and thus the value of instantaneous clearness index (Eq. 6.2).

The periodic pattern of the clearness index is removed when daily values are calculated with Eq. 4.2. This is shown in Fig. 6.11 where the daily global clearness index in Timisoara is plotted for the years 2009–2010. The graph outlines the aspect of a random time series.

The instantaneous clearness index database has been divided into two parts: one corresponding to the first 9 days (i.e. 12–20 June, the learning period) and the other corresponding to the 10th day (i.e. 21 June, the forecasting period). The clearness index time series from the learning period has been used to fit the ARIMA model.

A question is whether removing the k_t values associated to $h < 5°$ changes the autocorrelation properties of the series. Figure 6.3 displays the autocorrelation coefficients (Eq. 3.2b) for the beam clearness index $k_{t,b}$ series measured on June 13. Also, Fig. 6.3 shows the autocorrelation coefficients for the longer $k_{t,b}$ series, consisting of the series of June 13, preceded by data related to the last hour

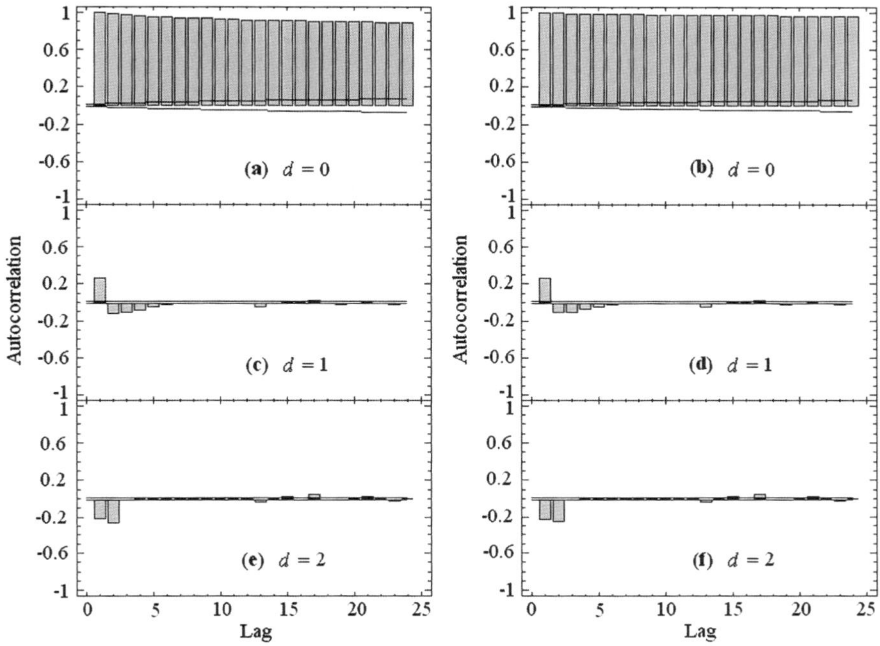

Fig. 6.4 Autocorrelation coefficients for: (**a**, **c**, **e**) the time series consisting of k_t values measured for 24 h during each of the 9 days, 12–20 June 2010; (**b**, **d**, **f**) the time series obtained by selecting and stacking for the same 9 days only data satisfying a sun elevation angle $h > 5°$. Differencing order d is indicated on the graphs. 51,840 measurements recorded during 12–20 June 2010 at Timisoara have been used

of the previous day (i.e., 12 June). It can be seen that there is a good agreement between the autocorrelation coefficients of the longer and shorter $k_{t,b}$ time series, respectively. Figure 6.4 displays the autocorrelation coefficients of the time series consisting of global clearness index k_t measurements for 24 h during each of the 9 days 12–20 June, on one hand, and the time series obtained by selecting and stacking for the same 9 days with only the data satisfying $h > 5°$, on the other hand. The difference between the appropriate autocorrelation coefficients for the two series is rather small and increases with the lag number. However, as seen further, only the first autocorrelation coefficient is relevant in most cases for ARIMA modeling of k_t. Thus, the procedure of data concatenation adopted in this chapter is expected to act reasonably well.

The day 21 June was used for checking the predictive performance of the model. When looking at the data, it is obvious that the day 21 June shows special peculiarities which do not allow expecting high model accuracy during the checking stage. Indeed, the daily mean clearness index is quite smaller on day 21 June than on the previous days (Fig. 6.5). The models would then be forced into predicting a somewhat different situation than it was the case in the learning stage. This is reasonable as long as the Markovian structure holds for days with

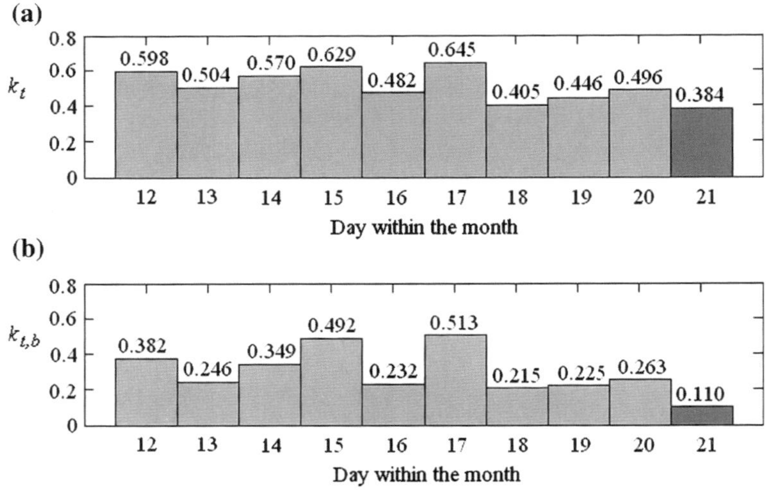

Fig. 6.5 Daily clearness index: **a** global k_t and **b** beam $k_{t,b}$ on the days 12–21 June 2010

dissimilar cloud cover amount, but in general a better performance is to be expected if the model is trained and used for forecasting on days with more similar radiative regimes.

6.1.2 ARIMA Models

In the learning period the computations were performed using the powerful data analysis tool Statgraphics (Statgrafics 2012), while MathCAD (PTC 2012) has been used to elaborate specific applications for the testing period.

The ARIMA models were estimated on days 12–20 June. The selection procedure of an ARIMA model followed the parsimony principle described in the preamble of Sect. 4.5. A set of values for the model parameters (i.e. the coefficients φ_i and θ_i in Eq. (4.11), the order of differencing d and the standard deviation σ_a of the white noise) is therefore obtained. These values are kept fixed when the ARIMA models are subsequently applied for forecasting on 21 June.

Some following comments are about the initial values of the autoregressive and moving average parts of Eq. (4.11). One can take estimates for k_t and white noise values from the last time moments of 20 June and associate them to the time index $t = 1$ on 21 June. Here we adopted a different procedure, which is explained now for the particular case of ARIMA(1,1,2) model. The first two measured k_t values on June 21 were used as a "seed" for $k_t^{(t-2)}$ and $k_t^{(t-1)}$, respectively. An *ad hoc* white noise a_t value has been generated. These three values $k_t^{(t-2)}$, $k_t^{(t-1)}$ and a_t allow for $k_t^{(t)}$ to be estimated from Eq. (6.4a) for the time index $t = 3$. The same procedure has been used to start using all other ARIMA models. The effect of the above starting procedure relaxes rapidly in time due to the Markovicity of the time series (note that the time series consists of more than three thousand values).

6.1.2.1 Fifteen Seconds Nowcasting Clearness Index

In this section, ARIMA models for both global and beam clearness index series are discussed. The following ARIMA models have been fitted:

$$\text{ARIMA}(1,1,2):\quad k_t^{(t)} = k_t^{(t-1)}(1+\varphi_1) - k_t^{(t-2)}\varphi_1 + a_t - \theta_1 a_{t-1} - \theta_2 a_{t-2} \tag{6.4a}$$

$$\text{ARIMA}(1,2,1):\quad k_t^{(t)} = k_t^{(t-1)}(2+\varphi_1) - k_t^{(t-2)}(1+2\varphi_1) + k_t^{(t-3)}\varphi_1 + a_t - \theta_1 a_{t-1} \tag{6.4b}$$

$$\text{ARIMA}(1,2,2):\quad \begin{aligned} k_t^{(t)} &= k_t^{(t-1)}(2+\varphi_1) - k_t^{(t-2)}(1+2\varphi_1) + k_t^{(t-3)}\varphi_1 \\ &\quad + a_t - \theta_1 a_{t-1} - \theta_2 a_{t-2} \end{aligned} \tag{6.4c}$$

where φ's and θ's are regression and moving coefficients, respectively. a_t is a white noise term with mean $\mu = 0$ and standard deviation σ_a, which are the only restrictions that the Box-Jenkins theory imposes on the white noise distribution. Uniform white noise distribution is used in practice and so it will be used next.

In addition to ARIMA models, several other models have been applied first to the clearness index series. Table 6.2 shows the main indicators of accuracy of these models. ARIMA(1,1,2) model performs the best and reaches *RMSE* of 9.09 % for beam series and 5.25 % for the global series. The percentage of *RMSE* refers to the average value of measured clearness index on June 21.

The fitted coefficients of the ARIMA models (6.4) and the estimated white noise standard deviation are given in Table 6.3, for both beam and global clearness index series.

Figure 6.6a displays the *beam* clearness index ($k_{t,b}$ series) measured and forecasted with the ARIMA(1,1,2) model for the day 21 June. Visual inspection shows that the forecasted points group as a cloud surrounding the curve produced by measurements. This picture gives a first conclusion: as a whole the forecasted points trace the measurements with noticeable accuracy. This graph also reveals a problem. At very low values of $k_{t,b}$ the model may forecast small negative values, which are unphysical ($k_{t,b}$ is a positive defined quantity) and unacceptable for further processing of forecasted data. These negative values of forecasted $k_{t,b}$ occurs in periods of time when the sun is covered by clouds. The simplest correction is to force negative values to zero, i.e., to cancel the beam component of solar irradiance when the sun is covered by clouds, which is reasonable. Thus, the outcome of the forecasting procedure $\hat{k}_{t,b}$ is specified by the following equation:

Table 6.2 Comparison of five different forecasting models in the learning period

Dataset	Model	RMSE	MAE	MBE
Beam	ARIMA(1,1,2)	0.02318	0.00639	0.00000
	ARIMA(1,2,1)	0.02480	0.00576	−0.00000
	Exponential smoothing	0.02475	0.00611	0.00000
	Moving average of three terms	0.03384	0.00906	0.00000
	Linear trend	0.19925	0.17986	0.00000
Global	ARIMA(1,1,2)	0.02379	0.00678	−0.00001
	ARIMA(1,2,2)	0.02429	0.00606	−0.00000
	Exponential smoothing	0.02559	0.00659	0.00000
	Moving average of three terms	0.03570	0.01027	−0.00001
	Linear trend	0.19415	0.16603	0.00000

Table 6.3 Coefficients of the selected ARIMA models and the estimated white noise standard deviation

Dataset	Model	φ_1	θ_1	θ_2	σ_a
Beam	ARIMA(1,1,2)	0.78389	0.53825	0.36410	0.02318
	ARIMA(1,2,1)	0.23949	0.99869	–	0.02408
Global	ARIMA(1,1,2)	0.83119	0.53517	0.38993	0.02379
	ARIMA(1,2,2)	−0.21984	0.44360	0.54819	0.02429

$$\hat{k}_{t,b} = \begin{cases} k_{t,b} & if \quad k_{t,b} > 0 \\ 0 & otherwise \end{cases} \tag{6.5}$$

Figure 6.6b shows the time series of forecasted beam clearness index $\hat{k}_{t,b}$ by processing the $k_{t,b}$ series with Eq. (6.5). The suppression of negative $k_{t,b}$ values is the only difference between the two representations in Fig. 6.7a, b.

Figure 6.7a displays the *global* clearness index (k_t series) measured and forecasted with the ARIMA(1,1,2) model for the day 21 June. Compared with Fig. 6.6a a smaller dispersion of the forecasted points can be noticed. The points are enveloping the curve of measurements. The problem of negative or very small positive values in heavily cloudy sky (k_t close to zero) has been solved again by forcing them to a constant value equal to the minimum $k_{t,\min}$ measured in the learning stage:

$$\hat{k}_t = \begin{cases} k_t & if \quad k_t > k_{t,\min} \\ k_{t,\min} & otherwise \end{cases} \tag{6.6}$$

In this case, $k_{t,\min}$ is set to a nonzero value since the global solar irradiance does not vanish even for the cloudiest sky. In this case, $k_{t,min} = 0.0039$. As Fig. (6.8b) illustrates, Eq. (6.6) is a reasonable solution for negative values in the k_t series.

To conclude, no significant differences can be observed between the predicted and measured series in Figs. 6.6b and 6.7b. Both graphs demonstrate that the forecasted clearness index, beam or global, mimics the measurements with reasonable accuracy.

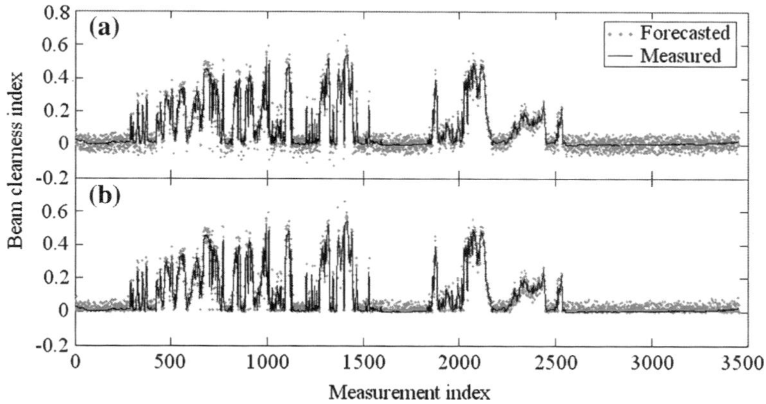

Fig. 6.6 Beam clearness index measured and forecasted on 21 June 2010 at Timisoara. **a** $k_{t,b}$ time series obtained by using Eq. (6.4a) and **b** $\hat{k}_{t,b}$ time series obtained by using Eq. (6.5)

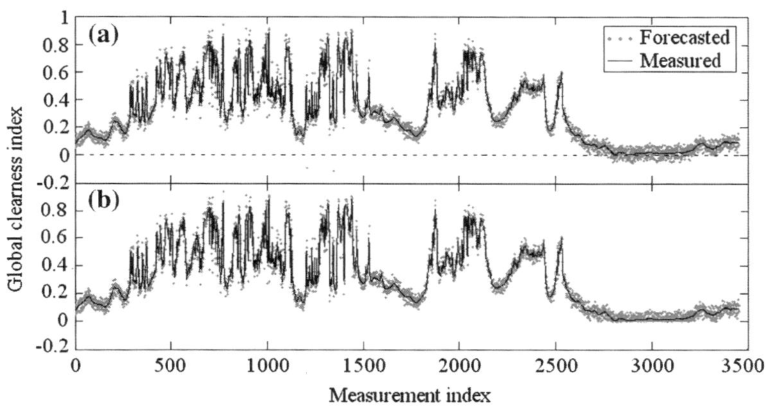

Fig. 6.7 Global clearness index measured and forecasted on 21 June 2010 at Timisoara. **a** k_t time series obtained by using Eq. (6.4a) and **b** $\hat{k}_t$ time series by using Eq. (6.6)

Table 6.4 shows the accuracy indicators for the ARIMA(1,1,2) and ARIMA(1,2,1) models applied to forecast the beam clearness index and the same applied to forecast the global clearness index on the testing day of 21 June. First, it can be seen that the processing of initial series with Eqs. (6.5) and (6.6) enhance the prediction accuracy. At this very short horizon of time the tested ARIMA models for forecasting global solar irradiance perform satisfactory even if, on the whole, the testing day exhibited solar radiative regime peculiarities different from those encountered in the learning period (see Sect. 6.1.1). The percentage of the average value of measured clearness index *RMSE* is 14.06 % for ARIMA(1,1,2) and 14.64 % for ARIMA(1,2,2). No bias in the forecasted series is noticed; *MBE* is less than 1 % after the corresponding transforming procedures were applied.

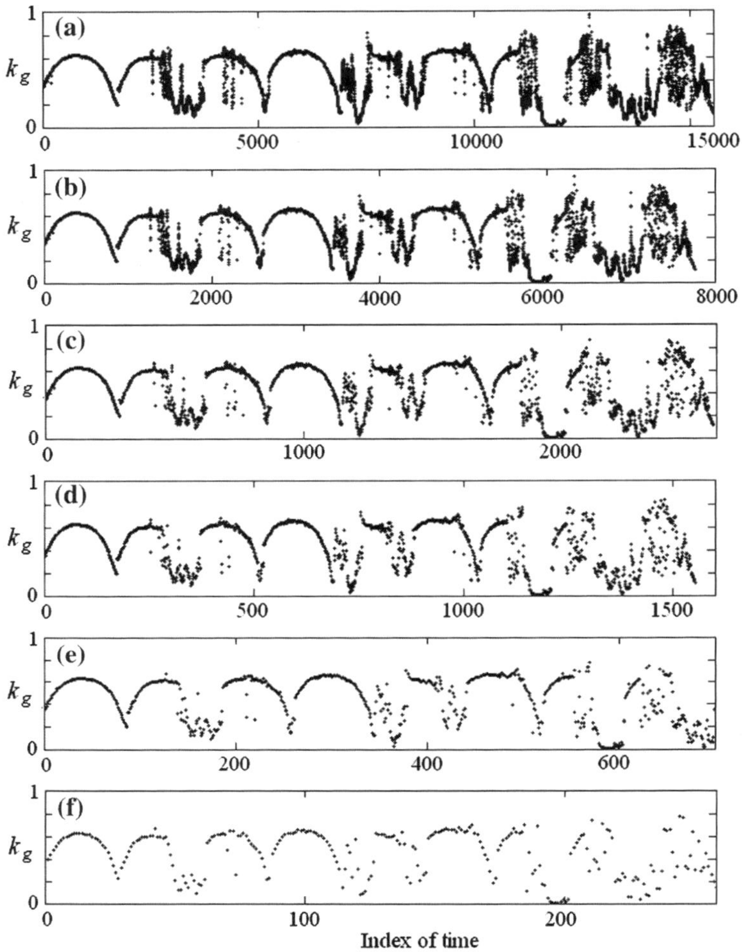

Fig. 6.8 Clearness index measured in Timisoara during 12–21 June 2010 at different time intervals: **a** 30 s; **b** 1 min; **c** 3 min; **d** 5 min; **e** 10 min; **f** 30 min concatenated daily series for an elevation angle greater than 5° are displayed

On the contrary, the performance of ARIMA models at forecasting beam solar irradiance is modest: *RMSE* is 44.54 % for ARIMA(1,1,2) and 51.21 % for ARIMA(1,2,1). To some extent, this is unexpected since the models are fitted with a good accuracy (Table 6.2). This poor behavior can be explained by the highly fluctuating solar regime on the day June 21 (Fig. 6.1). In such days, the variation of the beam clearness index is more abrupt than the variation of global solar irradiation. When the sun appears through the clouds, the beam clearness index suddenly changes between a value close to zero to a higher value (and vice versa, when the sun is covered by a cloud). This is very difficult to be predicted. Thus, for forecasting the beam solar irradiance, other models (e.g. logistic models) accepting more information at input (e.g. sunshine number) should be considered.

Table 6.4 Indicators of accuracy for ARIMA models in the testing period

Dataset	Model	Time series	C	RMSE	MBE	MAE
Beam	ARIMA(1,1,2)	$k_{t,b}$	4	0.0441	−0.0000	0.0315
		$\hat{k}_{t,b}$	4	0.0394	0.0082	0.0250
	ARIMA(1,2,1)	$k_{t,b}$	3	0.0527	−0.0000	0.0371
		$\hat{k}_{t,b}$	3	0.0453	0.0102	0.0286
Global	ARIMA(1,1,2)	k_t	4	0.0449	−0.0000	0.0318
		$\hat{k}_t$	4	0.0443	−0.0009	0.0308
	ARIMA(1,2,2)	k_t	4	0.0467	−0.0000	0.0330
		$\hat{k}_t$	4	0.0462	0.0010	0.0319

The number of regression coefficients C to be found in the learning period is also shown (it includes the standard deviation of the white noise)

Additional information about models' performance can be obtained by studying the distribution properties of the synthetic time series. Table 6.5 shows the four statistical moments of the beam and global clearness index time series measured on the day 21 June and various time series generated with the ARIMA models. At this level of approximation, there is good agreement between the distribution proprieties of the time series generated by all ARIMA models and the measured series. Processing the forecasted time series with Eqs. (6.5) or (6.6) alters to some extent the initial shape of the probability distribution of the synthetic time series, but does not change the data distribution around the mean.

6.1.2.2 Thirty Seconds to Thirty Minutes Nowcasting Clearness Index

ARIMA models for nowcasting global clearness index at higher time horizons were also fitted. For each time horizon Δt (of 30 s, 1, 3, 5, 10, and 30 min) the specific database in a day has been constructed starting from the 15 s-database. For a given Δt, the lines corresponding to index time $n\ \Delta t$, $n = 0, 1, 2, \ldots$ have been preserved and all other lines were filtered out. The index $n = 0$ corresponds to the first measurement line fulfilling the condition $h > 5°$. Figure 6.8 shows the clearness index for all Δt values. In clear sky days, one notices that the daily pattern of k_t is preserved regardless of the sampling Δt. As weather instability increases the clear shape of k_t in a day turns blurry and the random character of the series becomes more pregnant.

A summary of the results in the learning stage are presented in Table 6.6. The ARIMA models listed in Table 6.6 are the best found to fit the data. The specific equations of these models are:

$$\text{ARIMA}(0,1,2)\text{:}\ k_t^{(t)} = k_t^{(t-1)} + a_t - \theta_1 a_{t-1} - \theta_2 a_{t-2} \tag{6.7a}$$

$$\text{ARIMA}(0,1,1)\text{:}\quad k_t^{(t)} = k_t^{(t-1)} + a_t - \theta_1 a_{t-1} \tag{6.7b}$$

Table 6.5 First four statistical moments of the measured time series on 21 June and of various time series generated by the ARIMA models

Dataset	Model	Time series	Mean	Variance	Skewness	Kurtosis
Beam	–	Measured	0.0885	0.0171	1.6260	1.5003
	ARIMA(1,1,2)	$k_{t,b}$	0.0885	0.0192	1.5371	1.4664
		$\hat{k}_{t,b}$	0.0967	0.0175	1.6878	1.8192
	ARIMA(1,2,1)	$k_{t,b}$	0.0885	0.0204	1.4965	1.5285
		$\hat{k}_{t,b}$	0.0987	0.0181	1.7124	2.0020
Global	–	Measured	0.3155	0.0514	0.4662	−0.8004
	ARIMA(1,1,2)	k_t	0.3154	0.0538	0.5216	−0.6513
		$\hat{k}_t$	0.3165	0.0532	0.5439	−0.6511
	ARIMA(1,2,2)	k_t	0.3155	0.0542	0.5371	−0.6169
		$\hat{k}_t$	0.3165	0.0536	0.5601	−0.6159

Table 6.6 Coefficients of the selected ARIMA models, the estimated white noise standard deviation, and statistical indicators in the learning period

Dataset	N	ARIMA model	AR(1)	AR(2)	MA(1)	MA(2)	σ_a	MBE	MAE
30 s	15526	(2,1,2)	0.34005	0.28514	0.43656	0.42524	0.0399	−0.0000	0.0129
1 min	7763	(1,1,1)	0.53982	–	0.82476	–	0.0541	−0.0000	0.0202
3 min	2587	(1,1,1)	0.25722	–	0.67075	–	0.0733	−0.0000	0.0338
5 min	1552	(1,1,1)	0.23723	–	0.61734	–	0.0805	−0.0003	0.0402
10 min	776	(0,1,1)	–	–	0.38714	–	0.0964	−0.0004	0.0531
30 min	258	(0,1,2)	–	–	0.22328	0.13713	0.1297	−0.0028	0.0861

$$\text{ARIMA}(1,1,1):\quad k_t^{(t)} = k_t^{(t-1)}(1+\varphi_1) - \varphi_1 k_t^{(t-2)} + a_t - \theta_1 a_{t-1} \tag{6.7c}$$

$$\text{ARIMA}(2,1,2):\quad \begin{aligned} k_t^{(t)} &= k_t^{(t-1)}(1+\varphi_1) + k_t^{(t-2)}(\varphi_2 - \varphi_1) - \varphi_2 k_t^{(t-3)} \\ &+ a_t - \theta_1 a_{t-1} - \theta_2 a_{t-2} \end{aligned} \tag{6.7d}$$

The coefficients of Eqs. (6.7a–6.7d) are also given in Table 6.6. It can be seen that as Δt increases, the autoregressive terms vanish. On other hand, the increasing of the time interval Δt leads to an increase in the white noise standard deviation σ_a.

Table 6.7 displays the results of applying the ARIMA models to data in the testing stage, i.e., on June 21. For each time horizon, the specific database has been built using the same procedure as for the learning stage. The prediction accuracy decreases as the time horizon increases from 15 s to 30 min, from $RMSE = 0.0443$ (Table 6.3) to 0.2884 (Table 6.7), respectively.

The increasing of *RMSE* follows a nonlinear profile with respect to the forecasting time horizon. Let $k = \Delta t/\Delta t_0$, where $\Delta t_0 = 15$ s. The dependence *RMSE*(k) resulted from Table 6.6 is best fitted ($r^2 = 0.997$) by a power equation:

$$RMSE(k) = -37.899 + 42.303k^{0.094746} \tag{6.8}$$

Table 6.7 Indicators of accuracy of ARIMA models applied to nowcasting clearness index on June 21 at different time horizons Δt

Δt	N	RMSE	MAE	MBE	MEAN	VAR	SKEW	KURT	
30 s	1728	0.0673	0.044	0.0097	0.3163	0.0542	0.5627	−0.6061	**F**
					0.3154	0.0514	0.4695	−0.7912	M
1 min	864	0.1118	0.0780	0.0035	0.3191	0.00589	0.6442	−0.3675	F
					0.3156	0.0512	0.4672	−0.7964	M
3 min	288	0.1541	0.1070	0.0212	0.3244	0.0589	0.5432	−0.6436	F
					0.3183	0.052	0.4453	−0.8267	M
5 min	172	0.1835	0.1298	0.0055	0.3158	0.0591	0.7629	−0.1007	F
					0.3102	0.0492	0.5018	−0.6912	M
10 min	86	0.2179	0.1585	0.0038	0.3211	0.0570	0.5000	−0.4029	F
					0.3173	0.053	0.5263	−0.7499	M
30 min	28	0.2884	0.2161	0.0291	0.3717	0.0707	1.0525	0.5089	F
					0.3426	0.0606	0.4828	−0.8096	M

The first four statistical moments are displayed for both time series: predicted (Flag F) and measured (Flag M). N is the number of values in the clearness index series

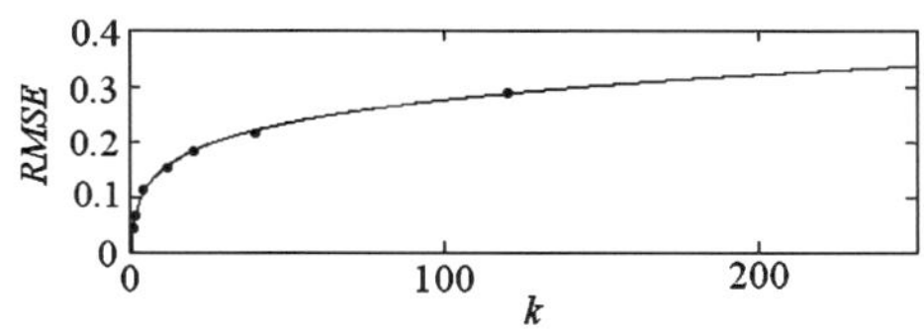

Fig. 6.9 Relative root mean square error as function of time horizon expressed as $k \times 15$ s, at forecasting clearness index. Data from Table 6.6 (*points*) and the fitted Eq. (6.8) (*line*) are displayed

The curve *RMSE*(*k*) is plotted in Fig. 6.9 and shows a rapid increase of the prediction errors for up to 5 min (*k* from 1 to 20), followed by a relaxation to a linear trend with a small slope.

At larger time horizons, the models' performance appears inadequate. This is a result of convergent actions from several assumptions on which this example was based. The fitted models, with knowledge accumulated in the learning stage, are put to predict on the day 21 which was rather different from the days encountered in the learning period. Moreover, as shown in Fig. 6.8, the difference between datasets in the learning and the testing periods is accentuated by enlarging the forecasting time horizon. Equation (6.7) shows that the estimation of clearness index series at a given time moment involves values of clearness index measured at previous time. The precise value of clearness index cannot be predicted. What ARIMA models can predict from the past is the conditional mean. In this example, increasing the forecasting time horizon forces the model to predict a different pattern than the one learned, severely affecting the prediction accuracy. This specific result can be easily extrapolated to the practice of forecasting solar irradiance via clearness index. It is obvious that in a given period, the solar

Table 6.8 Daily mean values of sunshine number $\bar{\xi}$ and sunshine stability number $\bar{\zeta}$in Timisoara during three periods of the year 2010

	January									
Days	22	23	24	25	26	27	28			
$\bar{\xi}$	0.0020	0.1385	0.7500	0.9928	0.5840	0.0056	0			
$\bar{\zeta}$	0.0010	0.0067	0.0005	0.0000	0.0040	0.0010	0			
	March									
Days	15	16	17	18	19	20	21	22	23	24
$\bar{\xi}$	0.3423	0.0054	0.7242	0.7446	0.9656	0.9223	0.3146	0.4970	0.0082	0.4061
$\bar{\zeta}$	0.0144	0.0023	0.0119	0.0141	0.0015	0.0034	0.0049	0.0090	0.0011	0.0040
	October									
Days	1	2	3	4	5	6	7	8		
$\bar{\xi}$	0.5697	0.3537	0.4846	0.2767	0.0769	0.1768	0.1343	0.7220		
$\bar{\zeta}$	0.0109	0.0035	0.0118	0.0035	0.0020	0.0100	0.0080	0.0097		

radiative regime may be stable or it may be very fluctuant as well. Ideally, the ARIMA model should accurately predict the clearness index for all regimes, stable and fluctuant, which is not easy to accomplish. ARIMA models' accuracy in predicting the clearness index is highly dependent on the radiative regime stability, as will be shown in the next section.

The results presented here prove that extrapolation of measurements is an appropriate method for forecasting solar irradiance on very short time intervals. Additional information on the weather pattern is required for obtaining a satisfactory accuracy at a larger time horizon than a few minutes.

6.1.2.3 Seasonal Effects

Seasonal stability of the ARIMA models is discussed here. Only the results for ARIMA(1,1,2) model applied to 15 s time series are shown since the other models yield similar results.

In addition to the above set of data (June, 12–21), three other different sets of data recorded at Timisoara in different seasons of 2010 are considered. The first set consists of 11,630 values of global instantaneous clearness index k_t measured during 7 days in winter (January 22–27); the second set consists of 26,287 k_t values measured during 10 days in spring (March 15–24); the third set consists of 20,072 values of k_t measured during 8 days in autumn (October 1–8). A summary on the selected day's radiative regime is given in Table 6.8 by means of the sunshine number and sunshine stability number. In each period, both sunny and cloudy days are present and the radiative regime varies from one day to another.

The dataset has been processed with the same procedure as for the 15 s lag dataset. The ARIMA(1,1,2) model has been fitted to data in each period, excepting those of the last day, which have been used to test the model. From Table 6.8 it

Table 6.9 Coefficients of the selected ARIMA models, the estimated white noise standard deviation, and statistical indicators in the learning period

Dataset	N	AR(1)	MA(1)	MA(2)	σ_a	MBE	MAE
January	11630	0.1928	−0.2364	−0.0834	0.0069	0.0000	0.0021
March	23599	0.1339	−0.1290	0.0863	0.0261	0.0000	0.0075
October	17609	0.2025	−0.1999	0.1699	0.0223	−0.0000	0.0069

can be seen that the solar radiative regime on every test day considered here is more stable than on June 21, used previously to test the models.

The coefficients of the ARIMA(1,1,2) model, the estimated white noise standard deviation, and statistical indicators in the learning stage for each period are listed in Table 6.9. It can be seen that the models are fitted with very high accuracy in each period. The percentage of the clearness index average *RMSE* is 1.63 % in January, 6.4 % in March, and 6.54 % in October. The results of testing the models are presented in Table 6.10. The same very good performance is noticed: *RMSE* is of 3.49 % on January 28, 7.35 % on March 24, and 7.11 % on October 8. For the same ARIMA(1,1,2), *RMSE* was 14.06 % on June 21 (see Tables 6.4 and 6.5).

At first sight, a seasonal dependence of the model's accuracy seems evident. In fact, the model's accuracy depends in a complex manner on the stability of the radiative regime. The particular radiative regime in the four test days is illustrated in Fig. 6.10 as instrumented by the sunshine number ξ. These days are clearly distinguished by different radiative regimes: January 28 was a fully stable overcast day; on March 24 the sky was overcast in the morning and later afternoon but variable around noon; day June 24 was fully unstable with high fluctuation rate in the state of the sky; on day October 8 the sky was variable at noon, overcast in the morning, but sunny in the afternoon.

Looking at Tables 6.1 and 6.8, the 4 days' hierarchy with respect to their stability of radiative regime, measured by the sunshine stability number $\bar{\zeta}$, is: (1) January 27 ($\bar{\zeta} = 0$), (2) March 24 ($\bar{\zeta} = 0.0040$), (3) October 8 ($\bar{\zeta} = 0.0097$), and (4) June 21 ($\bar{\zeta} = 0.0173$). The same ranking is noted for *RMSE*: (1) January 27 (*RMSE* = 0.0073), (2) March 24 (*RMSE* = 0.0269), (3) October (*RMSE* = 0.0350 %), and (4) June (*RMSE* = 0.0443). This verifies that the forecasting accuracy of the clearness index is conditioned by the stability of the respective radiative regime.

Relative to the mean, on the days March 24 and October 8 the model's accuracy is roughly the same, 7.35 and 7.11 %, respectively. The days differ by the value of sunshine number in the afternoon (see Fig. 6.10): $\xi = 0$ on March 24 and $\xi = 1$ on October 10, which is reflected in the daily mean sunshine number taking a lower value ($\bar{\xi} = 0.4061$) on March 24 and a higher value ($\bar{\xi} = 0.7220$) on October 10. In general, if several days are characterized by comparable fluctuations of their radiative regimes in a time interval, the magnitude of relative errors is conditioned by the mean values of sunshine number.

Table 6.10 Performance measures for the model ARIMA(1,1,2)

Dataset	N	RMSE	MAE	MBE	Mean	Var	Skew	Kurt	
January 28	1976	0.0073	0.0062	0.0000	0.2090	0.0017	0.0813	−1.0983	F
					0.2090	0.0017	0.0813	−1.1525	M
March 24	2687	0.0269	0.0231	0.0005	0.3655	0.0208	0.0262	−0.3755	F
					0.3650	0.0200	0.0451	−0.4108	M
October 8	2463	0.0350	0.0247	−0.0007	0.4921	0.0371	−0.5701	−0.9044	F
					0.4929	0.0362	−0.5825	−0.9197	M

N is the number of values in global clearness index series. The flags F and M stand for forecasted and measured series, respectively

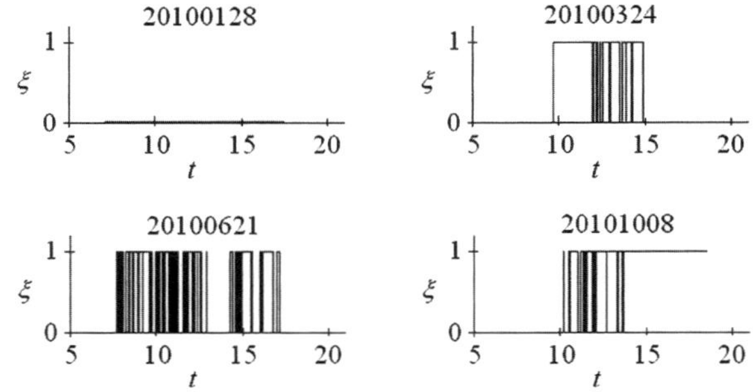

Fig. 6.10 Sunshine number ξ variations in Timisoara on the days of 2010 indicated at the *top* of each graph, as *yyyymmdd*

To conclude, the overall accuracy of nowcasting instantaneous clearness index in a given time is influenced by the state of the sky and its stability in this period. The more stable and sunnier the period is, the higher the nowcasting accuracy.

6.2 ARIMA Modeling of Solar Irradiation

In this section, an ARIMA model for forecasting global solar irradiation one day ahead is evaluated. Similar to the case of nowcasting solar irradiance, the forecasted quantity is the clearness index, this time daily clearness index.

The database consists of daily global solar irradiation measured during 2 years, 2009 and 2010, on the Solar Platform of the West University of Timisoara (see Sect. 9.4.1). From these data, the clearness index has been calculated with Eq. (4.2). This time series is plotted in Fig (6.11). The first 546 values (from January 1, 2009 to June 30, 2010) were used to construct the model, while the last 184 values (from July 1, 2010 to December 31, 2010) were used to test the model.

Using the procedures described in Sect. 4.5, the ARIMA(3,0,3) model has been selected. It is expressed by the equation:

$$k_t^{(t)} = \varphi_1 k_t^{(t-1)} + \varphi_2 k_t^{(t-2)} + \varphi_3 k_t^{(t-3)} + a_t - \theta_1 a_{t-1} - \theta_2 a_{t-2} - \theta_2 a_{t-2} \tag{6.9}$$

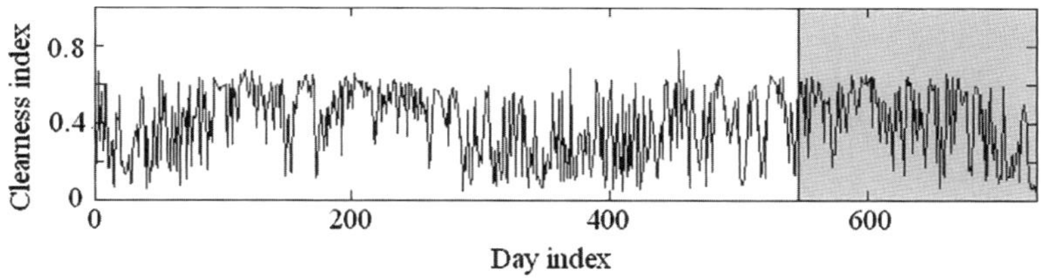

Fig. 6.11 Variations of daily clearness index measured in Timisoara during 2009 and 2010. Data from the region with *gray* background are used to test the model

Table 6.11 Comparison of ARIMA(3,0,3) model with the other three forecasting models in the learning period

Model	RMSE	MAE	MBE
ARIMA(3,0,3)	0.1525	0.1223	0.0041
Exponential smoothing	0.1564	0.1276	0.0006
Moving average of 3 terms	0.1665	0.1319	0.0005
Linear trend	0.1788	0.1556	0.0000

The fitted coefficients of the ARIMA(3,0,3) model are: $\varphi_1 = 0.8653$, $\varphi_2 = 0.5907$, $\varphi_3 = -0.4559$, $\theta_1 = 0.4838$, $\theta_2 = 0.7365$, $\theta_3 = -0.2976$ and the estimated white noise standard deviation is $\sigma_a = 0.15281$. Statistical indicators of ARIMA(3,0,3) in comparison with the other three forecasting models are shown in Fig. 6.11. Looking at the error statistics, the model with the smallest root mean squared error with the smallest mean absolute error during the estimation period is the ARIMA(3,0,3) model.

Using the Satgraphics tools, three tests: (1) Runs above and below median, (2) Runs up and down, (3) Box-Pierce test, have been run to determine whether or not the residuals form a random sequence of numbers. The first test counts the number of times the sequence was above or below the median. The number of such runs equals 252, as compared to an expected value of 274.0 if the sequence would be random. The p value for this test is 0.065. The second test counts the number of times the sequence rose or fell. The number of such runs equals 360, as compared to an expected value of 363.667 if the sequence would be random. The p value for this test is 0.747. The third test is based on the sum of squares of the first 24 autocorrelation coefficients. The p value for this test is 0.149. Since the p value for all three tests is greater than or equal to 0.05, the series is random at 95.0 % or higher confidence level. Thus, it is expected for the selected model to capture the structure in the data (Fig. 6.11).

The model has been applied to forecast daily clearness index in the second half of 2010. Due to the large dispersion of the white noise series ($\sigma_a = 0.15281$), values of clearness index out of the normal range may be generated by Eq. (6.9). In order to avoid such values, the series generated by Eq. (6.9) was limited to the minimum and maximum values occurred in the past. Thus, the final values of

Table 6.12 Performance measures for the model ARIMA(3,0,3)

RMSE	*MAE*	*MBE*	*MEAN*	*Var*	*Skew*	*Kurt*	
0.2485	0.1958	0.0073	0.4521	0.0576	−0.1259	−1.1997	F
			0.4179	0.0323	−0.5126	−1.0810	M

The flags F and M stand for forecasted and measured series, respectively

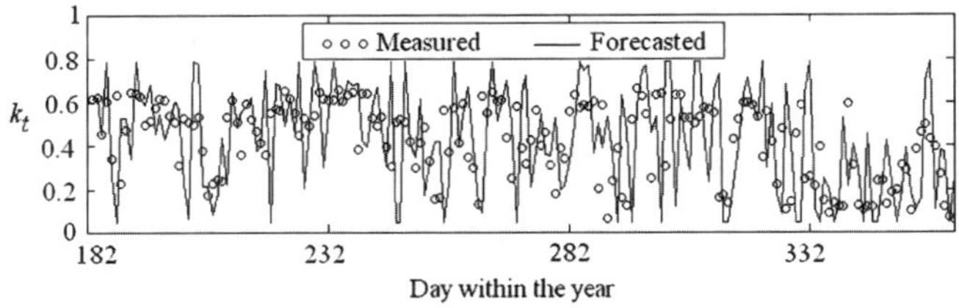

Fig. 6.12 Measured and forecasted daily global solar clearness index from July 1 to December 31, 2010

clearness indexes are obtained by confining the values generated by Eq. (6.9) with the conditional filter:

$$\hat{k}_t = \begin{cases} \min(k_t) & if \quad k_t < \min(k_{t,m}) \\ \max(k_t) & if \quad k_t > \max(k_{t,M}) \\ k_t & otherwise \end{cases} \tag{6.10}$$

where $k_{t,m}$ and $k_{t,M}$ are the minimum and maximum values in the measured time series.

Table 6.12 shows accuracy indicators for ARIMA(3,0,3) model after processing with Eq. (6.10). *RMSE* indicates a rather modest model performance.

Figure 6.12 shows the measured and forecasted daily global solar irradiation in the testing period. One can notice a certain inability of the model to grasp jumps from low to high values (or vice versa) of the measurements. These jumps of the clearness index are due to different states of the sky on consecutive days. Therefore, using a model that allows the entry of at least one additional variable (besides preceding values of the clearness index) could lead to increased forecast accuracy. This entry must be a parameter to measure the state of the sky, such as relative sunshine.

To conclude, this chapter was dedicated to short-time nowcasting of solar irradiance and one day ahead solar irradiation forecasting. We showed that a time series of instantaneous clearness index includes a daily pattern, which is visible if the measurement frequency is high. For forecasting instantaneous clearness index, ARIMA models with differencing order of at least one must be considered. As the forecast range increases, the instantaneous clearness index prediction accuracy decreases. We are not talking here about the generally lower accuracy for predicting values in a series at a growing number of lags. In the presented case, the prognosis is done for the next value in the series, but with the intervals between two measurements increasingly larger. Also, as the radiative regime is more stable, the prediction accuracy is better. In terms of daily solar irradiation forecast, even with an optimally selected model, the quality of the prediction accuracy in the test period is not the best. The results proved that ARIMA models can be used successfully to forecast solar irradiance on short intervals, i.e., minutes, tens of minutes. For longer intervals of time, better accuracy is possible by using models (e.g. logistic model) to integrate at the input other parameters, mainly associated with the state of the sky.

References

Box GEP, Jenkins GM (1970) Time series analysis. Forecasting and control. Holden-Day, San Francisco

Box GEP, Tiao GC (1992) Bayesian Inference in Statistical Analysis (Wiley Classics Library) Wiley-Interscience

Brockwell PJ, Davis RA (2002) Introduction to time series and forecasting. Springer, New York

Chen D, Peace KE (2010) Clinical trial data analysis using R. CRC Press, Boca Raton

De Gooijer JG, Hyndman RJ (2006) 25 years of time series forecasting. Int J Forecast 22:443–473

Gallego AJ, Camacho EF (2012) Estimation of effective solar irradiation using an unscented Kalman filter in a parabolic-trough field. Sol Energy (in press, corrected proofs). doi: 10.1016/j.solener.2011.11.012

Gillespie DT (1991) Markov processes: An introduction for physical scientists. Academic Press, London

Gottman JM (1982) Time-series analysis: a comprehensive introduction for social scientists. Cambridge University Press, Cambridge

Iizumi T, Nishimori M, Yokozawa M, Kotera A, Duy Khang N (2012) Statistical downscaling with bayesian inference: estimating global solar radiation from reanalysis and limited observed data. Int J Climatol 32:464–480

Kalman R (1960) A new approach to linear filtering and prediction problems. J Basic Eng 82:35–45

Palit AK, Popovic D (2005) Computational intelligence in time series forecasting: theory and engineering applications. Springer, Berlin

Paoli C, Voyant C, Muselli M, Nivet M-L (2010) Forecasting of preprocessed daily solar radiation time series using neural networks. Sol Energy 84(12):2146–2160

Pankratz A (1983) Forecasting with univariate Box-Jenkins models. Concepts and cases. Wiley, New York

Poggi P, Notton G, Muselli M, Louche A (2000) Stochastic study of hourly total solar radiation in Corsica using a markov model. Int J Climatol 20:1843–1860

PTC (2012)—MathCAD—Engineering calculations software. http://www.ptc.com/products/mathcad/

Sprott JC (2003) Chaos and time-series analysis. Oxford University Press, Oxford

Statgrafics (2012) Statgraphics centurion http://www.statlets.com/statgraphics_centurion.htm

Chapter 7
Fuzzy Logic Approaches

7.1 Artificial Intelligence Techniques

There are many different definitions, but in its broadest sense artificial intelligence (AI) is defined as the study and design of intelligent agents. An intelligent agent is an autonomous entity that perceives its environment, directs its activity toward achieving goals and takes actions that maximize its chances of success. The term *AI* has been introduced in 1956 by John McCarthy. *Programs with common sense* (McCarthy 1958) was probably the first paper on AI, in which logic is the method of representing information in computer memory and not just the subject matter of the program. Different from traditional algorithms, AI methods are highly successful in dealing with incomplete or uncertain input data. Most important, they are often capable to solve problems where the existing dependencies are too complex or insufficiently known to be programed in a traditional, rigid manner. These qualities are highly significant in the field of weather and solar energy modeling and forecasting. While AI systems are already used for some appliances, their use in photovoltaic field is just beginning. The book by Kalogirou (2007) includes a comprehensive review of AI techniques related renewable energy field while Mellit (2008) reviews AI systems related to modeling and forecasting solar radiation data.

There are several approaches developed inside AI such as: Genetic Algorithms (GA), Expert Systems (ES), Artificial Neural Networks (ANN), Fuzzy Logic (FL), and some hybrid systems that combine the above techniques.

A GA is a heuristic search that mimics the process of natural evolution. GAs generate solutions to optimization problems using techniques inspired by natural evolution, such as inheritance, mutation, selection, and crossover. The method is used in optimal sizing of power systems incorporated solar generators (Senjyu et al. 2006; Yang et al. 2008) but is rarely used in forecasting the output power of photovoltaic plant.

M. Paulescu et al., *Weather Modeling and Forecasting of PV Systems Operation*, Green Energy and Technology, DOI: 10.1007/978-1-4471-4649-0_7,

ES is a computer system that emulates the decision-making ability of a human expert. This AI technique is common in economic analysis but it is also seldom used in operating solar systems.

Several approaches for forecasting solar irradiance at different time horizon based on AI techniques have been reported. According to Mellit and Kalogirou (2008), these approaches can be categorized in three groups:

The models in the first group estimate solar irradiance based on meteorologic parameters such as air temperature (t_a), atmospheric pressure (p), relative humidity (RH), wind speed (v), cloud cover amount (C), or/and sunshine duration (σ) (Mohandes et al. 1998; Sen 1998; Donvlo et al. 2002; Reddy and Manish 2003; Lopez et al. 2005; Tymvios et al. 2005; Zarzalejo and Ramirez 2005; Tulcan-Paulescu and Paulescu 2008; Paulescu et al. 2008; Qin et al. 2011). In general, the equations of solar irradiance G can be expressed as:

$$G = f(t_a, p, RH, v, C, \sigma) \tag{7.1}$$

To write accurate correlations in the sense of Eq. (7.1), AI hybrid systems and FL stand out as fitting ways.

The second group includes models which predict the actual solar irradiance based on past observed data. Mathematically, the models can be formulated as:

$$G_t = f\left(G_{t-1}, G_{t-2}, \ldots, G_{t-p}\right) \tag{7.2}$$

Equation (7.2) is expressing the standard autocorrelation problem, but this time the solutions are developed by AI techniques (Kemmoku et al. 1999; Mihalakakou et al. 2000; Cao and Cao 2005; Mellit et al. 2006; Hocaoglu et al. 2008; Cao and Lin 2008; Paoli et al. 2010).

Models from the third group combine the defining approaches of the first and second group (Mellit et al. 2008):

$$G = f\left(G_{t-1}, G_{t-2}, \ldots, G_{t-p}, t_a, p, \mathrm{RH}, v, C, \sigma\right) \tag{7.3}$$

These approaches allow the forecasting of solar irradiance at different time horizons with different accuracies.

Since ANN and FL are used in many fields related to solar systems, including the estimation and forecasting of the available solar energy, these methods are briefly summarized in the following. In next sections, FL approaches proposed by our group at the West University of Timisoara (Romania) for estimating (Sect. 7.2) and forecasting (Sects. 7.3 and 7.4) solar irradiance and irradiation are detailed.

7.1.1 Artificial Neural Networks

An ANN is a mathematical model that is inspired by the structure and functional aspects of the biological neural networks. ANN consists of an interconnected

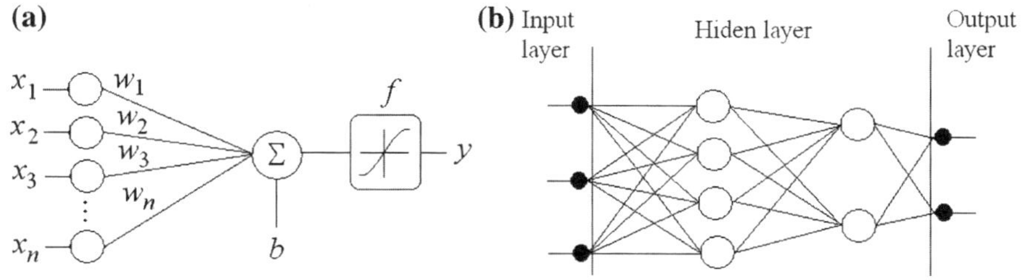

Fig. 7.1 **a** Schematic of an artificial neuron: x_i is the input amount from the synapse i, w_i is the weight for the specific input x_i, y is the output amount, and f stands for the activation function. **b** Example of a four layer ANN

group of artificial neurons organized in such a way that the network structure adapts itself to the requirements of the considered problem.

7.1.1.1 Introduction to ANN

The key element of any ANN is an artificial expression of the neuron, the fundamental cell of the brain. A schematic model of an artificial neuron is illustrated in Fig. 7.1a. The computer implementation resides in an algorithm whose operation can be summarized in four steps (Tymvios et al. 2008): (1) The inputs are multiplied by a predetermined weight and summed; (2) A bias is added to the result; (3) The sum is subjected to the activation (or transfer) function and is modulated accordingly; (4) The signal flows to the next neuron. The bias input to the neuron algorithm is an offset value that regulates the signal. Activation functions that are commonly used include the logistic function, the softmax function, and the gaussian function but in principle they may also take the form of other nonlinear functions, piecewise linear, sine, cosine or, step functions.

Mathematically, the output of an artificial neuron can be written as:

$$y = f\left(b + \sum_i w_i x_i\right) \tag{7.4}$$

where w_i is the weight for the specific input x_i, f is the activation function, and b is the bias for the neuron.

An artificial neural network is a collective set of such neural units, in which the individual neurons are connected through complex synaptic joints characterized by weight coefficients and every single neuron makes its contribution toward the computational properties of the whole system. A usual setup in solar energy applications has a three layer feedforward topology: input, hidden, and output layer. Occasionally, more than one hidden layer is used. An example of such a network, the multilayer perceptron (MLP) is given in Fig. 7.1b. When the term Artificial Network is used without any qualifications it refers to an MLP network.

It is demonstrated that in order for the hidden layer(s) to be useful, nonlinear activation functions are mandatory: a multilayer network using only linear activation functions is equivalent to some single layer, linear network.

An ANN is not a priori programed to perform specific tasks, but instead it is trained using a back propagation procedure until patterns in input data are learned. The ANN is typically initialized with random weights on its synapses. The learning process is similar to a gradient descent and requires that the activating functions are differentiable. It has the following steps: (1) forward propagation of the training pattern input, (2) calculating the error function Err by comparing the output o_j with the training pattern's target t_j:

$$\mathrm{Err} = \frac{1}{2}\sum_j \left|o_j - t_j\right|^2, \tag{7.5}$$

(3) back propagation of the error in order to generate the change in each weight Δw_{ij} for all output and hidden layer neurons, and (4) updating the weights w_{ij} using a rule like $w_{ij}^{\mathrm{new}} = w_{ij}^{\mathrm{old}} + \lambda \Delta w_{ij}$, where λ is a learning parameter. This way, the connections are adjusted so that the inputs are associated more strongly toward the expected answer. As the training proceeds, the network's response to the input data becomes better and better. Generally, more than one pattern is used, in which case the training is applied randomly and repeatedly for the set of examples until the network converges to a satisfactory local minima for the error function. Once the network is trained, all the synaptic weights are frozen (i.e. $\lambda = 0$, or, only step (1), forward propagation, remains enabled) and the network is ready to use.

ANN approach is an ongoing field in solar energy research. There are applications, such as for estimating solar irradiation at the ground (Reddy and Manish 2003; Tymvios et al. 2005; Mubiru and Banda 2008), forecasting solar radiation (Sfetsos and Coonick 2000; Mellit and Pavan 2010; Paoli et al. 2010), PV power output forecasting (Chen et al. 2011; Izgi et al. 2012).

7.1.1.2 Forecasting Solar Radiation with ANN

Results of two recent studies are summarized below. The presentation is focused on their forecasting accuracy rather than the model construction.

Mellit and Pavan (2010) developed an MLP for 24 h ahead forecasting solar irradiance. Mean daily solar irradiance and mean daily air temperature are the parameters used as entries in the model. The output is represented by the 24 values of solar irradiance in the next 24 h. The best performance was obtained with the following settings of the MLP: one input layer of three neurons ($\bar{T}_{j-1}$ daily mean of air temperature in the day $j-1$, $\bar{G}_{j-1}$ daily mean of global solar irradiance in the day $j-1$, and $j-1$ the day number within the year), two hidden layers containing 11 and 17 neurons, respectively, and an output layer with 24 neurons $\left(G_j^{(1)}, G_j^{(2)}, \ldots, G_j^{(24)}\right)$. This architecture has been successfully tested for forecasting solar irradiance at Trieste, Italy. This approach can be generalized to other areas by training using a measured dataset related to these areas.

The MLP output is related to the input by the equation:

$$\left(G_j^{(1)}, G_j^{(2)}, \ldots, G_j^{(24)}\right) = f\left(\bar{G}_{j-1}, \bar{T}_{j-1}, j\right) \tag{7.6}$$

where f is a nonlinear approximation function, which is estimated based on weights and the bias of the optimal MLP.

The model has been tested against measurements in four consecutive sunny days and other four consecutive cloudy days. For sunny days, *RMSE* (calculated for each day) varies between 18.9 and 67.0 % while for cloudy days *RMSE* varies between 54.6 and 85.7 %; *MBE* is between −6.4 and 32.0 % in sunny days and −45.5 and 53.6 % in cloudy days (Mellit and Pavan 2010). These results show that the algorithm performs better in the sunny days than in the cloudy day. However, the large values of *RMSE* indicate that there is room enough for increasing the model accuracy.

A comparison between the power produced by a 20 kWp grid connected photovoltaic plant and the one forecasted using the MLP predictor shows a rather modest performance of the model for the four sunny days (As percentage of the mean *RMSE* lies between 32.9 and 75.4 %). As indicated by the authors, this approach has many advantages with respect to other existing methods and it can be improved by adding more input parameters such as cloud cover or sunshine duration.

Another ANN algorithm for prediction daily global solar irradiation at daily horizon is reported in Paoli et al. (2010), constructed and validated with data collected at the meteorological station Ajaccio, Corsica Island, France. The algorithm uses an ad hoc time series preprocessing step and a time series prediction designed MLP. Without preprocessing step, verification against 2 years of measured data gives *RMSE* of 20.9 % for ANN and 21.1 % for AR(8). Interesting is that the authors calculated *RMSE* values for other methods (i.e. *K*—Nearest Neighbors, Bayesian inference, Markov chain) and found them to be greater than for multiyear (17) daily averages. Annual preprocessing ANN methods based on clearness index (ratio between ground and extraterrestrial daily solar irradiation) and clear sky index (ratio between ground actual and clear sky daily global solar irradiation) reduce forecasting errors with 5–6 % compared to classical predictors. The choice of preprocessing based on clearness index or clear sky index leads to comparable results.

The tool has been successfully validated on the DC energy prediction of a 1.175 kWp mono-Si PV grid connected system located at Vignola, near Ajaccio. By seasons, ANN with clear sky preprocessing represents an adequate solution for the winter months (*RMSE* ≈ 37 %). For summer months, ANN without preprocessing gives the best results (*RMSE* ≈ 15 %) (Paoli et al. 2010).

Results presented in both papers demonstrate that ANN is a viable option for forecasting solar irradiance, solar irradiation, and the output power of the PV systems. However, further studies are required to increase forecasting accuracy, for the method to be reliably implemented in practice. A discussion on this issue is inserted in Chap. 10, where the accuracy of various models developed for forecasting PV plant power output is assessed.

7.1.2 Fuzzy Logic

Since Aristotle, the theory of classical logic stated that every proposition must either be TRUE or FALSE, excluding the *middle*. In contrast, fuzzy logic is designed to allow computers to make use of the distinctions among data with shades of gray. It proposes making the values TRUE and FALSE operate over the range of real numbers [0, 1]. This should not lead to confusion between the degree of truth used in fuzzy theory and probabilities, which are conceptually distinct. Boolean logic can be seen as a subset of FL.

Fuzzy sets theory have been introduced in 1965 by Zadeh (1965) basically filling with real numbers the interval between 0 and 1. Because the fuzzy approach is quite different from classical sets theory, a short introduction to fuzzy sets theory is given. Additional references and examples can be found in many books (e.g. Zimmerman 1996; Passino and Yurkovich 1998).

7.1.2.1 Fuzzy Sets

The basis of Zadeh's logic theory is the fuzzy set concept, which is defined as follows. If X is a collection of objects, the associated fuzzy set A is defined as:

$$A \equiv \{ (x, m_A(x)) : x \in X\} \tag{7.7}$$

where $m_A(x)$ is the *membership function* showing the degree of affiliation of the element x to the fuzzy set A. Different subsets of A are separated by different membership functions. A physical variable is named *linguistic variable* and its values are not numbers (as in the case of deterministic variables), but linguistic values, called *attributes*, expressed by words or sentences. A membership function is associated to every attribute of a linguistic variable. It indicates the level of confidence with which that attribute characterizes a certain element from the set X. An intuitive example is presented below.

One assumes a sequence of measurements of daily relative sunshine σ (see Sect. 2.1 for definition). Let $X_\sigma \equiv [0, 1]$ corresponds to the X set from the definition (7.7). A value can be established (e.g., $\sigma_0 = 0.5$) to separate the set X_σ into two subsets named CLOUDY if $\sigma \in X_{\sigma 1} = [0, \sigma_0]$ and SUNNY if $\sigma \in X_{\sigma 2} = [\sigma_0, 1]$. We are accustomed to express that a day is sunny, i.e., σ belongs to $X_{\sigma 2}$ by an application f: $X_\sigma \to \{0,1\}$ defined as $f(\sigma) = 1$ if $\sigma \geq \sigma_0$ or else $f(\sigma) = 0$, as is shown in Fig. 7.2. It is what we call a crisp set. Dissimilar, fuzzy sets theory relaxes the crossing from $X_{\sigma 1}$ to $X_{\sigma 2}$ by replacing the abrupt boundary between

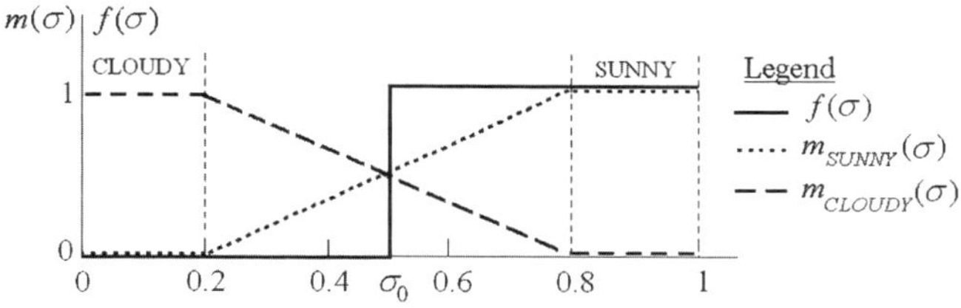

Fig. 7.2 The characteristic function $f(\sigma)$ defining the crisp sets CLOUDY and SUNNY and the membership functions m_{CLOUDY} and m_{SUNNY} of daily mean relative sunshine attributes CLOUDY and SUNNY, respectively

CLOUDY and SUNNY attributes with a slowly varying crossing in a finite interval around σ_0. The binary function $f(\sigma)$ is replaced by the membership functions $m_{\mathrm{SUNNY}}(\sigma)$ and $m_{\mathrm{CLOUDY}}(\sigma)$, which takes values in the interval [0,1].

From Fig. 7.2, a day characterized by relative sunshine below 0.2 certainly has the attribute CLOUDY, while a day characterized by relative sunshine above 0.8 certainly has the attribute SUNNY. A day with relative sunshine of 0.3 is assigned with a 5/6° to be CLOUDY and a 1/6° to be SUNNY. Effectively, the membership function reads out the *level of confidence* for σ to be member of the subsets SUNNY and CLOUDY. The fundamental difference with respect to the probability theory, which assigns for every day a probability to be SUNNY **OR** CLOUDY, is that fuzzy sets theory claims that every day has both the SUNNY **AND** the CLOUDY attributes simultaneously, the first with the confidence level $m_{\mathrm{SUNNY}}(\sigma)$ and the second with the confidence level $m_{\mathrm{CLOUDY}}(\sigma)$.

The number of attributes of a linguistic variable and the shape of the membership functions depend on the application, being specified in a heuristic way. Theoretically, the membership function can have any form; practically, due to the satisfactory results combined with the easy maneuverability, three symmetric forms are frequently used, namely triangular, trapezoidal, and Gaussian, as is shown in Fig. 7.3.

Usually, fuzzy models are constructed based on a set of measured data. There are some methods to determine more accurately the membership functions for a given set of data, for example by fuzzy c-mean clustering, as is shown at the end of this section.

7.1.2.2 Operations on Fuzzy Sets

Building a set theory requires defining operations between the elements of the set. In fuzzy sets theory, these operations are defined through the membership functions:

Fuzzy intersection (**AND**):

$$m_{A\cap B} = \min(m_A(x)\,,\; m_B(x)), \quad \forall x \in X \tag{7.8a}$$

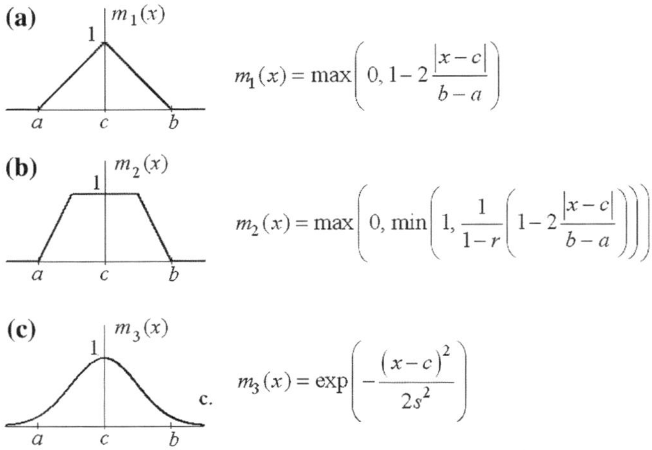

Fig. 7.3 Typical symmetric membership functions. **a** Triangular. **b** Trapezoidal (r is the trapeze bases ratio). **c** Gaussian (s is the Gaussian dispersion)

Fuzzy reunion (**OR**):

$$m_{A\cup B} = \max(m_A(x)\,,\ m_B(x)), \quad \forall x \in X \tag{7.8b}$$

Fuzzy complement:

$$m_{\bar{A}} = 1 - m_A(x)\,, \quad \forall x \in X \tag{7.8c}$$

Generally, a fuzzy logic model is a functional relation between two multidimensional spaces, $F : I^n \to O^p$, where I^n and O^p contain the fuzzy sets of n input and p output linguistic variables, respectively. The mapping between the input and the output spaces envelops the linguistic variables, the attributes, and the associative rules among different fuzzy sets. The rules are often expressed in the form:

IF (premises) **THEN** (conclusions) (7.9a)

Every premise or conclusion consists of an expression like:

(variable) **IS** (attribute) (7.9b)

This approach of the linguistically expressed rules is closer to human thinking than other mathematical approaches.

Thus, the information is carried out from the input to the output of a fuzzy model in three steps: fuzzification, inference, and defuzzification:

1. The *fuzzification* is a coding process in which each numerical input of a linguistic variable is transformed into the membership function values of its attributes.
2. The *inference* is a process consisting of two steps: (1) The computation of the degree in which a rule is fulfilled by the intersection of individual premises,

i.e., by applying the fuzzy operator **AND**. (2) Sometimes, some rules drive to the same conclusion, i.e., to the same attribute of the output linguistic variable. For finding the confidence level of this conclusion, the individual degrees of fulfilling the rules driving to this conclusion are joined by applying the fuzzy operator **OR**.

3. The d*efuzzification* is a decoding operation of the information contained in the output fuzzy sets resulted from the inference process, on the purpose of providing the most suitable output crisp value. There are many defuzzification methods (for details see Zimmerman 1996); in this chapter, we apply the center of gravity (COG) method, which is one of the most popular. According to COG, the suitable output crisp value is computed with the equation:

$$y_{\text{crisp}} = \frac{\sum_i c_i \int m_{y_i}(x)\mathrm{d}x}{\sum_i \int m_{y_i}(x)\mathrm{d}x} \tag{7.10}$$

In Eq. (7.10), c_i is the center of the membership function (generally, the value of the variable x where the membership function reaches its peak) and the integral $\int m_{y_i}(x)\mathrm{d}x$ represents the surface under the membership function $m_{yi}(x)$ corresponding to the attribute i of the output linguistic variable y.

7.1.2.3 Illustration of an FL Algorithm Operation

The problem of forecasting the output power of a PV system is considered in the following simplest example. One considers a fuzzy system with two input linguistic variables: cloud cover C and air temperature t, and one output linguistic variable, the PV system output power p expressed as percentage of the nominal power. The linguistic variable C has three attributes OVERCAST (O), VARIABLE (V), and SUNNY (S) and the linguistic variable t has four attributes, FREEZING (F), COOL (C), WARM (W), and HOT (H). The output linguistic variable p is characterized by three attributes SMALL (S), MEDIUM (M), and HIGH (H). Figure 7.4 displays the associated membership functions and the notations (which in a real application have to be elaborated heuristically or using a clustering procedure).

The associative rules between input and output are listed in Table 7.1. Every column is a rule in sense of Eq. (7.9a). For example rule #5 reads:

$$\text{Rule \#5}: \textbf{IF } C \text{ is VARIABLE and } t \text{ is Freezing } \textbf{THEN } p \text{ is SMALL} \tag{7.11}$$

Let us trace, using a numeric example, how the information is flowing through the fuzzy system from the input to the output.

1. *Fuzzification*: Let us assume at input the pair ($C = 0.85$; $t = 20$ °C). This means that the linguistic variable cloud cover is characterized by the attribute

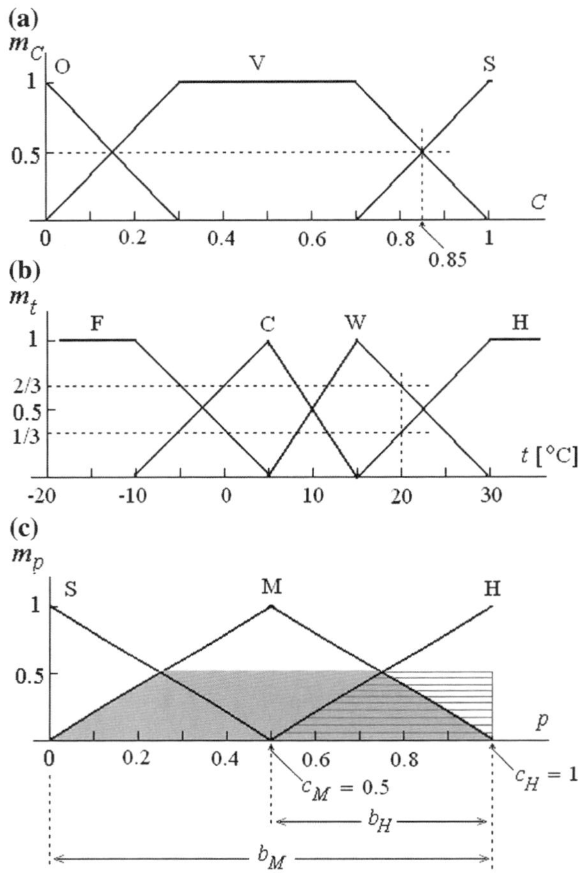

Fig. 7.4 Illustration of an FL model functioning. Membership functions associated to the attributes of the linguistic variables. **a** Cloud cover amount *C*. **b** Air temperature *t*. **c** Percentage of PV system power output *p*

Table 7.1 The associative rules connecting the input and output fuzzy sets defined in Fig. 7.4

Rule #	01	02	03	04	05	06	07	08	09	10	11	12
C attributes	**O**	**O**	**O**	**O**	**V**	**V**	**V**	**V**	**S**	**S**	**S**	**S**
t attributes	**F**	**C**	**W**	**H**	**F**	**C**	**W**	**H**	**F**	**C**	**W**	**H**
p attributes	**S**	**S**	**M**	**M**	**S**	**S**	**M**	**H**	**S**	**M**	**H**	**H**

VARIABLE with the confidence level $m_V(0.85) = 1/2$ *and* the attribute SUNNY with the confidence level $m_S(0.85) = 1/2$. The linguistic variable temperature for $t = 20$ °C has the attributes WARM *and* HOT with $m_W(20) = 2/3$ and $m_H(20) = 1/3$, respectively.

2. *Inference*: It can be seen that four rules, *R*#7, *R*#8, *R*#11, and *R*#12, are active. In the terms of membership functions, we can write:

   R#07: $m_{p,M} = \min(m_V, m_W) = 1/2$
   R#08: $m_{p,H} = \min(m_V, m_H) = 1/3$
   R#11: $m_{p,H} = \min(m_S, m_W) = 1/2$
   R#12: $m_{p,H} = \min(m_S, m_H) = 1/3$

   The rules *R*#8, *R*#11, and *R*#12 lead to the same conclusion, attribute HIGH for the variable *p*. To find the confidence level transferred to the output attribute *p*, the fuzzy operator **OR** should be applied:
   $m_{p,H} = \max(1/3,\ 0.5,\ 1/3) = 1/2$

3. *Defuzzification*: The output crisp value of *p* can be now calculated with the COG method (Eq. 7.10). The surfaces representing the integrals in Eq. (7.10) are sketched in Fig. (7.4c). Denoting the trapeze bases with $b_M = 1$ and $b_H = 0.5$ and the coordinate of the triangles peak with $c_M = 0.5$ and $c_H = 1$, the percentage of the output power is:

$$p = \frac{c_M b_M m_{p,M}\left(1 - \frac{m_{p,M}}{2}\right) + c_H b_H m_{p,H}\left(1 - \frac{m_{p,H}}{2}\right)}{b_M m_{p,M}\left(1 - \frac{m_{p,M}}{2}\right) + b_H m_{p,H}\left(1 - \frac{m_{p,H}}{2}\right)} = 0.666$$

7.1.2.4 Takagi–Sugeno Fuzzy Systems

In the Takagi–Sugeno (TS) approach (Takagi and Sugeno 1985), the premises and the map from the input to the output of the system resemble the fuzzy logic system, while the output membership attributes are replaced by mathematical functions. Thus, each rule, expressed as a sentence "**IF** (*premises*) **THEN** (*conclusion*)" (see Eq. 7.9a), drives to a crisp conclusion expressed as:

$$y = f(x_1, x_2, \ldots, x_N) \tag{7.12}$$

where x_i, x_2,…, x_N stand for the *N* input variables. In the standard TS model, the output y_k is calculated as a linear combination of the input variables:

$$y(x_1, x_2, \ldots, x_N) = a_0 + \sum_{k=1}^{N} a_k x_k \tag{7.13}$$

but any other function $f_k(x_1, x_2 \ldots x_N)$ can be considered.

In the TS model, the defuzzification procedure is simply carried out by taking up the weighted average outputs y_k, $k = 1, 2, \ldots, M$, where *M* is the total number of active rules. The suitable output crisp value is extracted with the relation:

$$y = \sum_{k=1}^{M} m_k y_k \bigg/ \sum_{k=1}^{M} m_k \tag{7.14}$$

7.1.2.5 Fuzzy c-Mean Clustering

Fuzzy c-mean clustering (Zimmermann 1996) is an algorithm parting the set X with N elements (x_k, $k = 1\ldots N$), into C fuzzy subsets (or clusters). The partition is represented by a matrix $M = m_{i,k}$ which consists of the values of the membership functions $m_{i,k}$ of all elements x_k, $k = 1\ldots N$, to every cluster i, $i = 1\ldots C$. In other words, $m_{i,k}$ is the membership function of the k element in the i cluster. The dimension of the matrix M is $C \times N$, where the number of clusters C meets the condition $2 \leq C \leq N$. The matrix M is a c-mean fuzzy partition on the set X if:

1. $m_{i,k} \in [0, 1], \quad \forall i = 1\ldots C, \quad \forall k = 1\ldots N$
2. $\sum_{k=1}^{N} m_{i,k} > 0, \quad \forall i = 1\ldots C$. This means that any cluster holds at least one element.
3. $\sum_{i=1}^{C} m_{i,k} = 1, \quad \forall k = 1\ldots N$. The membership functions should always be normalized to unity, i.e., the sum of the membership values of the element x_k over all clusters is equal to unity.

In order to find the fuzzy partition, an iterative algorithm is used for minimizing the function:

$$J_\mu(\mathrm{M}, \Gamma) = \sum_{k=1}^{N} \sum_{i=1}^{C} m_{i,k}^{\mu} d_{i,k}^{2}, \quad 1 \leq \mu \leq \infty \tag{7.15}$$

In Eq. (7.15) μ is the weighting parameter and $d_{i,k}$ are the elements of the similarity matrix:

$$d_{i,k} = |x_k - c_i| \tag{7.16}$$

The cluster centers $\{c_1, c_2, \ldots, c_C\}$ form the vector Γ and they are calculated with the following equation:

$$c_i = \sum_{k=1}^{N} m_{i,k}^{\mu} x_k \bigg/ \sum_{k=1}^{N} m_{i,k}^{\mu}, \quad i = 1\ldots C \tag{7.17}$$

The algorithm determines the matrix M (cluster matrix) in which every column is associated to a cluster. Each column contains the values of the membership functions for all elements of set X. This means that an element of the column i of M states the degree to which an element of the set X is belonging to the ith cluster. In the first step, the cluster matrix has to be initialized; a crisp partition may be

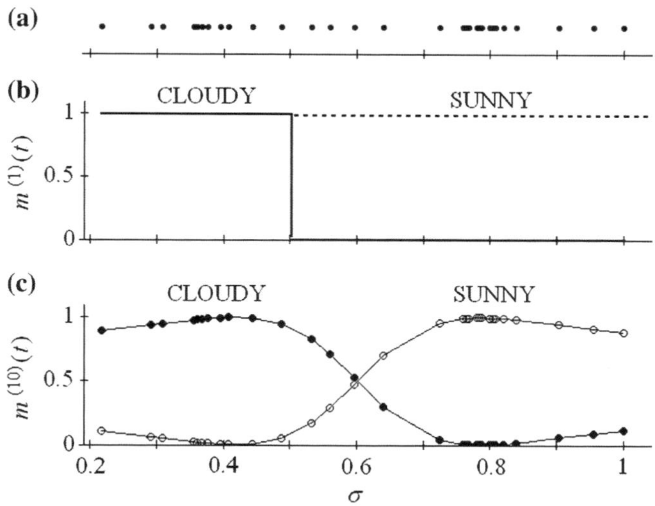

Fig. 7.5 Illustration of the fuzzy clustering procedure. **a** Distribution of the elements (relative sunshine, as is assumed) in the set X. **b** Crisp partition of the set X. **c** Fuzzy partition after nine steps of c-mean fuzzy algorithm

chosen, i.e., we have to fill M with 0 and 1. Then the cluster centers' vector is calculated with Eq. (7.17) and the partition M is updated. At the nth step, the membership function of the kth element to fall in the cluster i is (Zimmermann 1996):

$$m_{i,k}^{(n)} = \left[\sum_{j=1}^{C} \left(\frac{d_{i,k}^{(n-1)}}{d_{j,k}^{(n-1)}} \right)^{\frac{2}{\mu-1}} \right]^{-1} \tag{7.18}$$

The algorithm is stopped when the accuracy criterion is fulfilled:

$$\max\left(m_{i,k}^{(n)} - m_{i,k}^{(n-1)} \right) < \varepsilon \tag{7.19}$$

An illustrative example follows. Let X be a set with $N = 30$ elements consisting of relative sunshine σ recorded in a month. One assumes the distribution of σ as plotted in Fig. 7.5a. Visual inspection shows that two clusters can be identified in the proximity of the two data agglomeration. We will refer to them as CLOUDY and SUNNY. Again, one considers $\sigma_0 = 0.5$ the relative sunshine that separates the set X in the two subsets. Now, the initial matrix M_1 complying with the rule can be defined: for the CLOUDY subset $m_{C,k}^{(1)} = 1$ if $\sigma_k < \sigma_0$ and $m_{C,k}^{(1)} = 0$ otherwise; for the SUNNY subset $m_{S,k}^{(1)} = 0$ if $\sigma_k < \sigma_0$ and $m_{S,k}^{(1)} = 1$ otherwise, with $k = 1 \ldots N$. This is what is currently called a crisp partition. The initial membership functions $m_{\mathrm{CLOUDY},k}^{(1)}$ and $m_{\mathrm{SUNNY},k}^{(1)}$ are plotted in Fig. 7.5b. Now, the algorithm may

Table 7.2 Sequence of data from the set X and the membership function of initial M_1 and the final M_{10} matrices around the switching zone from CLOUDY to SUNNY initial clusters

k	σ_k	$m^{(1)}_{\text{CLOUDY},k}$	$m^{(1)}_{\text{SUNNY},k}$	$m^{(10)}_{\text{CLOUDY},k}$	$m^{(10)}_{\text{SUNNY},k}$
8	0.39	1	0	0.987	0.013
9	0.40	1	0	0.990	0.010
10	0.44	1	0	0.997	0.003
11	0.48	1	0	1.000	0
12	0.53	0	1	0.993	0.007
13	0.56	0	1	0.945	0.055
14	0.59	0	1	0.828	0.172
15	0.64	0	1	0.712	0.288
16	0.72	0	1	0.528	0.472

run as described above. After nine steps, the required precision $\varepsilon = 10^{-3}$ is reached and the fuzzy partition is encapsulated in M_{10}. The membership functions of the clusters CLOUDY $m^{(10)}_{\text{CLOUDY},k}$ and SUNNY $m^{(10)}_{\text{SUNNY},k}$ are plotted in Fig. 7.5c. It can be seen that a given day does not belong solely to a well-defined cluster, but it has both the CLOUDY *and* SUNNY fuzzy attributes at the same time.

The nine elements of the set X located around the switching zone from CLOUDY to SUNNY in the initial partition and the corresponding membership functions values included the initial and final matrices M_1 and M_{10} are listed in Table 7.2. Each matrix has two columns corresponding to $C = 2$ clusters.

Thus, in general, in fuzzy clustering every datum may belong to several clusters, with different values of confidence level. For each cluster, the discrete values of the membership function generated by a fuzzy clustering algorithm may be fitted in order to find the appropriate equations as premise in the FL algorithms.

7.2 Models for Estimating Solar Irradiance and Irradiation

In the last two decades, several models for estimating either solar irradiance or solar irradiation have been developed based on the FL theory. The basic ideas that started the construction of these models may be exploited for developing forecasting models. Three different models are reviewed next: the first evaluates the atmospheric transmittance using a typical fuzzy algorithm; the second model uses fuzzy clustering procedures to translate diffuse solar irradiance from a horizontal surface to an inclined surface; and the third model estimates solar irradiation from sunshine duration via FL.

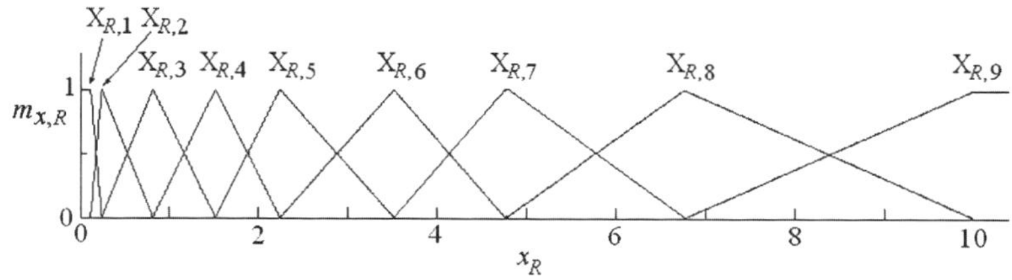

Fig. 7.6 Membership functions $m_{X,R}$ of the attributes $X_{R,i}$, $i = 1\ldots8$, in the case of Rayleigh scattering

7.2.1 Modeling Atmospheric Transmittance

In Paulescu et al. (2008), two models for solar radiation attenuation in the atmosphere are proposed. The distinctive feature of the models consists in using FL approach for evaluating the atmospheric transmittances associated to the main attenuators: Rayleigh scattering, aerosol extinction, ozone, water vapor, and trace gas absorption. The first model (I-ATM) include self-dependent fuzzy modeling of each characteristic transmittance, while the second (B-ATM) is a proper FL model for beam and diffuses atmospheric transmittances.

The starting point for the two fuzzy models described here is the Leckner's atmospheric transmittance (see Sect. 5.3.3.1), averaged with respect to wavelength (Eq. 5.48). Notations used in Sect. 5.3.3.1 are also used in the following.

In I-ATM model, every atmospheric transmittance is treated as an output linguistic variable while the product between corresponding surface meteorologic parameter and the atmospheric air mass is used as input linguistic variable: $x_R = m(p/p_0)$, $x_\beta = m\beta$, $x_w = ml_w$, $x_g = m$, $x_{O3} = ml_{O3}$. Fuzzy Associative Memory has been done independently for each attenuator, so that I-ATM entails five fuzzy models, which are formally similar. Every linguistic variable is characterized by nine attributes. The nonlinear behavior of input variables has been compensated by an appropriate progressive partition of attributes range. This is illustrated in Fig. 7.6 where the membership functions for the input variable x_R attributes are displayed. The geometry of all membership functions is triangular and all membership functions, either *in* or *out*, reads:

$$m_X(x) = \left| \begin{array}{ll} \max\left(0,\ \dfrac{x - a_i}{c_i - a_i}\right) & \text{if} \quad x < c_i \\ \max\left(0,\ 1 - \dfrac{x - c_i}{b_i - c_i}\right) & \text{otherwise} \end{array} \right. \tag{7.20}$$

where the constants a_i, b_i, and c_i have the signification as in Fig. 7.2a. Always, the vertex of a triangle coincides with the corresponding base extremities of adjacent triangles. Consequently, the vertices position c_i (Table 7.3) is enough to define all membership functions. The attributes of a linguistic variable are indexed with subscript number from 1 to 9, ranging from very low to very high. For every *in*

Table 7.3 c_i for the in and out membership functions in the I-ATM model

Process	Rayleigh scattering		Aerosol extinction		Water vapor absorption		Gas trace absorption		Ozone absorption	
Attribute	*In*	*Out*	*In*	*Out*	*In*	*Out*	*In*	*Out*	*In*	*Out*
c_1	0.1	0.59	0.005	0.2	0.0075	0.8	0.025	0.975	0.001	0.825
c_2	0.225	0.65	0.05	0.3	0.025	0.825	0.05	0.98	0.05	0.85
c_3	0.8	0.7	0.125	0.4	0.075	0.85	0.5	0.984	0.5	0.875
c_4	1.5	0.75	0.2	0.5	0.5	0.875	1.0	0.986	1.5	0.9
c_5	2.22	0.8	0.3	0.6	1.25	0.9	1.75	0.988	2.75	0.925
c_6	3.55	0.85	0.45	0.7	3.25	0.925	2.5	0.99	4.0	0.95
c_7	4.75	0.9	0.625	0.8	7.25	0.95	4.0	0.992	5.5	0.975
c_8	6.75	0.95	0.85	0.9	15.5	0.975	7.25	0.996	7.5	0.99
c_9	10.0	1.0	1.5	1.0	30.0	1.0	50.0	1	10	1.0

linguistic variable, the membership functions of the attributes indexed $\mathbf{X}_1$ and $\mathbf{X}_9$ have been saturated toward $x = 0$ and toward the upper domain limit, respectively.

Taking into account that each atmospheric transmittances $\tau(x)$ decrease with increasing x (See Sect. 5.3.4.2) the rules take a very simple form:

$$\text{IF } x = \mathbf{X}_\mathrm{i} \text{ THAN } y = \mathbf{Y}_{10-i} \tag{7.21}$$

Summarizing, to compute the atmospheric transmittance one starts with the air mass evaluation using a classical approach (Eq. 5.24). Then, transmittance is calculated for every atmospheric attenuator, using above fuzzy prescriptions. Finally, beam $\bar{\tau}_b$ and diffuse $\bar{\tau}_d$ transmittances are calculated with Eq. (5.52a, b). This swing between classic and fuzzy procedures is the model feature which suggests naming it *interlacing*.

The almost perfect way in which fuzzy algorithm reproduces the averaged transmittances is shown in Fig. 7.7. Note that the representation of fuzzy transmittances with dots was only made for convenience, to be distinct from transmittances calculated by Eq. (5.48). Otherwise, fuzzy transmittances are continuous and *de facto* can be integrated for solar irradiation.

In a different way, B-ATM enables an FL procedure which directly evaluates the mean beam atmospheric transmittance. The fuzzy algorithm uses at input five linguistic variables associated to the same atmospheric attenuators as in I-ATM. Practically, apart from I-ATM, the model B-ATM adds Eqs. (5.52a, b) inside Fuzzy Associative Memory. Every input linguistic variable is also characterized by 9 attributes, while the output linguistic variable is characterized by 20 attributes. As in previous case all membership functions are triangularly, being specified in Tables 7.4 and 7.5. In this case, FAM contains $9^5 = 59{,}049$ rules, written as:

IF

x_R is $\mathbf{X_R}$ AND x_β is $\mathbf{X}_\beta$ AND x_w is $\mathbf{X_w}$ AND x_{O3} is $\mathbf{X}_{O3}$ AND x_g is $\mathbf{X_g}$

THEN

τ *is* $\mathbf{T}$

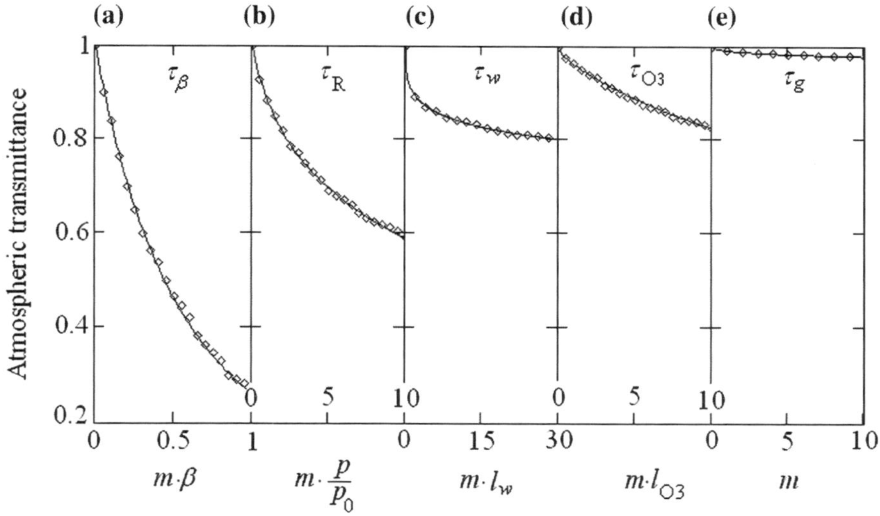

Fig. 7.7 Comparison between averaged Leckner's transmittance (*line*) and the fuzzy transmittance (*dots*): **a** Aerosol extinction τ_β; **b** Rayleigh scattering τ_R. **c** Water vapor absorption τ_w. **d** Ozone absorption τ_{O3}. **e** Trace gas absorption τ_g. From Paulescu et al. (2008), with permission from Elsevier

Table 7.4 c_i for the *in* membership functions of the B-ATM model

Process	Rayleigh scattering	Aerosol extinction	Water vapor absorption	Gas trace absorption	Ozone absorption
c_1	0.075	0.0005	0.0001	0.05	0.001
c_2	0.125	0.0025	0.001	1.0	0.05
c_3	0.325	0.02	0.01	2.5	0.25
c_4	1.0	0.075	0.075	3.5	1.55
c_5	1.5	0.175	0.75	5.5	2.15
c_6	2.25	0.275	2.5	12.5	3.25
c_7	3.75	0.75	10.0	30.0	4.5
c_8	6.25	1.55	22.5	50.0	8.25
c_9	9.0	2.25	75.0	75.0	11

Table 7.5 c_i for the *out* membership functions of the B-ATM model

c_1	0.15	c_6	0.575	c_{11}	0.8125	c_{16}	0.9375
c_2	0.25	c_7	0.625	c_{12}	0.8375	c_{17}	0.9575
c_3	0.35	c_8	0.675	c_{13}	0.8625	c_{18}	0.9725
c_4	0.45	c_9	0.725	c_{14}	0.8875	c_{19}	0.985
c_5	0.525	c_{10}	0.775	c_{15}	0.9125	c_{20}	1.0

Table 7.6 Input parameters for generation the rule base of B-ATM model

i	1	2	3	4	5	6	7	8	9
$x_{r,i}$	0	0.1	0.225	0.8	1.5	2.225	4.0	5.5	10.0
$x_{\beta,i}$	0	0.001	0.01	0.075	0.175	0.275	0.75	1.75	1.95
$x_{w,i}$	0	0.005	0.05	0.1	0.75	2.5	10.0	25.0	50.0
$x_{g,i}$	0	0.125	0.25	0.5	1.0	1.5	2.0	2.5	3.5
$x_{O3,i}$	0	0.001	0.05	0.5	1.5	2.75	5.5	7.5	12.5

Table 7.7 Parameters v_i for generating the rule base of the B-ATM model

v_1	0.15	v_6	0.575	v_{11}	0.8125	v_{16}	0.9375
v_2	0.25	v_7	0.625	v_{12}	0.8375	v_{17}	0.9575
v_3	0.35	v_8	0.675	v_{13}	0.8625	v_{18}	0.9725
v_4	0.45	v_9	0.725	v_{14}	0.8875	v_{19}	0.985
v_5	0.525	v_{10}	0.775	v_{15}	0.9125	v_{20}	1.0

Obviously, such a large rule base could be generated only using computer programs. In this case, the rules were established using Eq. (5.51) and the inputs from Table 7.6. The following algorithm was employed for sequentially calculating the attributes for every variable (*R#* denotes the rule number):

R# = 0
for i_1 =1 to 9
  for i_2 =1 to 9
    for i_3 =1 to 9
      for i_4 =1 to 9
        for i_5 =1 to 9
          *in*_attribute = $(i_1, i_2, i_3, i_4, i_5)$
          R# = R# + 1
          for k =1 to 18
            if $\tau\left(x_{r,i1}, x_{\beta,i2}, x_{w,i3}, x_{O3,i4}, x_{g,i5}\right) < v_1$ then *out*_attribute = 1
            if $v_i \le \tau\left(x_{r,i1}, x_{\beta,i2}, x_{w,i3}, x_{O3,i4}, x_{g,i5}\right) < v_{i+1}$ then *out*_attribute = i
            if $v_{19} \le \tau\left(x_{r,i1}, x_{\beta,i2}, x_{w,i3}, x_{O_3,i4}, x_{g,i5}\right)$ then *out*_attribute = 20
end

In order to run this procedure, the coefficients v_i are listed in Table 7.7. The *in* attributes and the corresponding *out* attribute are stored in a matrix with 9^5 rows and 7 columns. For example, rule $R\# = 923$ is written as:

00923 1 2 3 4 5 14

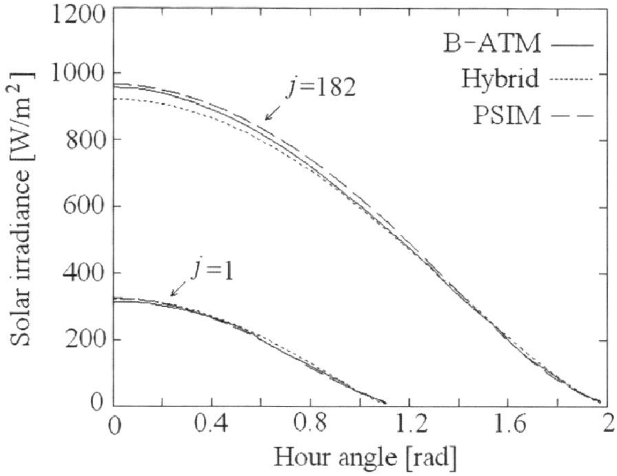

Fig. 7.8 Comparison between the models: B-ATM, Hybrid (Eq. 5.49a, b) and PSIM (Eq. 5.53b). From Paulescu et al. (2008), with permission from Elsevier

which reads:

IF

x_R is $\mathbf{X_{R,1}}$ AND x_β is $\mathbf{X_{\beta,2}}$ AND x_w is $\mathbf{X_{w,3}}$ AND x_{O3} is $\mathbf{X_{O3,4}}$ AND x_g is $\mathbf{X_{g,5}}$

THEN

τ *is* $\mathbf{T}_{14}$

Figure 7.8 shows the global solar irradiance calculated with B-ATM as functions of hour angle in two clear sky days: a winter one (Julian day $j = 1$) and a summer one (Julian day $j = 182$). The calculations have been done for 45° northern latitude and assuming the following atmospheric parameters: $l_{O3} = 0.35$ cm·atm, $\beta = 0.077$, $p = p_0$, $w = 2.3$ g/cm^2. It can be seen that the B-ATM curve is placed between the two curves generated with two well-known parametric models: the hybrid model (Eqs. 5.49a, b) and the PSIM model (Eq. 5.53b). It is remarkable to note that the model accuracy is preserved in winter days. This suggests that the accuracy of B-ATM model is at least comparable with the two above-mentioned models, demonstrating that FL is a viable alternative to classical parameterization. This conclusion is also supported by verification of the model against measured data at five meteorologic stations located in Romania, reported in Paulescu et al. (2008). Root mean square error at estimating daily global solar irradiation under clear sky has been found between 4.5 and 12.2 %.

7.2.2 Modeling Diffuse Irradiance on Inclined Surface

A model of estimating solar irradiance on arbitrarily oriented inclined surfaces based on hourly solar radiation average is proposed in Gomez and Casanovas (2003). Like the majority of the models for computation of solar irradiance on inclined surface, this model also evaluates the diffuse solar irradiation calculation. The characteristics that individualize the model are: (1) The sky categories (differentiate by the sky opacity) are defined by fuzzy clustering. (2) The optimum number of categories is fitted by means of competitive learning ANN. The optimum number of clusters is given by evaluating the partition entropy. (3) Global solar irradiance is evaluated taking into account that an input date may belong to different sky categories simultaneously, at different degrees of confidence.

The diffuse solar irradiance $G_d(\beta)$ on an surface tilted by angle β in respect to the horizontal plane, is considered to be a function of diffuse solar irradiance incident on horizontal surface $G_d(0)$ and the atmosphere clearness and brightness encapsulated in the function F_β:

$$G_d(\beta) = F_\beta G_d(0) \tag{7.22}$$

F_β is expressed with the equation (Perez et al. 1987, 1990):

$$F_\beta = \frac{0.5(1 + \cos\beta) + f\max(0, \cos\theta_z) + g\sin\beta}{1 + f\cos\theta_z} \tag{7.23}$$

where f is the relative increment in radiance associated to the circumsolar zone while g is the relative increment in radiance associated to the horizon zone of the sky. For $f = g = 0$, Eq. (7.23) reduces to the isotropic model (Eq. 5.101).

Two indices are used to measure the atmospheric conditions: (1) Atmospheric clearness index $k_n = G_n/G_{SC}$, defined as the ratio between solar irradiance at normal incidence G_n and solar constant G_{SC} and (2) Atmospheric brightness index $\Delta = mG_d(0)/G_{SC}$ where m is the optical air mass.

Four circular clusters (sky categories) are defined by means of indices k_n and Δ. Each cluster has the radius $r = 0.21$ and the centers: $k_n^c = 0.015$ and $\Delta^c = 0.161$ for overcast sky, $k_n^c = 0.135$ and $\Delta^c = 0.323$ for partly cloudy sky, $k_n^c = 0.404$ and $\Delta^c = 0.226$ for clear sky with high aerosol content in atmosphere and $k_n^c = 0.540$ and $\Delta^c = 0.119$ for clear sky and clean atmosphere. The membership functions indicated the degree of an input datum fitting in a cluster is inverse proportional to the Euclidian distance to those of the center of the class considered, i.e., $\left((k_n - k_n^c)^2 + (\Delta - \Delta^c)^2\right)^{-1/2}$. The relative radiance enhancements f and g in Eq. (7.23) are regarded as constant for a given sky category: $f = 0.122$ and $g = -0.063$ for overcast sky, $f = 0.994$ and $g = 0.078$ for partly cloudy sky, $f = 1.1363$ and $g = 0.358$ for clear sky with high aerosol content in atmosphere and $f = 1.765$ and $g = 0.921$ for clear sky and clean atmosphere (Gomez and Casanovas 2003). The factor F_β is computed by the weighted average procedure:

$$F_\beta = \sum_{i=1}^{4} \mu_i F_\beta^i \tag{7.24}$$

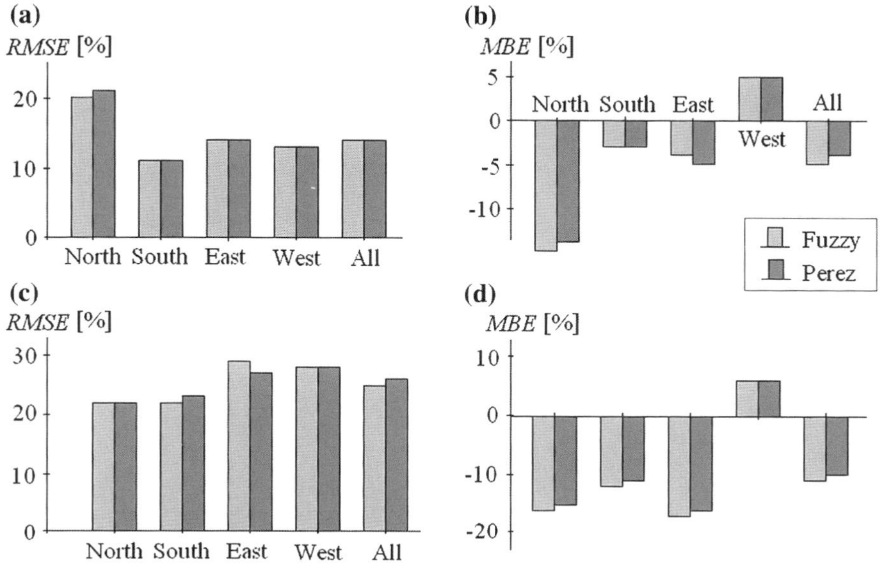

Fig. 7.9 Indicators of accuracy of the fuzzy and Perez et al. (1990) models estimating solar irradiance on vertical surfaces with different orientations: (**a**, **b**) total and (**c**, **d**) diffuse. Diagrams are build with data from Gomez and Casanovas (2003)

where μ_i are the membership functions of an input datum associated to each cluster. F_{β}^{i} are calculated with Eq. (7.23) using the coefficients f and g of each category.

Results of testing the fuzzy and Perez et al. (1990) models against measured data on vertical surfaces in Valencia, Spain, during the years 1992–1995, are summarized in Fig. 7.9. It can be seen that the performance of the fuzzy and Perez et al. models are comparable. As the author claims, the advantage of the fuzzy model is that it involves far fewer adjustable parameters and a smaller number of sky categories (while Perez et al. model operates with eight sky categories, the fuzzy model needs only four). The model proposed in Gomez and Casanovas (2003), allows physical interpretation of the model parameters obtaining through fitting.

7.2.3 Solar Irradiation From Sunshine Duration

In Reference Sen (1998), the theory of fuzzy sets is employed to represent the intimate relation of solar irradiation and sunshine duration as a set of fuzzy rules. Monthly averages of daily global solar irradiation and sunshine duration measured since 1982–1993 at the stations Istanbul (41°00′N, 28°58′E, 288 m), Ankara (39°55′N, 32°51′, 938 m), and Adana (37°0′N, 35°19′E, 20 m), Turkey, have

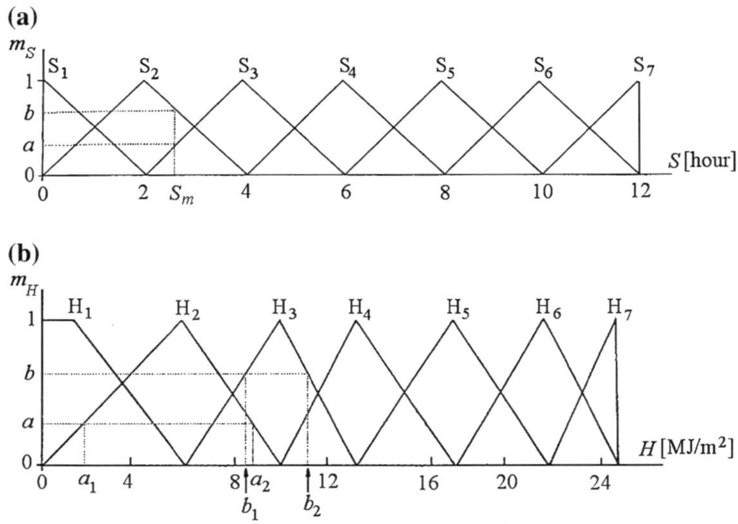

Fig. 7.10 Membership functions of the linguistic variable attributes. **a** Input—monthly mean of sunshine duration $\bar{S}$. **b** Output—monthly mean of daily global solar irradiation $\bar{H}$

been used to build the model. Sunshine duration, considered at maximum 12 h length, is the single input variable in the model and is characterized by seven attributes: $\mathbf{S_1}$, $\mathbf{S_2}$, $\mathbf{S_3}$, $\mathbf{S_4}$, $\mathbf{S_5}$, $\mathbf{S_6}$, and $\mathbf{S_7}$. The membership functions are considered triangular and are represented in Fig. 7.10a. This partition may be considered valid for the latitudes in the range 37–40°N. For the output variable, global solar irradiation, the author considered also seven attributes: $\mathbf{H_1}$, $\mathbf{H_2}$, $\mathbf{H_3}$, $\mathbf{H_4}$, $\mathbf{H_5}$, $\mathbf{H_6}$, and $\mathbf{H_7}$. However, because solar irradiation is dependent on location and altitude, thus the membership functions will be slightly different for different sites. Figure 7.10b shows the attributes of the variable monthly mean global solar irradiation for Istanbul. The following set of rules has been established:

$$\text{IF } S \text{ IS } \mathbf{S}_i \text{ AND } \mathbf{S}_{i+1} \quad \text{THEN } H \text{ IS } \mathbf{H}_i \text{ AND } \mathbf{H}_{i+1} \tag{7.25}$$

This is an unusual set of rules, because the consequent part is expressed in terms of two successive fuzzy subsets of the solar irradiation range.

The algorithm works as follows. For a given sunshine duration measurement S_m *always* there are two attributes of sunshine duration (see Fig. 7.10a) that characterize S_m. Let a and b to be the confidence level of these attributes, in this example $\mathbf{S}_2$ and $\mathbf{S}_3$ as is shown in Fig. 7.10a. According to the expression of rules (7.25) *always* two rules will be set up. As it can be seen in Fig. 7.10b, in this particular case, the rules drive to the attributes $\mathbf{H}_2$ and $\mathbf{H}_3$ of the output variable. The output value, i.e., the estimated value of solar irradiation H_e, is computed as the weighted average VB:

$$H_e = a\bar{a} + b\bar{b} \tag{7.26}$$

where $\bar{a} = (a_1 + a_2)/2$ and $\bar{b} = (b_1 + b_2)/2$, which is an innovative defuziffication approach.

Solar irradiation estimations from sunshine duration have been obtained so far either through the linear Ångström–Prescott equation (5.83) or via its modifications. The fuzzy algorithm developed in Sen (1998) does not provide an equation but can adjust itself to any type of linear or nonlinear form through fuzzy characterization of linguistic variables sunshine duration and solar irradiation. It is also possible to augment the conditional statements in the fuzzy implications used in Sen's paper, by including additional relevant meteorologic variables that might increase the precision of solar irradiation estimation. The author concluded that the application of this fuzzy algorithm is straightforward for any sunshine duration measurements in any part of the world.

7.3 A model for Nowcasting Solar Irradiance

In this section, a model for nowcasting solar irradiance is proposed. The model is constructed using the same set of data measured at 15 s interval as the ARIMA model in Sect. 6.1.2, where these data are completely described.

The presentation is focused on the model structure and functioning. The algorithm forecasts the instantaneous clearness index k_t, from which global solar irradiance is calculated with Eq. 4.1. According to the FL procedures, first the input variable should be chosen. The first input variable is k_t^{t-1}, the clearness index measured at time $t - 1$ (the prediction will be made for the time t). Since clearness index is a measure of the atmospheric transparency it encapsulates information on the state of the sky. Tests regarding sunshine number forecasting demonstrate that it may be correlated with the sun elevation angle (Brabec et al. 2012). Since sunshine number is also a measure for the state of the sky, a relation may exist also between k_t and sun elevation angle. Reference Paulescu et al. (2012) demonstrated that the accuracy of forecasting the sunshine number decreases with the increasing of radiative regime stability. Thus, in addition to k_{t-1}, two other variables have been considered at input: relative sunshine $\bar{\sigma}$ and sun elevation angle h. Here, $\bar{\sigma}$ represents the relative sunshine over 5 min prior to the moment $t - 1$ at which the prediction is made. Both h and $\bar{\sigma}$ are included in the list of input variables for a complementary quantification of the state of the sky, and thus for enhancing prediction in case of variable sky.

The attributes of each input variable and the attributes of the output variable are schematic represented in Fig. 7.11, where the notations are indicated. The membership functions are triangular, described by typical linear equations (see Fig. 7.3), with the coefficients specified in Fig. 7.11. The mapping of the inputs to the output of the fuzzy system, materialized in the rules base, is listed in Table 7.8 as a matrix. There are 64 rules, the each rule ($i = 1, 2$; $k = 1, 2, 3, 4$; $j = 1 \ldots 8$; $m = 1 \ldots 8$) reads:

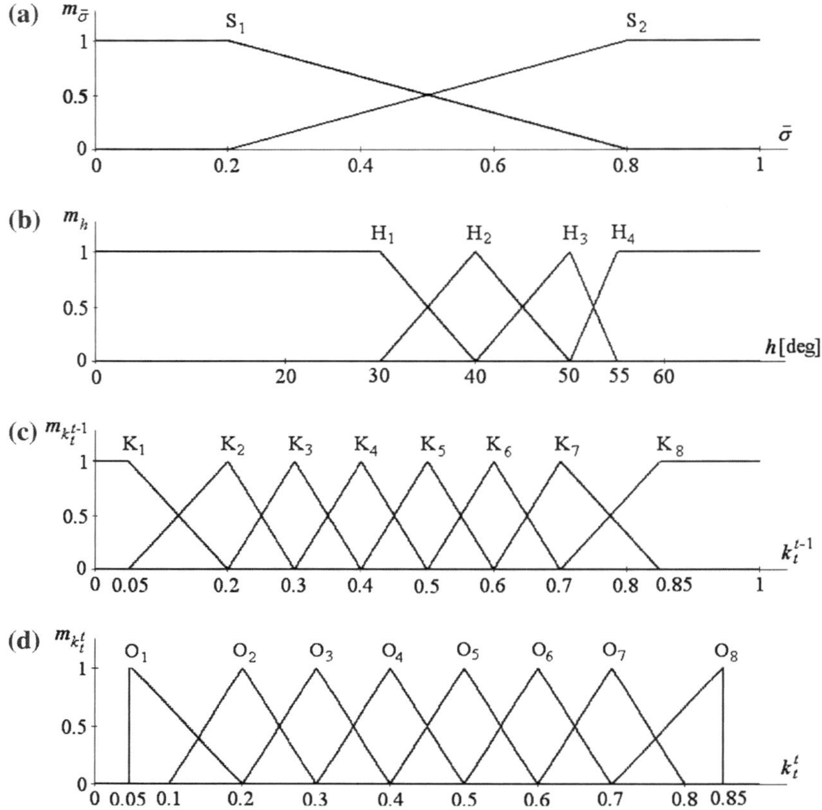

Fig. 7.11 The membership functions of the attributes characterizing the variables. **a** Relative sunshine. **b** Sun elevation angle. **c** Input clearness index at the moment $t-1$; output clearness index at the moment t

Table 7.8 Matrix of the system rules base. Each rule is a fuzzy implication in sense of Eq. (7.27)

$\bar{\sigma}$		S_1				S_2			
h		H_1	H_2	H_3	H_4	H_1	H_2	H_3	H_4
k_t^{t-1}	K_1	O_1	O_1	O_1	O_1	O_1	O_1	O_1	O_1
	K_2	O_2	O_2	O_2	O_2	O_2	O_2	O_2	O_2
	K_3	O_3	O_3	O_3	O_3	O_3	O_4	O_4	O_4
	K_4	O_4	O_4	O_4	O_4	O_4	O_4	O_5	O_5
	K_5	O_5	O_5	O_5	O_5	O_5	O_5	O_6	O_5
	K_6	O_6	O_6	O_6	O_6	O_6	O_6	O_6	O_6
	K_7	O_6	O_7	O_7	O_5	O_7	O_7	O_7	O_5
	K_8	O_6	O_7	O_6	O_4	O_8	O_8	O_6	O_4

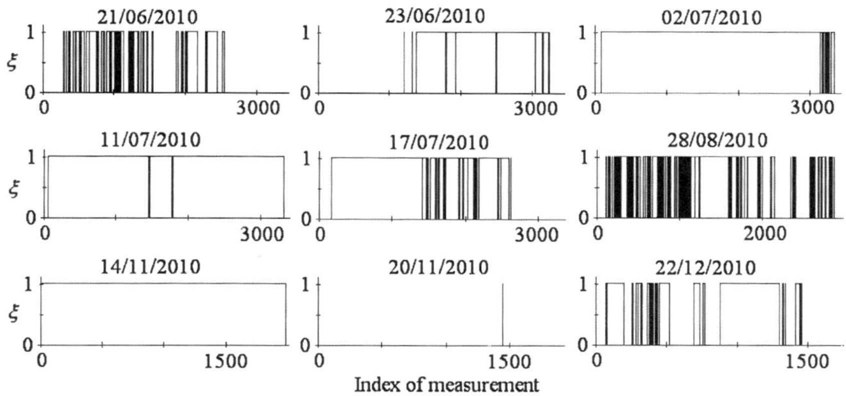

Fig. 7.12 Sunshine number ξ in 9 days used to test the fuzzy model

$$\text{IF } \bar{\sigma} \text{ IS } \mathbf{S}_i \text{ AND } h \text{ IS } \mathbf{H}_j \text{ AND } k_t^{t-1} \text{ IS } \mathbf{K}_k \text{ THEN } k_t^t \text{ IS } \mathbf{O}_m \quad (7.27)$$

With this information the model can be run. The inputs are the relative sunshine averaged over 5 min prior to the moment $t-1$ when the prediction is made, the instantaneous clearness index at the moment $t-1$ and the sun elevation angle at moment t (sun elevation angle is exactly calculated from astronomic consideration with Eq. 5.3). The output of the algorithm is the instantaneous clearness index k_t^t at the moment t.

The model has been tested against data measured in 9 days of the year 2010 at the station of Timisoara, characterized by different radiative regimes, as shown in Fig. 7.12. In this figure, sunshine number ξ measured at 15 s interval is plotted with respect to time (counted by the index of measurement) for every day. It can be seen that the day 28/08 is characterized by a fully unstable radiative regime, while the days 14/11 and 20/11 are characterized by a fully stable radiative regime. However, the days 14/11 and 20/11 are essentially different: 14/11 is a complete sunny day ($\xi = 1$) while 20/11 is an overcast day ($\xi = 0$).

For each day, a scatter plot of the estimated and measured instantaneous clearness index is included in Fig. 7.13. By comparison with Fig. 7.12, this sequence reveals an intimate relation between estimation accuracy of k_t and the stability of the radiative regime.

In the sunny day 14/11, the points are clustered close to the first diagonal showing that the model forecasts the clearness index with very high accuracy. This observation is also valid for the days 02/07 and 20/11 demonstrating that the forecasting accuracy does not depend on the season. Conversely, a large scatter of model outputs relative to the measured data occur in the days 21/06 and 28/08 characterized by a highly fluctuating solar radiative regime.

Statistical indicators of forecasting accuracy of the clearness index for each among the 9 days are collected in Table 7.9. In addition to the statistical indicators, for each day, the relative sunshine $\bar{\sigma}_d$ as an indirect measure for the cloud

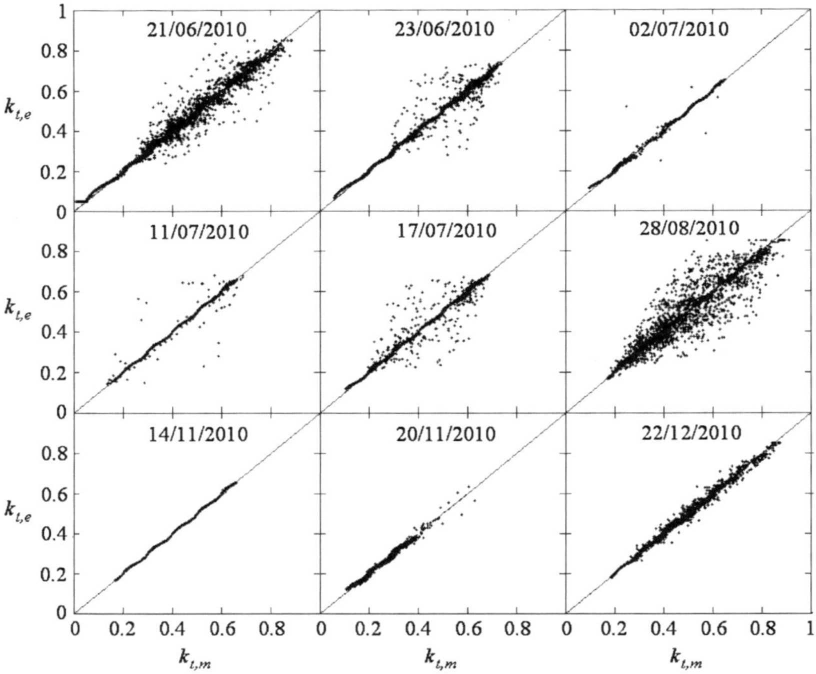

Fig. 7.13 Scatter plots of measured $k_{t,m}$ and estimated $k_{t,e}$ clearness index for different 9 days of 2010

Table 7.9 Statistical indicators of accuracy for forecasting instantaneous clearness index at 15 s

Date	$\bar{\sigma}_d$	$\bar{\zeta}_d$	N	*RMSE* [%]	*MBE* [%]
14/11/2010	0.9930	0.0005	1,982	0.81	0.26
20/11/2010	0.0020	0.0005	1,918	3.61	0.68
11/07/2010	0.9481	0.0011	3,398	2.78	0.12
02/07/2010	0.9250	0.0020	3,433	1.62	0.38
23/06/2010	0.5301	0.0040	3,446	5.84	0.42
17/07/2010	0.6531	0.0068	3,366	5.58	0.17
22/12/2010	0.4644	0.0114	1,734	3.24	−0.18
21/06/2010	0.3035	0.0173	3,446	10.94	1.69
28/08/2010	0.5205	0.0369	2,969	12.48	−0.23

lag in 9 days of the year 2010. $\bar{\sigma}_d$ is the daily relative sunshine, $\bar{\zeta}_d$ is the daily mean of sunshine stability number and N is the number of forecasted values. The days are sorted by ascending $\bar{\zeta}_d$

cover amount and the daily mean sunshine stability number $\bar{\zeta}_d$ as an indirect measure of the cloud cover variability (see Sect. 4.2), are included.

The first conclusion from Table 7.9 is the very good performance of the model, *RMSE* ranging from 0.81 % in perfect steady clear sky day ($\bar{\sigma}_d = 0.993$ and $\bar{\zeta}_d = 0.0005$) to 12.48 % in the day with highest variability of the sky ($\bar{\sigma}_d = 0.525$

Table 7.10 Statistical indicators of accuracy for forecasting instantaneous clearness index at 5 min lag in 9 days of the year 2010

Date	*N*	*RMSE* [%]	*MBE* [%]
14/11/2010	92	1.6	0.9
20/11/2010	89	22.8	0.9
11/07/2010	160	4.6	0.2
02/07/2010	162	4.5	0.5
23/06/2010	162	11.1	0.4
17/07/2010	158	23.1	0.3
22/12/2010	81	25.5	−1.1
21/06/2010	162	39.6	0.5
28/08/2010	139	37.0	−1.4

N is the number of forecasted values

and $\bar{\zeta}_d = 0.0369$). Second, numerical results from Table 7.9 not only confirm the above remark that there is a relation between the prediction accuracy and solar regime stability (here measured by $\bar{\zeta}_d$), but also complete it. *RMSE* clearly increase with increasing $\bar{\zeta}_d$ and is influenced by $\bar{\sigma}_d$.

To conclude, the accuracy increase of the forecasting procedure determined by a more stable solar radiative regime found in ARIMA modeling of sunshine number (Paulescu et al. 2012) is exhibited by the fuzzy model too. In order to increase the fuzzy model accuracy, further developments should include in the list of the linguistic variables a measure for the fluctuation of the radiative regime.

7.3.1 Five Minutes Forecasting of k_t

A question arising is whether the above model can be used for forecasting solar irradiance at different time intervals. Various tests have been performed to answer this question. Results from testing the model for 5 min ahead forecasting solar irradiance follows. The same dataset as in the previous case has been used. For each day, the database was constructed by staking lines of data measured every 5 min. This time the relative sunshine has been calculated over the period between two prediction moments.

An inspection of the output to input scattering shows that the membership functions can be preserved. Only one rule has been modified, that is:

$$\text{IF } \bar{\sigma} \text{ IS } \mathbf{S}_2 \text{ AND } h \text{ IS } \mathbf{H}_2 \text{ AND } k_t^{t-1} \text{ IS } \mathbf{K}_8 \text{ THEN } k_t^t \text{ IS } \mathbf{O}_7 \tag{7.28}$$

Comparing the rule (7.28) with the corresponding one from Eq. (7.27), it can be noted that the change is $\mathbf{O_8} \rightarrow \mathbf{O_7}$.

Results of testing the model against measured data are inserted in Table 7.10. The days are ordered according to sunshine stability number as in Table 7.9.

Table 7.10 shows a decreasing of the forecasting accuracy compared to the previous case, with *RMSE* ranging between 1.6 and 39 %. This was expected because it is well known that as the forecast time interval increases, the forecast accuracy decreases. Generally, the model accuracy is correlated with the daily solar radiative regime, but this relation may be affected by the larger interval between the two moments of forecasting (see results in 20/11). In a time interval of 5 min, it is possible to have a significant instability while the samples at the beginning and at the end of this time interval indicate stability, i.e., the same values for sunshine number. This risk increases with a longer forecasting interval, thus moving average should be adopted to maintain the prediction accuracy at a reasonable level. Including other quantities in the input list can also be considered. To conclude, a well-working fuzzy model at a certain timescale needs recalibration to perform at a different time scale.

7.4 A Model for Forecasting Solar Irradiation

A fuzzy model to forecast daily global solar irradiation at ground level is summarized here. This is based on the algorithm reported by Boata and Gravila (2012) where the stochastic component of the solar irradiation is quantified by means of the clearness index. Basically the model forecasts daily clearness index k_t (Eq. 4.2), tracking the rules of an autoregressive fuzzy algorithm. Fuzzy c-means clustering is used to establish the membership functions while the overall algorithm is developed in the frame of functional fuzzy systems.

The model proposed here is constructed with daily global solar irradiation data measured for 2 years, from January 1, 2008 to December 31, 2009, at the station of Madrid University (WMO index 08220, 40°27′N, 3°43′W, 664 m altitude above mean sea level), Spain. The source of data is World Radiation Data Center, St. Petersburg, Russia (see Sect. 2.2.2). Data recorded during 2010 have been used to test the approach. This is standard procedure for model validation, given that it is a fuzzy autoregressive algorithm.

In these autoregressive procedures, the input and output variables belong to the same set k_t; the difference between the variables is given by the time parameter. First, we discuss the simplest situation when the predicted value of clearness index depends only on the preceding value. Further, the results are compared with ones reported in Boata and Gravila (2012), where the model is constructed using two antecedent values at input.

The input variable $k_t^{(t-1)}$ is measured at the time $t-1$ while the output variable $k_t^{(t)}$ is forecasted for the time t. Four attributes have been considered for $k_t^{(t-1)}$, cloudy C, mostly cloudy MC, mostly sunny MS, and sunny S. The membership functions of these attributes have been evaluated with a fuzzy clustering procedure (see Sect. 7.1.2.5). Then, the discrete points have been fitted with various functions, a very good approximation ($r^2 = 0.999$) resulting for a Chebyshev equation:

Table 7.11 The coefficients c_i in Eq. (7.29)

i	C	MC	MS	S
1	0.69893	0.13628	0.07446	0.09009
2	−1.62317	−1.62410	−1.62418	−1.62418
3	−1.26493	−0.24525	−0.11655	0.00293
4	1.40488	1.40475	1.40475	1.40472
5	0.94046	0.20506	0.11588	0.14313
6	−0.77658	−0.77692	−0.77696	−0.77692
7	−0.55815	−0.14657	−0.07324	0.00108
8	0.48583	0.48572	0.48570	0.48568
9	0.25810	0.10010	0.05801	0.06950
10	−0.15387	−0.15390	−0.15390	−0.15390
11	−0.08345	−0.04784	−0.02216	−0.00046
12	0.06564	0.06558	0.06556	0.06557
13	0.01641	0.01639	0.01639	0.01639

$$m_j(k_t) \cong \frac{c_1 + \sum_{i=1}^{n} c_{2i+1} T_i[x'(k_t)]}{1 + \sum_{i=1}^{n} c_{2i} T_i[x'(k_t)]}, \quad j = \text{C, MC, MS, S} \tag{7.29}$$

with the coefficients c_i listed in Table 7.11. A few details on Chebyshev polynomials and the meaning of $T_i[x(k_t)]$ follow.

Chebyshev univariate polynomials are popular in approximations theory (Mason and Handscomb 2003). A Chebyshev rational–polynomial approximation of a function $f(x') : [-1, 1] \rightarrow \Re$ has been used here:

$$f(x') \cong \frac{c_1 + \sum_{i=1}^{n} c_{2i+1} T_i(x')}{1 + \sum_{i=1}^{n} c_{2i} T_i(x')} \tag{7.30}$$

where n is the polynomial order (in our case $n = 6$) and

$$T_i(x') = \cos[i \cdot \operatorname{acos} x'], \quad i = 1 \ldots n \tag{7.31}$$

In all four fitting processes (there are four attributes), $n = 6$ was the minimum polynomial order that gives $r^2 > 0.999$. The clearness index k_t domain [0.062, 0.788] has been linearly transformed into [− 1, 1]:

$$x'(k_t) = -1 + (k_t - 0.0628)\frac{2}{0.7880 - 0.0628} \tag{7.32}$$

Figure 7.14 displays the discrete points and the fitted functions $m_j(k_t)$, $j =$ C, MC, MS, and S. Such higher approximation ($r^2 = 0.999$) is required in order to preserve the normalization of the membership functions, i.e.,

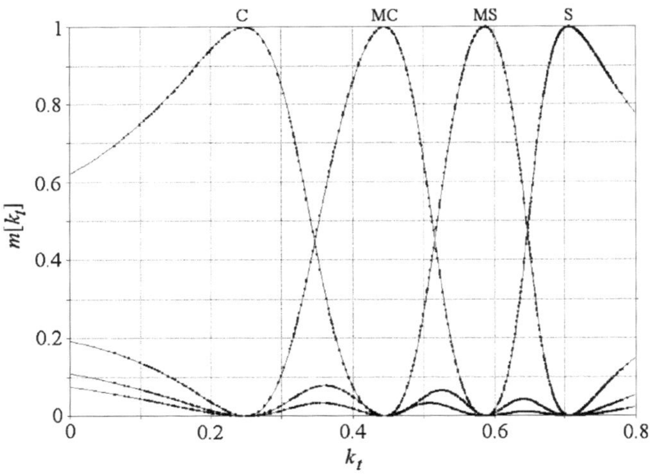

Fig. 7.14 The membership functions $m(k_t)$ for the attributes cloudy (C), mostly cloudy (MC), mostly sunny (MS), and sunny (S) associated of the input linguistic variable k_t

$m_C(k_t) + m_{MC}(k_t) + m_{MS}(k_t) + m_S(k_t) = 1$. The clusters centers are: $c_C = 0.249$, $c_{MC} = 0.447$, $c_{MS} = 0.587$, and $c_S = 0.706$.

The shape of the membership functions of the periphery attributes are not saturated to unit, as in the usual representation of the membership functions. These membership functions are exactly the outcomes of the c-mean clustering algorithm and are not artificially saturated. It is a more realistic picture of the common experience: a small k_t not always guarantees an entirely overcast day; it may be a day with fluctuating solar radiative regime which takes to some degree even the sunny day attribute (S). The above statement is substantiated by the comprehensive study on classification of the days in respect to the state of the sky and solar regime fluctuation reported in Sect. 4.4.

The mapping of the inputs to the output of our fuzzy system, materialized in the rules base, is the simplest possible:

$$\text{IF} \quad m_j\left(k_t^{(t-1)}\right) \text{ IS } j \quad \text{THEN} \quad k_t^{(t)} \in F_j\,, \quad j = \text{C}, \text{MC}, \text{MS}, \text{S} \tag{7.33}$$

where F_j is the subset which define the output function f_j. Thus, each attribute of the input variable drives to a distinct output function f_i.

Figure 7.15 displays the scattering of output variable, measured clearness index in the current day $t\left(k_t^{(t)}\right)$, to the input variable, clearness index measured in the previous day $\left(k_t^{(t-1)}\right)$. We notice a large scattering of output to input data, the output appears rather uncorrelated to the input, raising the question: has fuzzy theory strength enough to find the rules that govern the series of data? In the following, we answer this question.

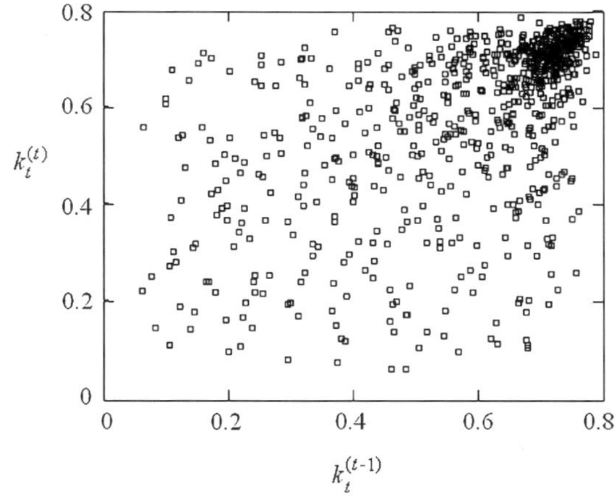

Fig. 7.15 Clearness index in the current day $k_t^{(t)}$ versus clearness index in the previous day $k_t^{(t-1)}$

In order to fit the four output functions, the database has been stratified into four classes F_j, $j = 1 \ldots 4$. We assume that a pair $\left(k_t^{(t-1)}, k_t^{(t)}\right)$ belongs to the F_i class if both input membership functions exceed 0.75:

$$\text{IF} \quad m_j^{(t-1)} \geq 0.75 \quad \text{THAN} \quad \left(k_t^{(t-1)},\ k_t^{(t)}\right) \in F_j \tag{7.34}$$

Therefore, the inputs with considerable uncertain attributes have been removed from the future fitting process of the output functions. Figure 7.16 shows the points $\left(k_t^{(t-1)}, k_t^{(t)}\right)$ in each class F_j, $j = 1 \ldots 4$. Visual inspection reveals a typical clustering image: in each class the majority of the points are gathering in a well-defined region; but in each cluster the points are dispersed in a large region. We choose a linear function to fit the points in each cluster:

$$f_j(\ k_t^{(t-1)}) = c_0 + c_1 k_t^{(t-1)}, \quad j = 1 \ldots 4 \tag{7.35}$$

The coefficients c_0 and c_1 are listed in Table 7.12. The lines $f_j(\ k_t^{(t-1)})$ are also plotted in Fig. 7.16. Additional inputs are required in order to do a better fit since all information encapsulated in the input variable $k_t^{(t-1)}$ has been exploited. Because it is very important for f_i to capture the essential feature encapsulated in the cluster F_i, the clearness index $k_t^{(t-2)}$ were introduced in the algorithm reported in Boata and Gravila (2012).

Now, the fuzzy algorithm is operational and for every input $k_t^{(t-1)}$ the output $k_t^{(t)}$ can be computed using Eq. (7.13).

The model was tested against data measured during 2010 at the same station, Madrid University. The performance indicators of the model are shown in Table 7.13. An *RMSE* value of 31.4 % points out that the accuracy of this fuzzy autoregressive model for forecasting solar irradiation is in the range of the fuzzy models for nowcasting solar irradiance at very short time horizon (see Table 7.10). The distribution properties of the synthetic series are rather insufficient; the

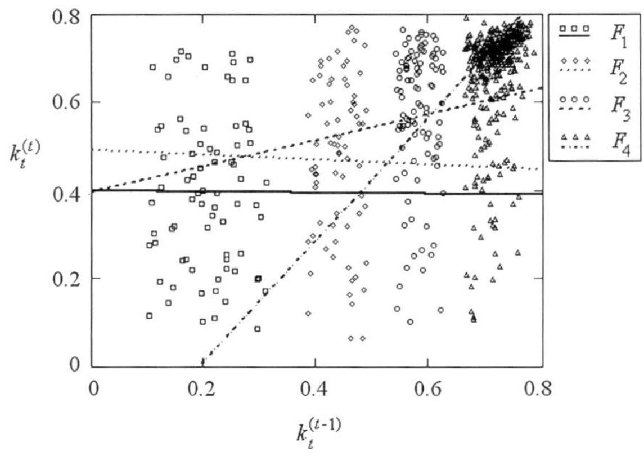

Fig. 7.16 Clusters $F_j j = 1...4$ defined in respect to Eq. (7.34) (*points*); superimposed are the *lines* given by Eq. 7.35

Table 7.12 The coefficients in Eq. 7.35

Cluster	F_1	F_2	F_3	F_4
c_0	−0.011	−0.055	0.292	1.394
c_1	0.400	0.492	0.398	−0.268

Table 7.13 Statistical indicators of accuracy of forecasting daily clearness index in 2010 at the station Madrid University

Series	Input	*RMSE*	*MBE*	*MAE*	Mean	Variance	Skewness	Kurtosis
Measured	–	–	–	–	0.557	0.036	−0.922	−0.314
Forecasted	1	0.175	0.055	0.127	0.611	0.012	0.118	−1.399
Forecasted	2	0.172	0.042	0.123	0.559	0.009	−0.706	−0.866

difference between the mean of the measured and synthetic time series is roughly 10 %, the distribution of the synthetic series is skewed to the right while the measured series is skewed to the left.

Table 7.13 also contains statistical indicators resulted after testing a fuzzy autoregressive model at the same station and in the same period as reported in Boata and Gravila (2012). Both models are constructed following the same procedures, the only difference being the fact that the model reported in Boata and Gravila (2012) are using recordings from the previous 2 days at input. Evaluating the data from Table 7.13, it can be figured out that adding to the input the clearness index measured 2 days ago (the day before yesterday), the model performance is only slightly improved while the distribution of the synthetic series much better replicates the ones of the measured series.

Reference Boata and Gravila (2012) also reports fuzzy autoregressive models for forecasting clearness index using at input values measured in the previous

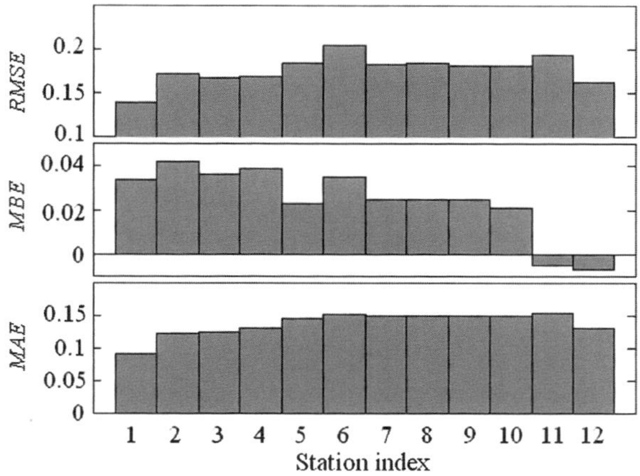

Fig. 7.17 Statistical accuracy indicators of forecasting daily clearness index with fuzzy autoregressive models applied at the stations: *1* Valencia/Spain (39°29′N, 0°23′W, 23 m altitude), *2* Madrid Univ/Spain (40°27′N, 3°43′W, 664 m), *3* Bucharest/Romania (44°30′N, 26°13′E, 91 m), *4* Timisoara/Romania (45°47′N, 21°17′E, 90 m), *5* Zagreb/Croatia (45°50′N, 16°00′E, 182 m), *6* Locarno-Monti/Switzerland (46°10′N, 8°47′E, 366 m), *7* Innsbruck Arpt/ Austria (47°15′N, 11°21′E, 579 m), *8* Sopron/Hungary (47°41′N, 16°36′E, 233 m), *9* Bratislava-Koliba/Slovakia (48°10′N, 17°07′E, 304 m), *10* Wien/Austria (48°15′N, 16°21′E, 203 m), *11* Strbske-Pleso/Slovakia (49°07′N, 20°04′E, 1,387 m), and *12* Uccle/Belgium (50°48′N, 4°21′E, 100 m). Data from Boata and Gravila (2012) are used

2 days for other 11 European localities. Data measured during 1 year have been used to test the model. A summary of models testing results is presented in Fig. 7.17. *RMSE* ranges in a short interval between 0.139 and 0.204, indicating that the model accuracy is not critically sensible to location. The best performance occurs at stations located at low altitude (Valencia, Uccle, Bucharest).

The ending point of the forecasting procedure is the daily global solar irradiation. A conclusion of Reference Boata and Gravila (2012) is that the accuracy of forecasting daily global solar irradiation trails the accuracy of forecasting the daily clearness index. The ranges of the statistical indicators listed in Fig. 7.17 are comparable with other results reported in the literature.

To conclude, a model forecasting daily global solar irradiation was presented in this section. It is constructed via TS approach (Sect. 7.1.2.4). Being a measure of the stochastic component of global solar irradiation, the daily clearness index is the effective forecasted quantity. In principle, the actual model can be applied only in sites where measurement of daily global solar irradiation is currently performed. However, the algorithm is general and can be adapted by potential users to fit their own location. For sites where measurements are not available, the algorithm may be adapted by subsequently using forecasted meteorologic parameters in models for estimating daily global solar irradiation. The algorithm has the strength to translate the information enclosed in previous two days of measurements into an actual prediction of the clearness index with an acceptable accuracy. This accuracy is preserved when the forecasted clearness index is used to calculate daily global solar irradiation.

The results presented in this chapter are indicative of FL as a feasible approach in nowcasting solar irradiance or forecasting daily solar irradiation. Further studies should be focused to enhance the prediction accuracy when large fluctuations are present in the solar irradiance/irradiation time series.

References

Boata St R, Gravila P (2012) Functional fuzzy approach for forecasting daily global solar irradiation. Atmos Res 112:78–88

Brabec M, Badescu V, Paulescu M (2012) Nowcasting sunshine number by using logistic modeling (submitted)

Cao S, Cao J (2005) Forecast of solar irradiance using recurrent neural networks combined with wavelet analysis. Appl Therm Eng 25:161–172

Cao J, Lin X (2008) Application of the diagonal recurrent wavelet neural network to solar irradiation forecast assisted with fuzzy technique. Eng Appl Artif Intell 21:1255–1263

Chen C, Duan S, Cai T, Liu B (2011) Online 24 h solar power forecasting based on weather type classification using artificial neural network. Sol Energy 85(11):2856–2870

Donvlo ASS, Jervase JA, Al-Lawati A (2002) Solar radiation estimation using artificial neural networks. Appl Energy 74:307–319

Gomez V, Casanova A (2003) Fuzzy modeling of solar irradiance on inclined surfaces. Sol Energy 75:307–315

Hocaoglu FO, Gerek ON, Kurban M (2008) Hourly solar radiation forecasting using optimal coefficient 2-D linear filters and feed-forward neural networks. Sol Energy 82:714–726

Izgi E, Oztopal A, Yerli B, Kaymak MK, Sahin AD (2012) Short–mid-term solar power prediction by using artificial neural networks. Sol Energy 86(2):725–733

Kalogirou S (ed) (2007) Artificial intelligence in energy and renewable energy systems. Nova Science, New York

Kemmoku Y, Orita S, Nakagawa S, Sakakibara T (1999) Daily insolation forecasting using a multi-stage neural network. Sol Energy 66:193–199

Lopez G, Batlles FJ, Tovar-Pescador J (2005) Selection of input parameters to model direct solar irradiance by using artificial neural networks. Energy 30:1675–1684

McCarthy J (1958) Programs with Common Sense. In: Proceedings of teddington conference on the mechanization of thought processes. Available at http://library.thinkquest.org/05aug/01158/mccarthy.html

Mason JC, Handscomb DC (2003) Chebyshev polynomials. Chapman & Hall, CRC Boca Raton

Mellit A (2008) Artificial intelligence technique for modelling and forecasting of solar radiation data: a review. Int J Artif Intel Soft Comput 1(1):52–76

Mellit A, Kalogirou SA (2008) Artificial intelligence techniques for photovoltaic applications: a review. Prog Energy Combust Sci 34:547–632

Mellit A, Benghanem M, Kalogirou SA (2006) An adaptive wavelet network model for forecasting daily total solar radiation. Appl Energy 83:705–722

Mellit A, Kalogirou SA, Shaari S, Salhi H, Hadj Arab A (2008) Methodology for predicting sequences of mean monthly clearness index and daily solar radiation data in remote areas: application for sizing a stand-alone PV system. Renewable Energy 33:1570–1590

Mellit A, Pavan AM (2010) A 24-h forecast of solar irradiance using artificial neural network: application for performance prediction of a grid-connected PV plant at trieste. Italy Sol Energy 84(5):807–821

Mihalakakou G, Santamouris M, Asimakopoulos DN (2000) The total solar radiation time series simulation in Athens, using neural networks. Theor Appl Climatol 66:185–197
Mohandes M, Rehman S, Halawani TO (1998) Estimation of global solar radiation using artificial neural networks. Renewable Energy 14:179–184
Mubiru J, Banda EJKB (2008) Estimation of monthly average daily global solar irradiation using artificial neural networks. Sol Energy 82(2):181–187
Paoli C, Voyant C, Muselli M, Nivet M-L (2010) Forecasting of preprocessed daily solar radiation time series using neural networks. Sol Energy 84(12):2146–2160
Passino KM, Yurkovich S (1998) Fuzzy Control. Adison Wesley Longman. Available at www.ece.osu.edu/~passino/books.html
Paulescu M, Gravila P, Tulcan-Paulescu E (2008) Fuzzy logic algorithms for atmospheric transmittances of use in solar energy estimations. Energy Convers Manage 49:3691
Paulescu M, Badescu V, Brabec M (2012) Tools for grid and pv plant operators: nowcasting of passing clouds. Submitted to Journal
Perez R, Ineichen P, Seals R, Stewart R, Meniccucci D (1987) A new simplified version of the of the Perrez diffuse irradiance model for tilted surfaces. Sol Energy 39:221–231
Perez R, Ineichen P, Seals R, Michalsky J, Stewart R (1990) Modeling daylight availability and irradiance components from direct and global irradiance. Sol Energy 44:271–289
Qin J, Chen Z, Yang K, Liang S, Tang W (2011) Estimation of monthly-mean daily global solar radiation based on MODIS and TRMM products. Appl Energy 88:2480–2489
Reddy KS, Manish R (2003) Solar resource estimation using artificial neural networks and comparison with other correlation models. Energy Convers Manage 44:2519–2530
Sen Z (1998) Fuzzy algorithm for estimation of solar irradiation from sunshine duration. Sol Energy 63:39–49
Senjyu T, Hayashi D, Urasaki N, Funabashi T (2006) Optimum configuration for renewable generating systems in residence using genetic algorithm. IEEE Trans Energy Convers 21(1):459–467
Sfetsos A, Coonick AH (2000) Univariate and multivariate forecasting of hourly solar radiation with artificial intelligence techniques. Sol Energy 68(2):169–178
Takagi T, Sugeno M (1985) Fuzzy identification of systems and its applications to modeling and control. IEEE Trans Syst Man Cybern 15:116–132
Tulcan-Paulescu E, Paulescu M (2008) Fuzzy modeling of solar irradiation using air temperature data. Theor Appl Climatol 91:181–192
Tymvios FS, Jacovides CP, Michaelides SC, Scouteli C (2005) Comparative study of Ångström's and artificial neural networks' methodologies in estimating global solar radiation. Sol Energy 78(6):752–762
Tymvios FS, Michaelides SC, Skouteli CS (2008) Estimation of surface solar radiation with artificial neural networks. In: Badescu V (ed) Modeling solar radiation at the Earth surface. Springer, Berlin
Yang HX, Zhou W, Lu L, Fang ZH (2008) Optimal sizing method for stand-alone hybrid solar–wind system with LPSP technology by using genetic algorithm. Sol Energy 82(4):354–367
Zadeh LA (1965) Fuzzy sets. Inf Control 8:338–353
Zarzalejo LF, Ramirez LJ (2005) Artificial intelligence techniques applied to hourly global irradiance estimation from satellite-derived cloud index. Energy 30:1685–1697
Zimmermann HJ (1996) Fuzzy set theory and its application, 3rd edn. Kluwer Academic; Norwell, MA

Chapter 8
Air Temperature-Based Models

8.1 Introduction

A brief introduction to modeling global solar irradiation via air temperature data has been given in Sect. 5.5.3. Two models have been presented: (1) Supit and Van Kappel (1998) Eq. (5.94) at which daily air temperature amplitude is taken as a refinement next to the cloud cover amount in an Ångström equation and (2) Bristow and Campbell (1984) (Eq. 5.96) which build an Ångström equation based only on daily air temperature amplitude. Each of these equations are representative for an individual group of models: the first group takes into account air temperature besides other indicators (total cloud amount, relative sunshine) to quantify the state of the sky while the second group uses only air temperature-based estimates for the state of the sky. In the next sections of this chapter, in addition to the presentation from Sect. 5.5.3, more details on both classes are given and other models are introduced.

With respect to the subject of this book, the study of air temperature-based models is important because these models can be included in algorithms for forecasting solar radiation. The procedure's flowchart is illustrated in Fig. 8.1. Its construction is based on two observations: air temperature is a common forecast parameter everywhere and air temperature-based models for estimating solar irradiation achieve an acceptable level of accuracy. Thus, the forecasted air temperature may be used as entry in the mentioned models for forecasting solar irradiation.

Forecasting of air temperature is a component of weather models and daily minimum and maximum air temperature forecasts in the horizon of 24–72 h are usually provided by meteorological services. At present, the accuracy of forecasting air temperature is very high. For example, MetOffice, the UK's National Weather Service, compares forecasts for both maximum and minimum temperatures to the actual values observed at 45 stations across the UK. The stations used for verification are those where MetOffice have quality-controlled data. The early

M. Paulescu et al., *Weather Modeling and Forecasting of PV Systems Operation*, Green Energy and Technology, DOI: 10.1007/978-1-4471-4649-0_8,

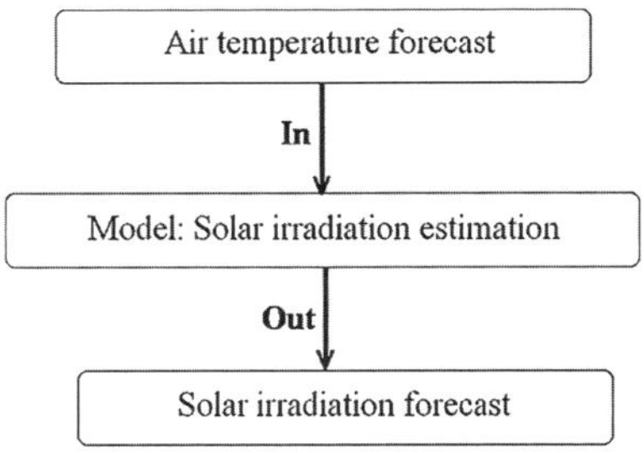

Fig. 8.1 Flowchart of the general procedure for forecasting solar irradiation using air temperature data

morning forecast is used to produce a percentage number of the times when the forecast is accurate to within ±2 °C. The final result is based over a rolling 36 month period to smooth out extremes and give a representative average. In March 2012, the following data has been reported: 87.8 % of maximum temperature forecasts are accurate to within ±2 °C on the current day and 79.3 % of minimum temperature forecasts are accurate to within ±2 °C on the first night of the forecast period (MetOffice 2012).

The accuracy of models for estimating solar irradiation via air temperature data does not reach the accuracy of air temperature forecasting. In the sequential chain (Fig. 8.1), the temperature prediction error is propagated to the output, through the model for estimating solar irradiation. Therefore, it is logical that the estimation used as input in the algorithm's second step should be as good as possible. In Sect. 8.5, a critical assessment of the accuracy of temperature-based models for solar irradiation is conducted. Results discussed in Sect. 8.5 demonstrate that, even if present models seem adequate enough for many applications, future efforts are required for improving their accuracy. The chapter ends with the presentation of a simulation exercise of forecasting daily global solar irradiation by air temperature-based models.

8.2 Solar Irradiance Modeling

When solar irradiance is computed it is obvious that replacing the total cloud amount with an air temperature function is impossible. Common experience shows that instant cloud amount and air temperature cannot be correlated: there are sunny days in warm summer as are sunny days in frosty winter. Contrasting with this, when solar irradiation is evaluated by summing up the irradiance, the goal of replacing the cloudiness with an air temperature equation becomes feasible. SEAT (Solar Energy—Air Temperature), introduced in Paulescu et al. (2011), is a model for global solar irradiance and its application to the computation of daily solar irradiation is presented in the following. The model runs in a standard way: first, global solar irradiance is computed, then a cloud cover correction is applied, ending up in summing the results for the daily irradiation.

8.2.1 SEAT Equations

Global solar irradiance at ground level $G(h, C, t)$ is calculated with the equation:

$$G(\theta_z, C, t) = G_{SC}\varepsilon f(\cos\theta_z, C, t) \tag{8.1}$$

where $G_{SC} = 1366.1$ W/m^2 is the solar constant, ε is the solar constant adjustment in respect to the earth trajectory eccentricity (Eq. 5.1), θ_z is the solar zenithal angle and t is the actual air temperature in degree Celsius. $f(C, t, \cos\theta_z)$ is parameterized as:

$$f(C, t, \cos\theta_z) \equiv f(C, t_r \cdot \cos\theta_z) = \left[f_1(C) + f_2(C)C^{f_3(C)}\right](t_r \cos\theta_z)^{1.128} \tag{8.2}$$

where $t_r = 1 + t/273$. $f_i(C)$, $i = 1, 2, 3$, are given by the equations:

$$f_1(C) = \frac{0.61337518 + 1.5678507C - 2.229234C^2}{1 + 9.0637694C - 12.161647C^2 + 4.3582103C^3} \tag{8.3a}$$

$$f_2(C) = 0.69356389 - 0.47655C \tag{8.3b}$$

$$f_3(C) = \exp\left(-0.65499162 + 1.939823C^3\right) \tag{8.3c}$$

SEAT can be applied in practice *as it is*. But bearing in mind the mathematical integration as the next step for solar irradiation, it is desirable to substitute the instantaneous value of air temperature t in Eq. (8.2) with a model consisting of continuous equations. An appropriate substitution which relates $t_j(t_{\max}, t_{\min}, \omega)$ with daily air temperature maximum $t_{\max}$, minimum $t_{\min}$, and hour angle ω is also given in Paulescu et al. (2011):

$$t(j, \omega) = a \cdot t_0(j, \omega) + b \tag{8.4}$$

where $t_0(j, \omega)$ is a sine–cosine model:

$$t_0(j,\omega) = \begin{cases} t_{\max}(j) - (t_{\max}(j) - t_{\min}(j)) \cdot \left(1 - \cos\left(\dfrac{\pi}{2}\dfrac{\omega_m - \omega}{\omega_m + \omega_0(j)}\right)\right), & \omega \leq \omega_m \\ t_{\max}(j) - (t_{\max}(j) - t_{\min}(j+1)) \cdot \sin\left(\dfrac{\pi}{2}\dfrac{\omega_m - \omega}{\omega_m - \omega_0(j+1)}\right), & \omega > \omega_m \end{cases} \tag{8.5}$$

The coefficients a and b empirically adjust Eq. (8.4) to a local temperature regime. For Timisoara (see Fig. 3.1 for localization) in Paulescu et al. (2011) the fitted values are $a = 0.99$ and $b = -0.41$. $t(j, \omega)$, in Celsius, is the estimated air temperature in the Julian day j at hour angle ω. $\omega_0(j)$ represents the sunset hour angle while ω_m represents the hour angle at which the maximum air temperature is reached. In this model, ω_m is assumed to be the same in everyday. By using Eq. (8.5) in the evaluation of daily air temperature, SEAT is turned into an integrable

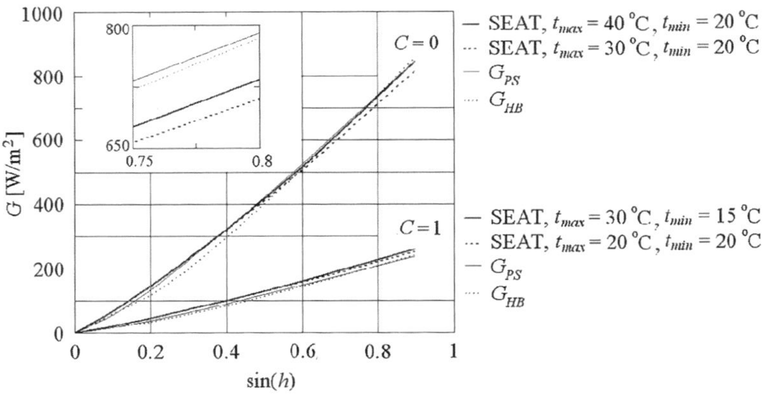

Fig. 8.2 SEAT, G_{PS} (Eq. 5.51) and G_{HB} (Eqs. 5.66 and 5.70.) as function of sinus of elevation angle in the Julian day 182. Inset are details for the case $C = 0$. The graphs cut-off at $\sin(h) = 0.93$ is a consequence of maximum elevation of the sun in the respective days at 45°N latitude where the models had been run. From Paulescu et al. (2011), with permission from Wiley

model. For computing solar irradiance, in addition to the geographical coordinates and temporal reference, the cloud cover amount and air temperature are required.

8.2.2 SEAT Accuracy to the Computation of Solar Irradiance

Figure 8.2 shows a graphical comparison between SEAT running with different daily air temperature extremes as input and two global irradiance models without air temperature as input: G_{HB} an empirical model built by combining the Hottel (1976) model for the direct component (Eq. 5.66) and the Bugler (1977) model for the diffuse component (Eq. 5.70), and G_{PS} (Eq. 5.51), which is a parametric model proved highly accurate in the Romanian climate (Paulescu and Schlett 2003). Since both models operate under clear sky, the actual total cloud amount has been considered using the Kasten and Czeplak (1980) equation (Eq. 5.93). All models have been run assuming 45°N latitude. The curves displayed in Fig. 8.2 are calculated for July 1 ($j = 182$) for a sunny ($C = 0$) and an overcast day ($C = 1$). It can be concluded that SEAT performs at least as accurate as the models which do not use air temperature at input.

Table 8.1 shows the statistical indicators of accuracy when the empirical models SEAT and HB are applied against data (global solar irradiance) measured of 15 s lag in the first 10 days of July 2010 at the station of the West University of Timisoara (SRMS 2012). Only data for an elevation angle greater than five degrees have been kept. The data recorded in a day have been separated in two classes according to the sunshine number $\xi = 0$ and $\xi = 1$. Then, in each day, for the entire dataset and separately for each class the statistical indicators *RMSE* and *MBE* have been calculated. It can be seen that in mostly sunny days ($0.5 < \bar{\xi} < 1$),

Table 8.1 Statistical indicators of accuracy for SEAT

Date	N	$\bar{\xi}$	$\bar{\zeta}$	ξ	SEAT		HB	
					RMSE	MBE	RMSE	MBE
09	3418	0.955	0.0011	0, 1	0.090	0.006	0.079	0.021
	154			0	0.574	0.042	0.615	−0.126
	3264			1	0.088	−0.007	0.077	0.024
10	3414	0.946	0.0020	0, 1	0.088	0.013	0.087	0.033
	188			0	0.528	−0.265	0.721	−0.520
	3226			1	0.085	0.014	0.084	0.035
02	3443	0.925	0.0020	0, 1	0.083	0.042	0.089	0.061
	261			0	0.379	−0.282	0.879	−0.776
	3182			1	0.080	0.043	0.085	0.064
06	3430	0.903	0.0087	0, 1	0.187	−0.013	0.170	0.016
	338			0	0.289	−0.102	0.391	−0.252
	3092			1	0.180	−0.012	0.162	0.020
01	3445	0.866	0.0014	0, 1	0.085	0.033	0.091	0.053
	466			0	0.340	−0.089	0.586	−0.422
	2979			1	0.079	0.035	0.084	0.058
04	3437	0.816	0.0096	0, 1	0.182	−0.031	0.189	−0.003
	636			0	0.880	−0.656	1.062	−0.829
	2801			1	0.143	0.002	0.145	0.035
03	3440	0.571	0.0026	0, 1	0.172	0.061	0.170	0.072
	1475			0	0.627	0.376	0.634	0.299
	1965			1	0.072	0.009	0.089	0.035
08	3422	0.355	0.0216	0, 1	0.323	−0.131	0.328	−0.107
	2207			0	0.730	0.418	0.891	−0.588
	1215			1	0.182	0.036	0.169	0.027
05	3434	0.243	0.0099	0, 1	0.441	−0.079	0.468	−0.102
	2598			0	0.714	−0.155	0.808	−0.267
	836			1	0.188	−0.007	0.193	0.031
07	3426	0.022	0.0067	0, 1	0.626	−0.068	0.689	−0.132
	3348			0	0.662	−0.074	0.735	−0.150
	78			1	0.197	0.006	0.202	0.067

(Eq. 8.1) and HB (Eqs. 5.66, 5.70 and 5.93) models. $\bar{\xi}$and $\bar{\zeta}$ are daily mean sunshine number and sunshine stability number. N represents the number of measurements in each class of sunshine number ξ. Data recorded during the first 10 days of July 2010 at the station of the West University of Timisoara (SRMS 2012) have been used. Days are classified according to the magnitude of mean sunshine number

when the sun is shining ($\xi = 1$), both models estimate global solar irradiance with reasonable accuracy (for SEAT, *RMSE* takes values between 8.3 and 18.2 %). The estimation accuracy decreases when the sun is behind clouds ($\xi = 0$). Moreover, the estimation accuracy decreases dramatically (for SEAT, *RMSE* takes values between 32.3 and 62.6 %) in mostly cloudy days ($0 < \bar{\xi} < 0.5$), keeping the better estimation for moments when the sun is shining ($\xi = 1$). Results from Table 8.1 shows that the statistical indicators are not set in order by the daily mean sunshine number. For example, the day 09 was the most "clear" ($\bar{\xi} = 0.955$) while in the day 03 the sun was shining little over half of daylight ($\bar{\xi} = 0.571$). For the class $\xi = 1$ *RMSE* is 8.8 % in the day 09 and 7.2 % in the day 03. But, these days are characterized by almost the same sunshine stability number $\bar{\zeta} = 0.0011$ in 09 and $\bar{\zeta} = 0.0026$ in 03. In the sunny day 06, with $\bar{\xi} = 0.903$ and $\bar{\zeta} = 0.0087$, even for class $\xi = 1$ the estimation accuracy is low, $RMSE = 18.0$ %.

What we learn from Table 8.1 is that the SEAT (as well as HB) model accuracy depends on the radiative regime: as the radiative regime instability increases, the model accuracy decrease. Future effort should be dedicated to increase the accuracy of estimating solar irradiance in days when the state of the sky frequently changes.

8.2.3 Daily Irradiation Computation

Daily solar irradiation may be straightforwardly computed by integrating the SEAT irradiance module (Eq. 8.1) between sunrise and sunset:

$$H_j = \frac{12}{\pi} \int_{-\omega_0}^{\omega_0} G\left[\bar{C}_j, \omega, t_j(t_{\max}, t_{\min}, \omega)\right] d\omega \tag{8.6}$$

where $\bar{C}_j$ is the daily average of cloudiness which can be related to daily air temperature extremes by a third order polynomial (Paulescu et al. 2011):

$$\begin{aligned} C(\Delta t, \bar{t}_w) = & 2.797 \cdot 10^{-5} \bar{t}_w^2 \Delta t - 5.137 \cdot 10^{-5} \bar{t}_w^3 + 4.3949 \cdot 10^{-4} \bar{t}_w^2 + 5.2192 \cdot 10^{-3} \bar{t}_w \\ & + 2.1554 \cdot 10^{-3} \bar{t}_w \Delta t - 1.4196 \cdot 10^{-4} \bar{t}_w \cdot (\Delta t)^2 + 1.0558 - 1.6282 \cdot 10^{-2} \Delta t \\ & - 4.8898 \cdot 10^{-3} (\Delta t)^2 + 2.0761 \cdot 10^{-4} (\Delta t)^3 \end{aligned} \tag{8.7}$$

In Eq. (8.7), $\bar{t}_w$ stands for the weekly average of air temperature and for a Julian day j it is computed simply as an arithmetical mean:

$$\bar{t}_w(j) = \frac{1}{7} \sum_{j-3}^{j+3} \frac{t_{\max}(j) + t_{\min}(j)}{2} \tag{8.8}$$

Since data have been collected in temperate climate (South-Eastern Europe), at some extent the correlation $C(\Delta t, \bar{t}_w)$ emerges naturally: at small Δt_j there is a small probability to be a sunny day, thus cloudiness takes only values close to 1; as Δt increases, the range of cloudiness increases, $\bar{t}_w$ acquiring more weight. High Δt and high $\bar{t}_w$ could be associated certainly with a sunny day. As $\bar{t}_w$ decreases, the days pass toward winter and the model counts a higher probability to be a cloudy day at moderate Δt.

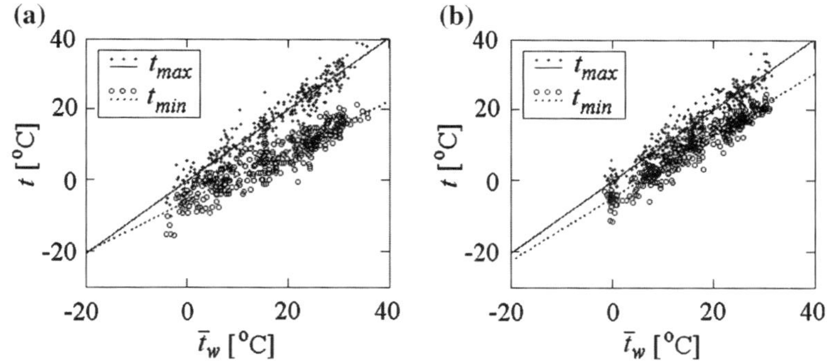

Fig. 8.3 Daily maximum and minimum air temperature as function of its weekly average and the corresponding linear regression at the stations **a** Iasi and **b** Constanta, during year 2000

By using Eq. (8.7) SEAT is able to compute the daily global solar irradiation using as entry only the daily air temperatures extremes.

8.2.4 Extending the Application Area

Because SEAT parameters have been fitted with data coming from only one location, it is evident that SEAT is sensitive to the origin location. A method for extending the application area of SEAT, based on findings reported in Paulescu et al. (2006), follows.

The daily amplitude of air temperature and daily mean air temperature are parameters influenced in a complex manner by the local meteorological regime. Conceivably, SEAT performance decreases in other sites due to a different behavior of $\Delta t = \Delta t(\bar{t}_w)$. An example is presented in Figs. 8.3 and 8.4. Figure 8.3 displays a graphic comparison of linear regressions of maximum and minimum air temperatures depending on weekly average air temperatures recorded during the year 2000 at two Romanian meteorological stations in Iasi and Constanta (see Fig. 3.1 for localization):

$$\text{Iasi: } t_{\max}(\bar{t}_w) = 1.018\bar{t}_w - 0.292\ ; \quad t_{\min}(\bar{t}_w) = 0.71\bar{t}_w - 6.032 \tag{8.9}$$

$$\text{Constanta: } t_{\max}(\bar{t}_w) = 1.011\bar{t}_w - 0.182\ ; \quad t_{\min}(\bar{t}_w) = 0.883\bar{t}_w - 4.911 \tag{8.10}$$

It can be noted a clear difference between them. While at both station $t_{\max}(\bar{t}_w)$ is roughly the same, $t_{\min}(\bar{t}_w)$ is different. The daily amplitude at the station Constanta is smaller than at station Iasi. This difference is more visible in Fig. 8.4, where $\Delta t(\bar{t}_w) = \Delta t = t_{\max}(\bar{t}_w) - t_{\min}(\bar{t}_w)$ is plotted for the two stations and for other two, Timisoara and Galati (see Fig. 3.1 for localization).

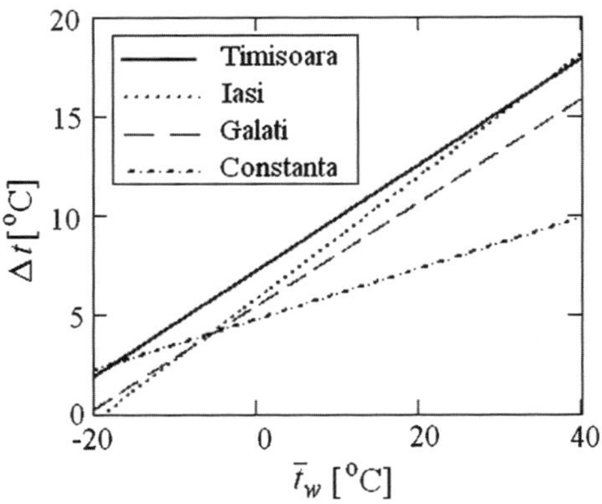

Fig. 8.4 Daily air temperature amplitude estimated by linear regression with respect to $\bar{t}_w$ at four Romanian stations. Data recorded during the year 2000 have been used

These figures suggest an idea of adapting the procedure for other sites with different temperature regimes. This may be done by introducing a correction χ, which acclimatize the variables Δt to the origin location temperature regime. This means that in Eq. (8.7), Δt is replaced by $\chi \cdot \Delta t$. A practical implementation of the method is reported in Paulescu et al. (2011), where χ is calculated as the ratio:

$$\chi = \chi_{\Delta t} / \chi_{\bar{t}_w} \tag{8.11}$$

$\chi_{\Delta t}$ is the ratio of yearly mean air temperature amplitude in the origin location Δt_{TM} (Timisoara for SEAT) and the actual location Δt_l :

$$\chi_{\Delta t} = \Delta t_{TM} / \Delta t_l \tag{8.12}$$

and $\chi_{\bar{t}_w}$ is the ratio of yearly mean air temperature in the origin location $\bar{t}_{w,TM}$ and actual location $\bar{t}_{w,l}$:

$$\chi_{\bar{t}_w} = \bar{t}_{w,TM} / \bar{t}_{w,l} \tag{8.13}$$

SEAT has been build with data recorded in Timisoara in the years 1998–2000 (details on this matter are given in Paulescu et al. 2011), for which $\Delta t_{TM} = 11.4\,°\mathrm{C}$ and $\bar{t}_{w,TM} = 11.8\,°\mathrm{C}$. Consequently:

$$\chi = 0.9661 \cdot \frac{\bar{t}_{w,l}}{\Delta t_l} \tag{8.14}$$

8.2.5 Model Application

SEAT can be applied for the estimation of global solar irradiance on a horizontal surface and the estimation of global solar irradiation in a certain time period.

The computation of global solar irradiance is obtained in three steps. In addition to geographical coordinates and temporal information, instantaneous air temperature t_a and total cloud amount C has to be known. The steps are:

(a) Using Eqs. (8.3a, 8.3b, 8.3c), the approximation functions f_i, $i = 1, 2, 3$, are computed;
(b) $f(C, t_r \cdot \cos\theta_z)$ is computed with Eq. (8.2);
(c) The global solar irradiance is computed with Eq. (8.1).

The computation of global solar irradiation runs in four steps. Daily maximum and minimum air temperature in a week around the current day and the yearly mean of daily air temperature amplitude and mean are needed as input.

(d) The adaptation factor of the daily temperature amplitude is calculated with Eq. (8.14);
(e) Daily mean cloudiness is estimated with Eq. (8.7);
(f) $G(\bar{C}_j, \omega, t_j(\omega))$ is built by replacing into Eq. (8.1) the cloudiness with its daily mean value computed at step 2 and the instantaneous air temperature with the approximation Eq. (8.5);
(g) Daily global solar irradiation is calculated with Eq. (8.6) by the integration of $G(\bar{C}_j, \omega, t_j(\omega))$ between sunrise and sunset hour angle.

A discussion on SEAT accuracy applied to calculating the daily global solar irradiation is included in Sect. 8.5.

8.3 Ångström-Type Equations

Ångström-type equations (Ångström 1924) correlate the ground solar irradiation H with the corresponding value at the extraterrestrial level H_{ext} by means of simple measures of the state of the sky (commonly, relative sunshine or total cloud cover amount). The Ångström equation has been introduced in Sect. 5.5 where it is shown that daily air temperature, maximum, minimum and amplitude, encapsulate information about the state of the sky and can be used as entry in Ångström equations. Two models representing historical milestones in the field have been summarized in Sect. 5.5.3: (1) The equation of Supit and Van Kappel (1998) which takes daily air temperature extremes besides cloud cover amount to compute the daily clearness index (Eq. 5.94) and (2) the Bristow and Campbell (1984) equation which uses daily air temperature amplitude as indicator of the state of the sky (Eq. 5.96).

Here, three other models are discussed: from the first category the polynomials correlations reported in El Metwally (2003) and from the second category the models developed by Donatelli and Bellocchi (2001) and Paulescu et al. (2006).

8.3.1 El Metwally' Models

The coefficients of the El Metwally (2003) models have been fitted with data provided by seven meteorological stations distributed over Egypt territory from the North to South. Three models are proposed:

Table 8.2 Regression coefficients in Eq. (8.15) for three cloud cover class (El Metwally 2003)

Coefficient	*a*	*b*	*c*	*d*	*e*
5PP					
CLEAR SKY	0.8027	−0.0211	−0.0369	−0.1448	−2.2660
PARTLY CLOUDY	0.7927	−0.0218	−0.0293	−0.7237	−0.8569
HEAVY CLOUDY	0.6630	0.2939	−0.2387	−2.1343	4.9700
4PP					
CLEAR SKY	0.7544	−0.0679	0.0170	−0.3358	–
PARTLY CLOUDY	0.7819	−0.0368	−0.0168	−0.8190	–
HEAVY CLOUDY	0.6647	0.2953	−0.2377	−1.3184	–
PEP					
CLEAR SKY	0.0353	−0.0005	−0.0014	−0.0076	1.9818
PARTLY CLOUDY	0.0387	−0.0005	−0.0019	−0.0338	1.9279
HEAVY CLOUDY	0.0452	0.0147	−0.0151	−0.1455	2.0678

$$\text{Five Parameter polynomial (5PP): } H = aH_{\text{ext}} + bT_{\max} + cT_{\min} + d\bar{C} + e \tag{8.15a}$$

$$\text{Four Parameter Polynomial (4PP): } H = aH_{\text{ext}} + bT_{\max} + cT_{\min} + d\bar{C} \tag{8.15b}$$

Parameter exponential polynomial (PEP):

$$H = \exp(aH_{\text{ext}} + bT_{\max} + cT_{\min} + d\bar{C} + e) \tag{8.15c}$$

In Eq. (8.15a) H and H_{ext} stand for ground and extraterrestrial daily global solar irradiation in MJ/m^2 and $\bar{C}$ is the daily mean of total cloud cover amount in octas ($0 \le \bar{C} \le 8$). The coefficients in Eq. (8.15a) are given in Table 8.2 for three classes of daily mean cloud cover amount, corresponding to the state of the sky: CLEAR SKY ($\bar{C}/8 < 0.2$), PARTLY CLOUDY ($0.2 \le \bar{C}/8 \le 0.6$), and HEAVY CLOUDY ($0.6 < \bar{C}/8$).

El Metwally noted that local climatology has a visible effect on the model performance. Splitting the model into three equations associated to different classes of total cloud amount substantially increase the accuracy. The same conclusion has been previously noticed in Badescu (2002).

8.3.2 RadEst Tool

The Bristow and Campbell (1984) model has been constantly improved in the 1990s by a team of Research Institute for Industrial Crops (ISCI), Bologna, Italy. In the models reported in Donatelli and Marletto (1994) and Donatelli and Campbell (1998) a correction factor for seasonality occurring in mid-latitude areas was introduced. A further improvement was proposed in Donatelli and Bellocchi (2001), to better account for seasonality at a wide variety of sites. The last model is summarized in the following.

The Donatelli-Bellocchi (DB) model estimates global clearness in the Julian day j index with the equation:

$$H_j/H_{ext} = \tau(1 + f_j)\left[1 - \exp\left(\frac{-b \cdot (\Delta t_j)^2}{\Delta t_w}\right)\right] \tag{8.16}$$

where τ and b are empirical coefficients. In Eq. (8.16), τ has the meaning of atmospheric transmittance. Δt_j is the daily air temperature amplitude calculated as $\Delta t_j = t_{\max,j} - 0.5(t_{\min,j} + t_{\min,j+1})$ and Δt_w is the mean of daily temperature amplitude over seven days around the current day. f_j is the seasonality function given by:

$$f_j = c_1\left[\sin\left(j_r c_2 \frac{\pi}{180}\right) + \cos\left(j_r f(c_2)\frac{\pi}{180}\right)\right] \tag{8.17}$$

In Eq. (8.17), c_1 and c_2 are the seasonality factors, varying from 0 to 0.5 and from 1 to 1.5, respectively, and j_r is a the so called "reverse option" $j_r = 361 - j$. $f(c_2)$ is expressed as:

$$f(c_2) = 1 - 1.9[c_2 - \mathrm{int}(c_2)] + 3.83[c_2 - \mathrm{int}(c_2)]^2 \tag{8.18}$$

with int(c_2) the integer part of c_2.

DB model is part of a suite of models implemented in the RadEst global solar radiation estimation tool (Donatelli et al. 2003), available through the ISCI website, ISCI (2012). The data input requirements are daily values of maximum and minimum air temperature and location-specific parameters. A sample of model parameters at about 200 worldwide sites can be downloaded from the RadEst web page (ISCI 2012), as obtained through optimization procedures over multiple year solar irradiation data sets.

8.3.3 AEAT Models

Two empirical global solar irradiation models using daily air temperature extremes at input build with data from Timisoara, Romania, have been reported in Paulescu et al. (2006). The models are recognized by the acronym AEAT, Ångström Equation by Air Temperature.

The first model AEAT-1 consists of a typical Ångström equation in which daily minimum $t_{\min}$ and maximum $t_{\max}$ temperatures together play the role of relative sunshine duration σ. Daily global solar irradiation H is related to its maximum possible (clear sky) H_0 with the equation:

$$H = H_0 \cdot f_a(\Delta t, \bar{t}_5) \tag{8.19}$$

where $\Delta t = t_{\max} - t_{\min}$ is the daily amplitude of air temperature and $\bar{t}_5$ is the five days average of $\bar{t} = (t_{\max} + t_{\min})/2$. The Ångström equation is expressed as:

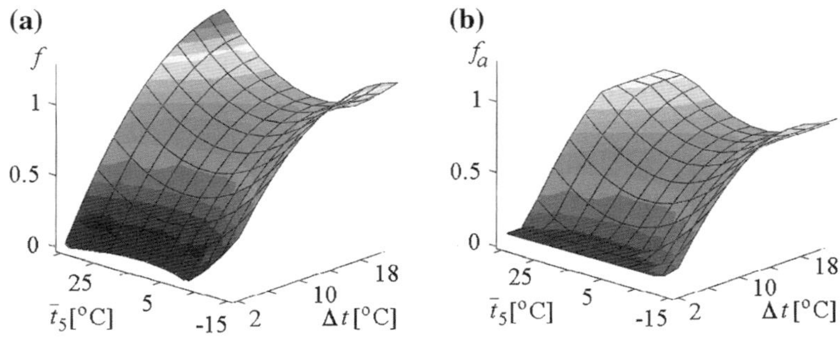

Fig. 8.5 3D plot of (**a**) $f(\Delta t, \bar{t}_5)$ given by Eq. (8.21) and (**b**) $f_a(\Delta t, \bar{t}_5)$ given by Eq. (8.20)

$$f_a(\Delta t, \bar{t}_5) = \begin{cases} 0.288 & \text{if} \quad f(\Delta t, \bar{t}_5) \leq 0.288 \\ f(\Delta t, \bar{t}_5) & \text{if} \quad 0.288 < f(\Delta t, \bar{t}_5) < 1.031 \\ 1.031 & \text{if} \quad f(\Delta t, \bar{t}_5) \geq 1.031 \end{cases} \tag{8.20}$$

$f(\Delta t, \bar{t}_5)$ is a two degree polynomial fitted with $r^2 = 0.948$:

$$\begin{aligned} f(\Delta t, \bar{t}_5) = \; & 5.29 \cdot 10^{-4} \bar{t}_5 \Delta t + 5.49 \cdot 10^{-4} \bar{t}_5^2 - 0.01832 \bar{t}_5 - 0.0095 \\ & + 0.099237 \Delta t - 0.002838 (\Delta t)^2 \end{aligned} \tag{8.21}$$

The two surfaces f and f_a are plotted in Fig. 8.5. Equation (8.19) needs to cover a wide range of daily air temperature extremes that can be met in temperate climate, 2 °C < Δt < 20 °C and $-15° < \bar{t}_5 < 35$ °C. It can be seen from Fig. 8.5a that there are pairs (Δt, $\bar{t}_5$) that are producing unphysical values of $f(\Delta t, \bar{t}_5)$, i.e., below the low range limit of any Ångström correlation, $H/H_0 \approx 0.2$, reached in the overcast days ($\sigma = 0$) or over the upper limit $H/H_0 \approx 1$, reached in perfect clear sky days ($\sigma = 1$). Thus, the limitation Eq. (8.20) has been imposed to adjust this behavior, through a comparison with the nonlinear variant of Ångström equation Eq. (5.89).

The second model, AEAT-2, considers a linear dependence of H in respect to H_0, with slope and intercept depending on daily extreme air temperatures. The AEAT-2 equations are Paulescu et al. (2006):

$$H(\Delta t, \bar{t}_5) = H_0 f_1(\Delta t) + f_2(\bar{t}_5) \tag{8.22a}$$

$$f_1(\Delta t) = -0.32414 + 0.36689 (\Delta t)^{0.42449} \tag{8.22b}$$

$$f_2(\bar{t}_5) = 0.00576 + 0.37256 \sin\left(\frac{2\pi \bar{t}_5}{26.35} + 1.83278 \right) \tag{8.22c}$$

Figure 8.6 displays the way in which $H(\Delta t, \bar{t}_5)$ acts on H_0 in the usual ranges of Δt and $\bar{t}_5$ in two days, one during the winter (H_0 = 2 kWh/m^2) and the other one during the summer (H_0 = 8 kWh/m^2). The graphs clearly reveals the role of the sine function $f_2(\bar{t}_5)$ in Eq. (8.22a): being independent of Δt and depending only on

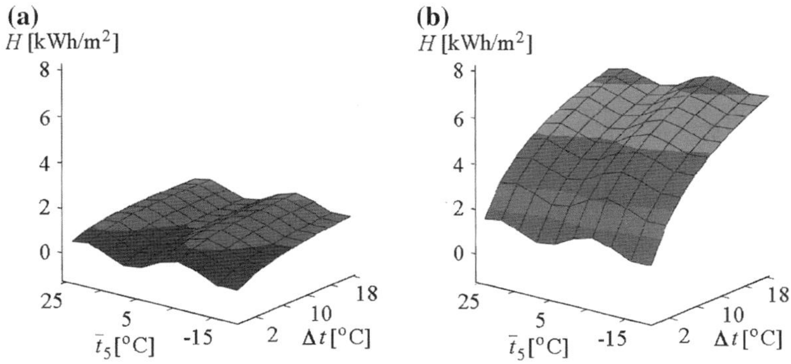

Fig. 8.6 Graph of $H(\Delta t, \bar{t}_5)$ given by Eq. (8.22a) in the usual diurnal air temperature range for (**a**) $H_0 = 2$ kWh/m^2 and (**b**) $H_0 = 8$ kWh/m^2

$\bar{t}_5$, it operates in a natural way as a seasonal adaptor. On other hand, $f_2(\bar{t}_5)$ acts as a refinement for the typical Ångström equation $H(\Delta t, \bar{t}_5) = H_0 f_1(\Delta t)$.

In sites other than the origin location of AEAT models, an improving of the estimation accuracy of the daily global solar irradiation can be achieved by introducing an adjustment parameter, as in the case of SEAT model. For the AEAT-1 model, Eq. (8.20) is rewritten as:

$$f_{\chi'}(\Delta t, \bar{t}_5) = \begin{cases} 0.288 & \chi' f(\Delta t, \bar{t}_5) < 0.288 \\ 1.031 & \chi' f(\Delta t, \bar{t}_5) > 1.031 \\ \chi' f(\Delta t, \bar{t}_5) & \text{otherwise} \end{cases} \tag{8.23}$$

For a given location, the parameter χ' is constant and depends on mean values characterizing the entire databases of air temperature in the origin location and the new location. In Eq. (8.23), the parameter χ' acts as a multiplier of polynomial coefficients resulting from the fitting process at the origin location. χ' is calculated as follows (Paulescu et al. 2006):

$$\chi' = \frac{\Delta\bar{t}_5|_{TM}}{\Delta\bar{t}_5|_{\text{site}}} \tag{8.24}$$

where $\Delta\bar{t}_5 = t_{\max}(\bar{t}_5) - t_{\min}(\bar{t}_5)$, with $t_{\max}(\bar{t}_5)$, $t_{\min}(\bar{t}_5)$ the lines which better approximate the database $(t_{\max}, \bar{t}_5)$, $(t_{\min}, \bar{t}_5)$, respectively.

AEAT-2 is also close to the origin location. Since χ' is independent of the AEAT-1 and AEAT-2 approaches, it is expected that the action of χ' on AEAT-2 will be the same as on AEAT-1.

8.4 Fuzzy Models

In Tulcan-Paulescu and Paulescu (2008), a model for daily global solar irradiation based on fuzzy sets theory is reported (for an introduction see Sect. 7.1.2). The important feature of the model is the use of only air temperature as the input parameter. To account for the fact that air temperature-based models are sensitive to the origin, an adaptive algorithm for the membership functions has been developed.

The algorithm outlined next is a slightly modified version of that reported in (Tulcan-Paulescu and Paulescu 2008). The model is conducted with two input linguistic variables: daily amplitude of air temperature $\Delta t_j = t_{\max,j} – t_{\min,j}$ and Julian day j. The output variable is the daily global clearness index $k_t = H_j/H_{ext}$ where H_j represents the daily solar irradiation in the day j while H_{ext} is its extraterrestrial value.

Because of relative higher than average scattering of Δt to k_t, these variables are characterized by eight attributes, T_i and K_i $i = 1,\ldots,8$, respectively. The attributes range from VERY LOW to VERY HIGH with ascending i. For the linguistic variable Julian day only two attributes have been considered, WINTER (W) and SUMMER (S).

The membership functions of input attributes are plotted in Fig. 8.7, where the notation for every attribute is specified.

The membership functions for Δt_i $(i = 1\ldots8)$ attributes are triangular:

$$m_{\Delta t,i}(\Delta t,\chi) = \begin{cases} \max\left(0, \dfrac{\Delta t - a_i\chi}{c_i\chi - a_i\chi}\right) & \text{if } \Delta t < c_i\chi \\ \max\left(0, 1 - \dfrac{\Delta t - c_i\chi}{b_i\chi - c_i\chi}\right) & \text{otherwise} \end{cases} \tag{8.25}$$

The coefficients a_i,, b_i ,and c_i have the meaning depicted in Fig. 8.7. Numerical values for the coefficients a_i, b_i, and c_i, $i = 1\ldots8$, are listed in Table 8.3. The membership function of attributes T_1 and T_8 are saturated toward zero ($m_{\Delta t,1} = 1$ if $\Delta t < c_1$) and infinite ($m_{\Delta t,8} = 1$ if $\Delta t > c_8$), respectively. χ is an adjustment factor which fits the algorithm to the location. As introduced in Eq. (8.25), it compresses or expands the membership functions associated to Δt attributes to overlay the specific Δt range in a given location. A recipe for computing χ as a function of yearly mean of air temperature $\bar{t}$ and yearly mean of daily air temperature amplitude $\Delta\bar{t}$ is reported in (Tulcan-Paulescu and Paulescu 2008). $\chi(\Delta\bar{t},\bar{t})$ is expressed as a third-order bivariate polynomial:

$$\begin{aligned} \chi(\Delta\bar{t},\bar{t}) = {} & 0.00413\bar{t}^3 - 0.964\bar{t}^2 + 1.078\bar{t} - 0.00565\bar{t}^2\Delta\bar{t} - 0.023\bar{t}\Delta\bar{t} \\ & + 0.009476\bar{t}(\Delta\bar{t})^2 + 0.495\Delta\bar{t} - 0.0468(\Delta\bar{t})^2 - 0.002223(\Delta\bar{t})^3 - 3.581 \end{aligned} \tag{8.26}$$

Eqation (8.26) is not applicable everywhere; the required condition in a given location characterized by the pair $(\Delta\bar{t},\bar{t})$ is: $0.6 < \chi(\Delta\bar{t},\bar{t}) < 1.4$.

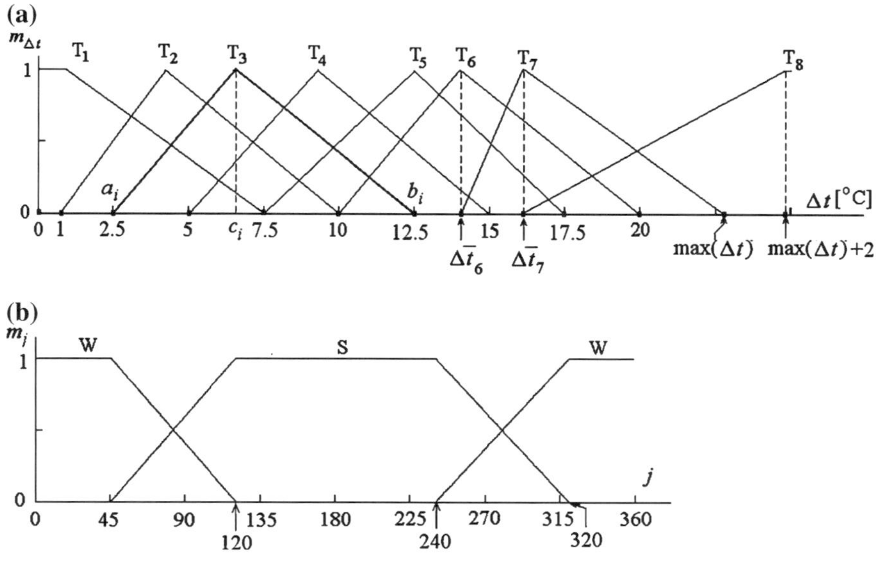

Fig. 8.7 The membership functions of the input linguistic variable: **a** Daily temperature amplitude and **b** Julian day. The notation for the triangular membership functions is indicated to the attribute T_3 ($i = 3$). $c_i = \Delta \overline{t_i}$ represents the mean of the elements of set the T_i

Table 8.3 Coefficients a_i, b_i and c_i, $i = 1 \ldots 8$, in Eqs. (8.25) and (8.29)

Linguistic variable	i	1	2	3	4	5	6	7	8
Δt	a_i	0.00	1.00	2.50	5.00	7.50	10.00	15.00	18.75
	b_i	7.50	10.00	12.50	15.00	17.50	20.00	22.50	–
	c_i	3.75	5.50	7.50	10.00	12.50	15.00	18.75	24.50
k_t	a_i	0.00	0.10	0.20	0.30	0.40	0.50	0.60	0.70
	b_i	0.10	0.20	0.30	0.40	0.50	0.60	0.70	0.80
	c_i	0.20	0.30	0.40	0.50	0.60	0.70	0.80	0.90

The role of the Julian day linguistic variable is to enhance the model prediction in cold season, when the irradiation models accuracy decays. Thus, it is allowed to enable specific rules for days characterized with WINTER attribute. On the other hand, everyone knows from routine observations that some spring or autumn days are sometimes closer to the summer one and other times to the winter ones; this behavior is well accounted for by the trapezoidal membership functions of the Julian day attributes:

$$m_{j,w} = \begin{cases} \max\left(0, 1 - \dfrac{j-45}{75}, \dfrac{j-240}{80}\right) & \text{if} \quad 45 < j < 320 \\ 1 & \text{otherwise} \end{cases} \tag{8.27}$$

Table 8.4 Input/output associative rules of the fuzzy algorithm

Rule#	1	2	3	4	5	6	7	8	9	10	11	12	13	14	15	16
Δt	T_1	T_2	T_3	T_4	T_5	T_6	T_7	T_8	T_1	T_2	T_3	T_4	T_5	T_6	T_7	T_8
j	S	S	S	S	S	S	S	S	W	W	W	W	W	W	W	W
k_t	K_1	K_2	K_3	K_4	K_5	K_6	K_7	K_8	K_1	K_3	K_4	K_5	K_5	K_5	K_6	K_7

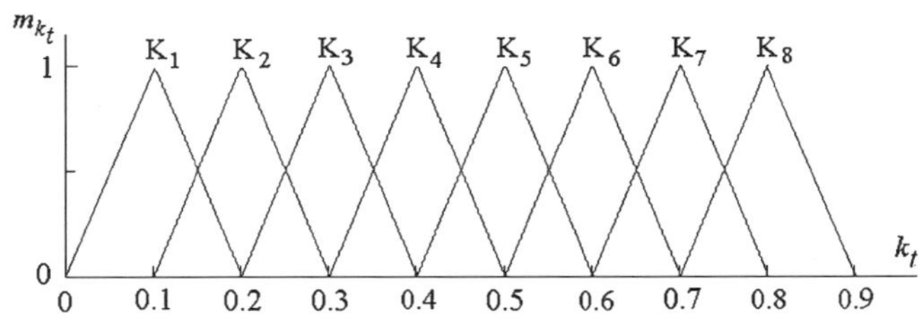

Fig. 8.8 The membership functions of the output linguistic variable k_t attributes

$$m_{j,S} = \begin{cases} \max\left(0\,,\ \frac{j-45}{75}\right) & \text{if} \quad j < 120 \\ 1 & \text{if} \quad j < 120 \\ \max\left(0\,,\ 1 - \frac{j-240}{80}\right) & \text{otherwise} \end{cases} \tag{8.28}$$

The membership function attributes of K_i, $i = 1 \ldots 8$, are fixed in shape as triangular, symmetric and equidistant:

$$m_{k_t,\,i}(k_t) = \begin{cases} \max\left(0,\ \dfrac{k_t - a_i}{c_i - a_i}\right) & \text{if} \quad k_t < c_i \\ \max\left(0,\ 1 - \dfrac{k_t - c_i}{b_i - c_i}\right) & \text{otherwise} \end{cases} \tag{8.29}$$

The coefficients in Eq. (8.29) a_i, b_i, and c_i have the meaning from Fig. 7.3a. Figure 8.8 displays the membership function attributes of $k_{t,i}$. Numerical values for the coefficients a_i, b_i, and c_i, $i = 1 \ldots 8$, are listed in Table 8.3.

The input/output mapping of the fuzzy system is presented in Table 8.4. Every rule is encompassed in a column meaning a fuzzy implication in the relation Eq. (8.30). For example, rule #7 is reading:

$$\text{IF } \Delta t \text{ IS } T_7 \text{ AND } j \text{ IS S THEN } k_t \text{ IS } K_7 \tag{8.30}$$

As a matter of fact, the rule Eq. (8.30) has to be understood as follows: If daily temperature amplitude is high in a summer day then also the clearness index is high, with the assumption that HIGH is associated to T_7 and K_7 attributes.

With the input/output mapping listed in Table 8.4, the fuzzy algorithm is ready for use. A handling example is presented in the following. Let the input be: $\Delta t = 18$ °C, $\chi = 1$ and $j = 100$. The process is illustrated graphically in Fig. 8.9.

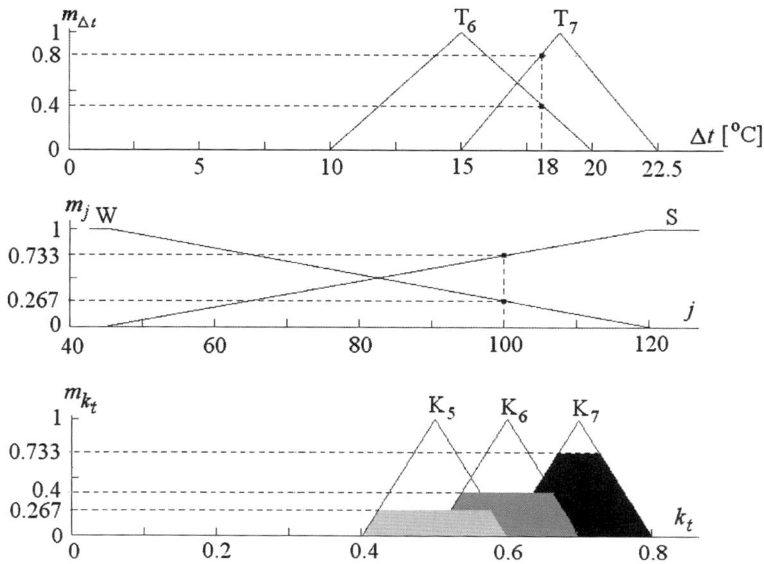

Fig. 8.9 Membership functions associated to the attributes of input linguistic variables $m_{\Delta t}$ and m_j and output linguistic variable m_{kt}. Only the attributes with nonzero confidence level are plotted. The area corresponding to the integral of output membership functions, truncated at the corresponding degree appears in *gray* shading

The three steps experienced by the information passing the fuzzy system from the input to the output are summarized below:

(1) *Fuzzyfication.* Crisp inputs are transformed into confidence levels of input linguistic variable attributes, being computed with Eqs. (8.27) and (8.28). For $\Delta t = 18$ °C, the linguistic variable air temperature amplitude is characterized by two attributes T_6 and T_7 with the confidence levels $m_{\Delta t,6} = 0.4$ and $m_{\Delta t,7} = 0.8$, respectively. Julian day $j = 100$ have both attributes SUMMER and WINTER with the confidence level $m_S = 0.733$ and $m_W = 0.267$, respectively.
(2) *Inference.* According to the rule-base from Table 8.4, four rules are set-up. First the fuzzy inputs are combined logically using the operator AND (see Sect. 7.1.2.2) to produce the output values:

$$\begin{array}{ll}
\text{Rule\#6} & m_{k_t,6} = \min\left(m_{\Delta t,6}, m_{j,S}\right) = \min(0.4, 0.733) = 0.4 \\
\text{Rule\#7} & m_{k_t,7} = \min\left(m_{\Delta t,7}, m_{j,S}\right) = \min(0.8, 0.733) = 0.733 \\
\text{Rule\#14} & m_{k_t,5} = \min\left(m_{\Delta t,6}, m_{j,W}\right) = \min(0.4, 0.267) = 0.267 \\
\text{Rule\#15} & m_{k_t,6} = \min\left(m_{\Delta t,7,}, m_{j,W}\right) = \min(0.8, 0.267) = 0.267
\end{array}$$

Each rule leads to an attribute of output linguistic variable clearness index. But the rules Rule#6 and Rule#15 sums up to the same conclusion: attribute K_6. The different degree of fulfillment K_6 needs to be summarized in just one conclusion, which is achieved by unifying the individual results with the fuzzy operator OR (see Sect. 7.1.2.2). Thus, the confidence level of output linguistic variable attribute K_6 is obtained as:

$$m_{k_t,6} = \max(0.4,\ 0.267) = 0.4$$

(3) *Defuzzyfication.* The result of the inference process is translated from fuzzy logic into a crisp value using the COG method (Eq. 7.10). After simple manipulations it reads:

$$k_t = \frac{c_5 m_{k_t,5}\left(1 - \frac{m_{k_t,5}}{2}\right) + c_6 m_{k_t,6}\left(1 - \frac{m_{k_t,6}}{2}\right) + c_7 m_{k_t,7}\left(1 - \frac{m_{k_t,7}}{2}\right)}{m_{k_t,5}\left(1 - \frac{m_{k_t,5}}{2}\right) + m_{k_t,6}\left(1 - \frac{m_{k_t,6}}{2}\right) + m_{k_t,7}\left(1 - \frac{m_{k_t,7}}{2}\right)}$$

and, using the numerical values from the inference task, the k_t predicted by the fuzzy algorithm is equal to 0.623.

A C source code that computes the daily global solar irradiation using the above fuzzy procedures can be downloaded from the website SRMS (2012). To compute the adjustment factor in Eq. (8.25), a data file "stationtemperatures.prn" is read from the disk. It should contain 365 rows with the daily air temperatures organized in 4 tab-delimited columns as follows:

Julian day Mean Maximum Minimum

This file (stationtemperatures.prn) should be prepared by the user. For this, a large on-line database, Global Surface Summary of Day Data, from National Climatic Data Center—NCDC, Asheville, USA, which contains surface meteorological parameters collected at over 8000 stations around the world, including air temperature mean maxima and minima, is available online NCDC (2012). The program has been elaborated aiming to compute the global solar irradiation in a given day and for a given air temperature amplitude. The user is asked to input the local latitude (in degrees), Julian day, air temperature maxima and minima (in Celsius). The program will return the global solar energy (in kWh/m^2). The C source file can be easily modified to meet user requirements. For example, the stationtemperatures.prn file can be extended for a better account of local metrological particularities by adding data of several years. One can build a loop to compute the solar irradiation in a given period, by reading input data from file instead of asking for keyboard input.

8.5 On the Temperature-Based Models Accuracy

Many of the air temperature-based models presented above have been elaborated with the aim to estimate daily global solar irradiation values, which are eventually summed up to produce monthly values. Thus, in the original articles the validation task is usually performed for the estimated daily and/or monthly values. A critical assessment on several temperature-based models accuracy following.

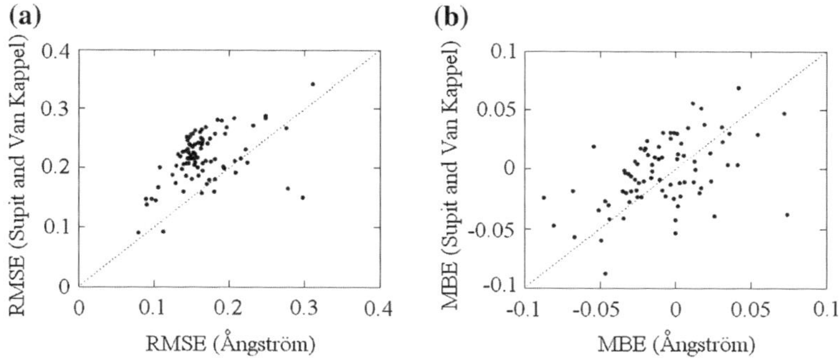

Fig. 8.10 Statistical indicators of accuracy for the Supit and Van Kappel model (Eq. 5.94) in comparison with the relative sunshine-based Angstrom equation: **a** Relative root mean square error (*RMSE*) and **b** Relative mean bias error. The graphs have been constructed with data from Supit and Van Kappel (1998)

8.5.1 SK Model

The SK model (Supit and Van Kappel 1998) has been tested against data recorded in 91 locations in Europe ranging from Italy to Finland. The study concentrates mainly on UK and Ireland (55 stations). To assess the accuracy, daily global solar irradiation values were estimated and compared with observed values by means of *RMSE* and *MBE*. Figure 8.10 summarizes the results reported in (Supit and Van Kappel 1998) for both temperature-based (Eq. 5.94) and Ångström-Prescott (Eq. 5.94) models. The *MBE*s for both methods fall generally inside ±5 %, indicating that for either method the systematic under- and over-estimation is small. The difference in *RMSE* between the two methods is as follows: at 6 stations (6.6 % from all testing sites), the SK model performs better than the Ångström-Prescott equation. At 9 stations (9.9 %) the performance is the same, at 20 stations (22 %) *RMSE* for SK estimation is 0.1 to 5.0 % greater and at 58 stations the *RMSE* for SK estimation is 5.1 to 10.8 % greater than for Ångström-Prescott equation. Generally, the Ångström-Prescott method provides better estimates, however the differences with SK method are small. As a requisite, the method should use cloud cover amount observations which can be retrieved from meteorological satellite data.

8.5.2 El Metwally' Models

The same behavior as with the SK model is encountered with El-Metwally models (Eqs. 8.15a, b, c), which also use the cloud cover amount beside daily air temperature extremes to estimate daily global solar irradiation. Figure 8.11 displays the statistical indicators for the three models reported by El-Metwally: 5PP (Eq. 8.15a), 4PP (Eq. 8.15b) PEP (Eq. 8.15c) and the S–K model (Eq. 5.94) versus Ångström-Prescot equation (Eq. 5.83). The models are tested against data measured in seven locations from Egypt.

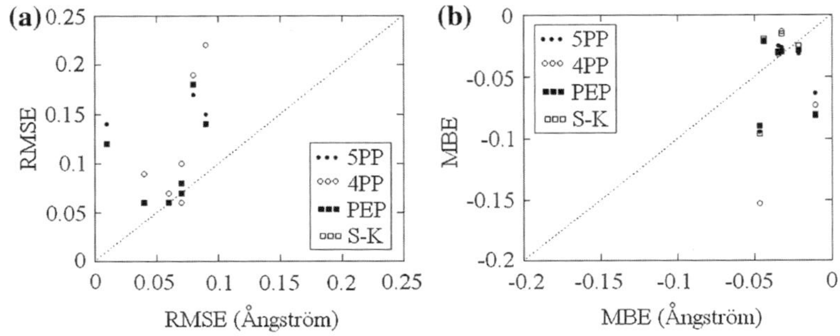

Fig. 8.11 Statistical indicators of accuracy for the El Metwally (Eq. 8.15a, b, c) and S–K model (Eq. 5.94) in comparison with the relative sunshine-based Ångström equation: **a** Relative root mean square error (*RMSE*) and **b** Relative mean bias error (*MBE*). The graphs have been constructed with data from El Metwally (2003)

Again, the Ångström-Prescott method provides better estimations but the difference with temperature-based method is very small: the average of the relative RMSE was found 9 % for the Ångström-Prescott equation, 10 % for the El-Metwally equations and 13 % for the S–K equation (El-Metwally 2004). Additional tests pointed out that local climatology has noticeable effects on the performance of the temperature-based methods. They provide low performance in both winter and spring, which is induced by the increased cloud amount and instability of the state of the sky, respectively. Low biasing is noted in the summer. Generally the errors depend on the sky condition, being low for clear sky, increasing with the cloud amount and reaching high values for overcast sky.

8.5.3 SEAT Model

The SEAT performance has been assessed against data recorded in three different years at 15 European stations located between 40 and 50°N latitudes (Paulescu et al. 2011). Statistical indicators of monthly mean global solar irradiation are collected in Table 8.5 and show that the estimation accuracy is reasonable and compares well with the estimation accuracy of other approaches. For example, *RMS*E and *MBE* for monthly mean daily global solar irradiation ranging between +3.7…+26.0% and −22.0 …+9.1 %, respectively, have been found in (Paulescu and Schlett 2004) after the verification of five models with cloud amount and sunshine duration at input.

Figure 8.12 presents SEAT estimations tracking with good accuracy the daily global solar irradiation measured in 2000 at station Bucharest (44.5°N; 22.2°E; 131 m). As most of the traditional models based on Ångström–Prescott equation, the estimation errors increase in winter months when there are many cloudy or overcast days (*RMSE* is 35.3 % in January and 12.7 % in July). As anyone can notice, in the winter months the amount of solar energy is three to five time smaller

Table 8.5 Statistical indicators of accuracy of monthly mean of daily global solar irradiation estimation with SEAT (Eq. 8.6)

Station	Latitude [deg.]	Longitude [deg.]	Altitude. [m]	Years	χ	*RMSE* [%]	*MBE* [%]
Ajaccio	41.91	8.80	9	2000	1.45	6.4	−1.6
Marseille	43.45	5.23	32	2000	1.59	9.1	−5.9
Bordeaux	44.83	−0.70	61	1998	1.29	6.3	1.7
				2000	1.44	12.9	8.1
Auxerre	47.80	3.50	212	1998	1.23	7.1	1.7
				2000	1.31	10.0	5.8
Strasbourg	48.55	7.63	153	1998	1.11	8.9	1.1

than in summer months. This encourages using air temperature models to predict solar irradiation one day ahead, because as we see in Sect. 8.1 the meteorological services forecast the next day maximum and minimum air temperature within an interval ±2 °C, with a probability close to 70 and 80 % respectively. Thus, it is instructing to see how the accuracy of forecasting air temperature influences the accuracy of the SEAT output. Results of such a study follow.

Figure 8.13 shows daily global solar irradiation estimated with SEAT as function of the measured one in 2000 at the station Bucharest. Two cases have been considered. First, SEAT was run with measured values of daily air temperature extremes (Fig. 8.13a). Second, the entries in SEAT have been modified to emulate inaccurate prediction of both air temperature minimum and maximum in every day: the minimum air temperature used as entry has been taken the actual value minus 2 °C and the maximum air temperature the actual value plus 2 °C. A visual inspection of Fig. 8.13 reveals no major difference between the two scatter plots. Compared with Fig. 8.13a, in Fig. 8.13b can be noted a slight increase of scattering but also a small correction of bias. This is confirmed by statistical indicators: *RMSE* equals 23.3 % and *MBE* −9.4 % with the measured values of air temperature extremes. Using the altered predictions of air temperature, *RMSE* increased slightly to 24.1 % while *MBE* decreased close to zero, i.e. 0.2 %. From these results it can be concluded that the actual level of accuracy reached in one day ahead forecasting of daily air temperature extremes is enabling the use of air temperature based-models to forecast global solar irradiation. The actual challenge lies in increasing the estimation accuracy of these models.

Three limitations are mentioned in Paulescu et al. (2011): (1) SEAT should be applied with care in mountain regions since the model originates and is validated with data recorded under 500 m altitude; (2) the accuracy of SEAT at seacoast may be smaller than for continental sites. Mainly, this is due to the peculiar seacoast air temperature regime, which is not represented in the SEAT algorithm; (3) SEAT was verified only in the latitude interval 40 and 50°N, so an application outside requires verification.

Based on the results of testing SEAT, it can be concluded that the model exhibits a level of accuracy comparable with that of traditional Ångström-Prescott

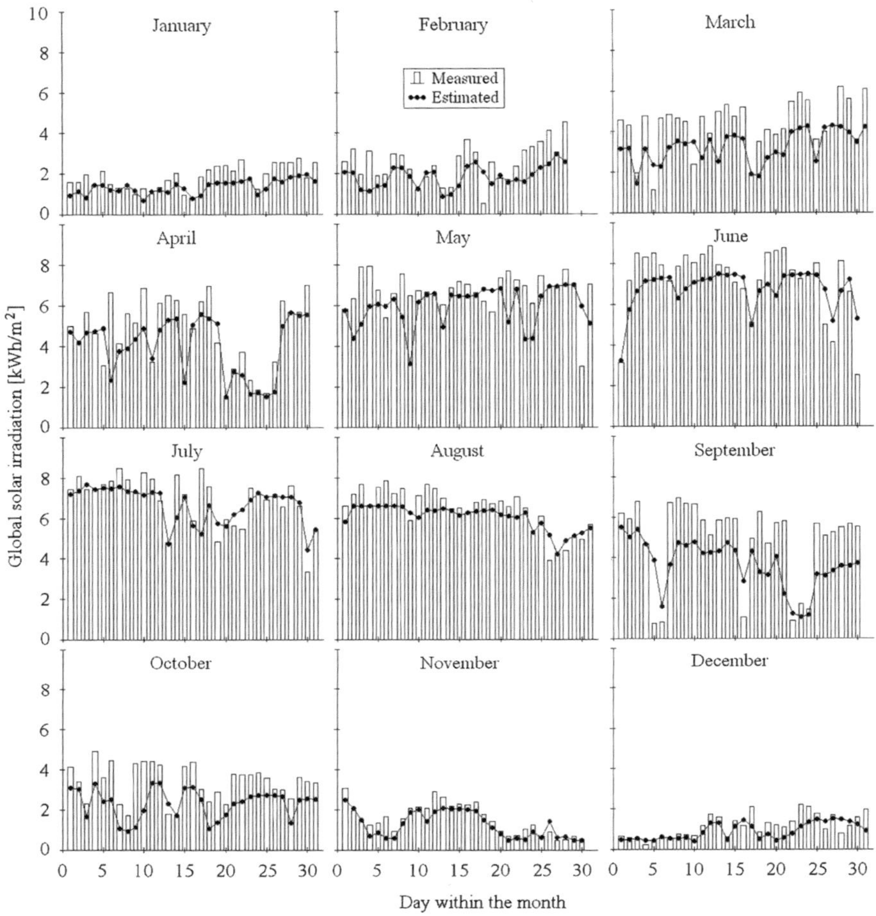

Fig. 8.12 Daily global solar irradiation estimated with SEAT and measured at the station Bucharest in 2000

models. To address the issue that air temperature-based models are sensitive to origin, a simplified adaptive algorithm has been established. The approach is presented in detail (Sect. 8.2.4) and is intended to provide the means to devise local and accurate models to be used for forecasting the daily global solar irradiation.

8.5.4 AEAT Models

A summary of the results of testing AEAT models reported in Paulescu et al. (2006) follows. Table 8.6 presents the statistical indicators of accuracy when

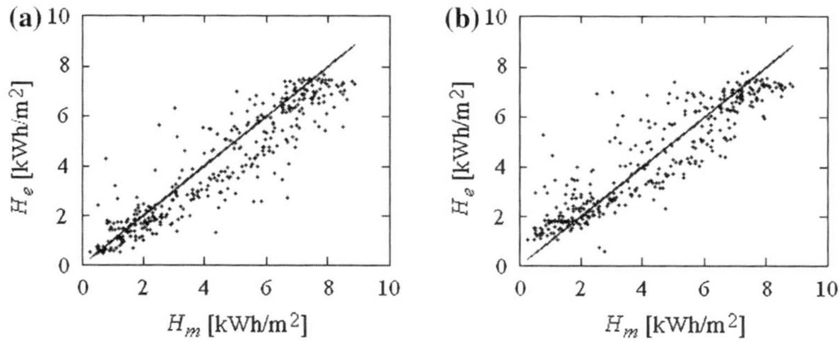

Fig. 8.13 Daily global solar irradiation estimated with SEAT vs. measured in Bucharest in all days of 2000: To run SEAT, the following values of daily extremes temperatures have been used: **a** Measured and **b** Measured minimum minus 2 °C and measured maximum plus 2 °C

Table 8.6 Statistical indicators of accuracy of monthly mean daily global solar irradiation estimation with the models AEAT-1 and AEAT-2, in the year 2000 (after Paulescu et al. 2006)

Location	Statistical indicator	AEAT-1 [%]	AEAT-2 [%]	D-B [%]	S–K [%]
Bucharest	*RMSE*	11.9	11.3	14.4	10.9
	MBE	−7.6	−7.0	−12.9	−5.8
Constanţa	*RMSE*	9.3	11.8	35.3	15.1
	MBE	−0.6	6.1	−29.6	−13.2
Craiova	*RMSE*	16.8	15.9	13.4	13.0
	MBE	11.0	10.8	−4.2	1.7
Iaşi	*RMSE*	8.7	7.9	16.0	21.1
	MBE	−1.3	−0.7	−12.5	−18.8
Timişoara	*RMSE*	19.8	19.9	12.9	14.6
	MBE	15.8	16.4	10.7	12.3

AEAT-1 (Eq. 8.19), AEAT-2 (Eq. 8.22a, b, c), DB (Eq. 8.16), and SK (Eq. 5.94) models were applied to estimate daily global solar irradiation at several Romanian stations in 2000 (see Fig. 3.1 for localization). For clear sky daily global solar irradiation H_0, the parametric model PS (Eq. 5.51) has been used. The model has been run using at input the following climatologic parameters: ozone column content 0.35 cm atm (Badescu 1997); water vapour column content 1.7 g/cm^2 (Gueymard 1995); Ångström turbidity coefficient $\beta = 0.077$ (as a mean value at the latitude 45°N, Leckner (1978)). The DB model (Eq. 8.16) has been run with the parameters: $\tau = 0.71$, $b = 0.112$, $c_1 = -6.72 \cdot 10^{-3}$, $c_2 = 1.135$, as mean values at the latitude 45°N provided by authors (Donatelli et al. 2003). The model S–K (Eq. 5.94) have been run with the parameters $a = 0.075$, $b = 0.428$, $c = -0.283$, computed by averaging of the parameters provided by authors (Supit and Kappel 1998) for stations of France, Germany and the Czech Republic, chosen as being of a latitude close to 45°N (parallel of 45°N crosses through Romania) and having a similar climate.

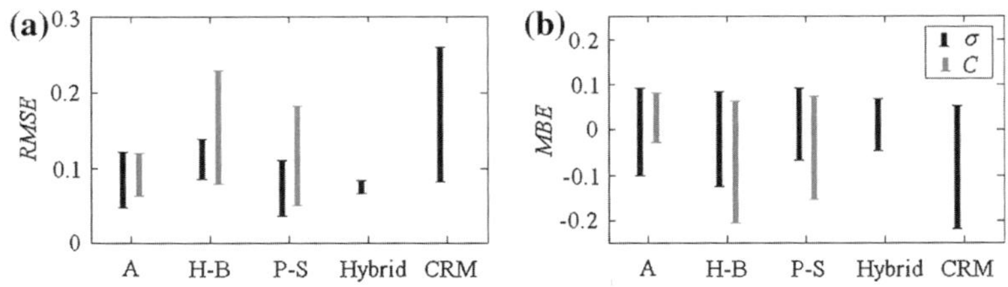

Fig. 8.14 Range of statistical indicators of accuracy of estimating monthly mean of daily global solar irradiation, with the models A (Eq. 5.71), H–B (Eqs. 5.66 and 5.70) P–S (Eq. 5.51) Hybrid (Eqs. 5.49a, b) and CRM (Gul et al. 1998) having been verified at the stations Bucharest, Constanta, Timisoara and Iasi. The listed models have been used to calculate the daily global solar irradiation under clear sky H_0. Ångström-Prescott equations based on relative sunshine (σ) and cloud cover amount (C) have been used to adjust H_0 to the actual state of the sky

The first conclusion is that AEAT, D-B, and S–K models performance are similar, even if S–K model uses in equations a direct indicator for the state of the sky, which is daily mean of total cloud amount. The second conclusion derived from Table 8.6 is that the estimation accuracy of monthly mean of daily global solar irradiation is comparable with the accuracy reached by using correlations with relative sunshine or total cloud amount at input. Results of testing five such *classical* solar irradiation models (Paulescu et al. 2006) using the same input parameters as for Table 8.6, are summarized in Fig. 8.14. It can be seen that RMSE varies between 3.7 and 13.8 % for relative sunshine-based correlations and between 5.0 and 26.0 % for total cloud amount-based correlations. The test proves again that daily air temperature extremes may be successfully used to construct specific measures for the state of the sky.

As all the above results show, the great benefit of air temperature-based models AEAT stem from the synergism of using simplified clear sky solar irradiation models, which require only geographical coordinates as entry, and a Ångström-Prescott type equation, which uses as entry only air temperature.

8.5.5 *Fuzzy Model*

The model performance has been assessed at 11 European stations (Tulcan-Paulescu and Paulescu 2008). Representative values of statistical indicators are collected in Table 8.7. It can be seen that the estimation accuracy is reasonable and compares well with the estimations using classical correlations (see Fig. 8.14). Results from Table 8.7 point out an excellent ability of the fuzzy model to fit a particular yearly air temperature regime: it is evident for the station of Auxerre, where *RMSE* about 0.07 is reached in 1998 and 2000 for very different adjustment factors $\chi = 0.929$ and $\chi = 0.857$, respectively.

From Table 8.7, two limitations of the model can be noticed. The adjustment factor χ given by Eq. (8.26) seems to be not applicable at seacoast locations. At the

Table 8.7 Statistical indicators of accuracy of monthly mean of daily global solar irradiation estimation with the fuzzy model

Station	Latitude [deg.]	Longitude [deg.]	Altitude [m]	Years	χ	*RMSE* [%]	*MBE* [%]
Constanta (RO)	44.20	28.63	17	2000	0.762	30.5	−25.3
Galati (RO)	45.48	28.01	72	2000	0.825	8.2	−2.0
Payerne (CH)	46.81	6.95	491	2000	1.012	10.2	−5.3
Nantes (FR)	47.15	−1.55	27	2000	0.778	22.9	6.7
Innsbruck (AT)	47.26	11.38	584	2000	0.996	8.6	−2.1
				2002	1.004	9.7	−7.5
Budapest (HU)	47.43	19.18	138	2000	0.871	8.3	−1.1
				2002	0.857	5.6	−1.3
Auxerre (FR)	47.80	3.55	212	1998	0.929	6.8	−1.4
				2000	0.857	7.0	0.4

stations Constanta, at the Black Sea coast and Nantes, at the Atlantic coast, the fuzzy model performance is considerably lower than at the continental sites. It is due to the peculiar seacoast air temperature regime, which has not been enclosed in the algorithm. Consequently, the adaptive mechanism is not able to compensate for. Figure 8.15a, a scatter-plot of daily global solar irradiation estimations versus measurements at the station of Constanta in the year 2000, shows that an underestimation occurs in days with high solar irradiation. But, it is well known that daily air temperature amplitude is lower than it is inside the Continent (see Fig. 3.1). At 150 km from the Black Sea coast, at Galati, at Danube River (see Fig. 3.1) the scatter-plot from Fig. 8.15b shows an acceptable accuracy of the estimation. Another limitation stems from the fact that the model originates in data collected up to 500 m (Tulcan-Paulescu and Paulescu 2008). Since at high altitudes the thermal regime is different from low altitudes, the model should be applied with care in the mountain locations. A third limitation is the latitude domain, between 40 and 50°N, where the stations used to derive the model are located.

The results presented in this section can be regarded as a starting point for future developments with increased generality level of temperature based models. The model universality and versatility is determined by the way in which the adjustment factor χ can be related to the local climate. In order to increase the accuracy of solar irradiation estimation, the fuzzy approach could be further developed either by including additional relevant meteorological variables, or by regional particularization.

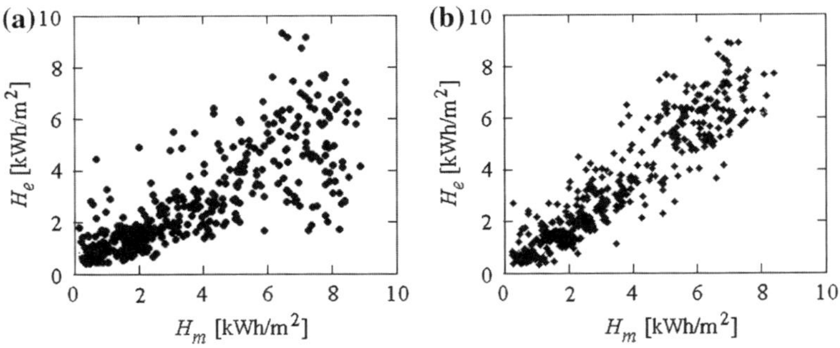

Fig. 8.15 Estimated versus measured daily global solar irradiation at the stations **a** Constanta and **b** Galati, in the year 2000

8.6 Simulation of Forecasting Daily Global Solar Irradiation

This section illustrates the procedure of forecasting daily global solar irradiation using the algorithm outlined in Fig. 8.1. The three tasks are presented at large: (1) forecasting air temperature; (2) fitting the solar irradiation temperature-based model, and (3) forecasting solar irradiation, followed by the assessment of overall procedure prediction accuracy. We are not interested here in the accuracy of the air temperature-based models (discussed in the previous section) but in the way in which the precision of forecasting air temperature influences the accuracy of forecasting solar irradiation. The very simple model of Bristow and Campbell (Eq. 5.96) has been selected to illustrate the algorithm. The model estimates daily global solar irradiation using at input only one parameter, the air temperature amplitude. A synthetic time series of daily air temperature amplitudes instead of one forecasted by a meteorological service is employed for the testing stage. These choices were made in order to facilitate the interpretation of the results.

Daily minimum and maximum air temperature and daily global solar irradiation recorded between 1997 and 2000 at the station Timisoara (see Fig. 3.1) have been used in this study. Data from the first three years have been used to fit the Bristow-Campbell model, while data from 2000 have been used to test the overall algorithm.

8.6.1 Generation of the Synthetic Daily Air Temperature Amplitude Time Series

The interpretation of the results of this simulation exercise in forecasting global solar irradiation starts with the way of generating the series of random daily air temperature amplitudes used in the testing period. It is therefore important to describe here in detail the properties of the synthetic series.

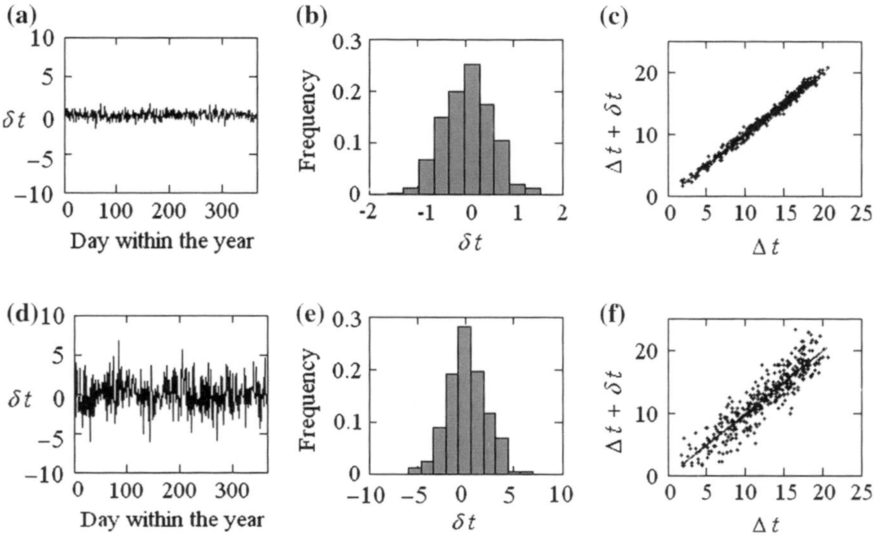

Fig. 8.16 Synthetic white noise series with normal distribution generated for simulating the errors δt in forecasting daily air temperature amplitude, frequency distribution for ten classes δt, and the scattering of simulated daily air temperature amplitude (Eq. 8.30) to the measured ones. The graphs were generated using: $\sigma_e = 0.5°\mathrm{C}$ (a, b, c) and $\sigma_e = 2°\mathrm{C}$(d, e, f)

As stated above, the model is tested against data measured in 365 days (the entire year 2000). For each day of the year, the air temperature amplitude Δt_m is known from measurements. In general, weather stations forecast the maximum and minimum temperature from one day to come. The amplitude of air temperature forecast is calculated simply as the difference of the two values and is accompanied by an error caused by the errors made in forecasting the two values, minimum and maximum. In this exercise, the error δt of air temperature amplitude forecasting is simulated in a *controlled* way. Four white noise time series were generated (using MathCAD specific functions), each consisting of 365 values, having zero mean and normal distribution. The difference between the four series is given by the standard deviation value σ_e. The experiment is controlled by choosing the following values for σ_e: 0.5 °C, 1.0 °C, 1.5 °C, and 2.0 °C. The time series δt with standard deviations $\sigma_e = 0.5$ °C and $\sigma_e = 2$ °C along with the corresponding histograms are plotted in Fig. 8.16.

In a day, the artificial value of the air temperature amplitude Δt used as entry in the procedure of forecasting global solar irradiation is obtained by adding the simulated error δt to the measured value:

$$\Delta t = \Delta t_m + \delta t \tag{8.30}$$

In case $\sigma_e = 0.5$ °C, a high accuracy prognosis is simulated: 65 % of the values are in the interval ±0.5 °C and 95 % in ±1 °C around measured values. In case

$\sigma_e = 2$ °C, a low accuracy prognosis of air temperature amplitude is simulated: 65 % of the values are in the interval ±2 °C and 95 % in ±4 °C around the measurement.

The values of the generated white noise series fall in the range −1.6 to +1.5 °C in the case of the first series ($\sigma_e = 0.5\,°\mathrm{C}$). This range increases to values of −5.9 to +6.9 °C in the second case ($\sigma_e = 2\,°\mathrm{C}$). In very few instances, the series δt with large dispersion Eq. (8.30) may lead to negative values of Δt in overcast conditions (exhibiting small Δt_m values). To overcome these artifacts, Eq. (8.30) has been replaced with the following:

$$\Delta t = \begin{cases} \Delta t_m + \delta t & \text{if} \quad \Delta t_m + \delta t > \Delta t_{\min} \\ \Delta t_{\min} & \text{otherwise} \end{cases} \tag{8.31}$$

Equation (8.31) limits the Δt series to the minimum value $\Delta t_{\min}$ that was met in the measured series Δt_m used to fit the model. In this case $\Delta t_{\min} = 1.0\,°\mathrm{C}$. Fortunately, from 365 values in the high dispersion simulated series ($\sigma_e = 1.5\,°\mathrm{C}$ and $\sigma_e = 2\,°\mathrm{C}$), very few (none or one) Δt's had to be slightly adjusted by Eq. 8.31, which means that the series distribution remains practically unaffected. The scattering of simulated daily air temperature amplitude (Eq. 8.31) to the measured ones is displayed in Fig. 8.16c, for the series with $\sigma_e = 0.5\,°\mathrm{C}$ and$\sigma_e = 2\,°\mathrm{C}$, respectively.

8.6.2 Air Temperature-Based Model

The Bristow-Campbell model (Eq. 5.96) basically relates the daily clearness index k_t (Eq. 4.2) to daily air temperature amplitude. The values of air temperature, minima and maxima, and the values of global solar irradiation measured in the 1095 days of period 1997–1999 have been used to fit the coefficients of Eq. (5.96). The resulted equation is ($r^2 = 0.606$):

$$k_t(\Delta t) = 0.584\left[1 - \exp\left(-0.014\Delta t^2\right)\right] \tag{8.32}$$

Equation (8.32) and the discrete points used in the fitting process are displayed in Fig. 8.17. A large scattering of $k_{t,m}$ versus Δt_m is evident; the approximation by a single curve is forced. To improve estimation accuracy, additional meteorological parameters could be added to refine Eq. (8.32).

8.6.3 Assessment of Results

To predict global solar irradiation H in a day, the forecasted air temperature amplitude for this day shall be inserted in Eq. (8.32), obtaining the daily clearness index. Then, the forecasted value H is determined by means of Eq. (4.2).

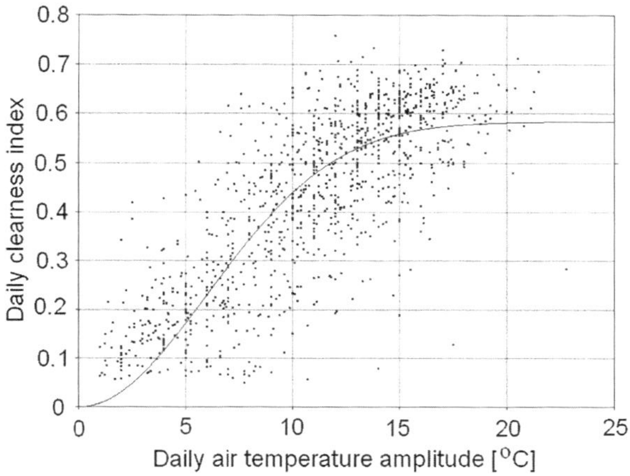

Fig. 8.17 Measured daily clearness index $k_{t,m}$ as function of the daily air temperature amplitude Δt_m (*points*) and the fitted curve given by Eq. (8.32) (*line*)

Table 8.8 Statistical indicators of accuracy of forecasting daily global solar irradiation. σ_e stands for the standard deviation of the synthetic time series

σ_e	*RMSE* [%]	*MBE* [%]	*MAE* [%]
Original	23.9	3.0	17.8
0.5	24.0	3.1	17.9
1.0	25.0	2.6	18.7
1.5	25.7	2.8	19.1
2.0	27.8	1.6	20.4

In this exercise, first the series consisting of 365 measured air temperature amplitudes were introduced in Eq. 8.32. Then, values of daily solar irradiation were estimated. Statistical indicators are placed on the first line in Table 8.8 and represent reference values, i.e., they would be the statistical indicators in the ideal situation when the amplitude of air temperature is projected exactly in proportion of 100 %. Basically this is a measure of the model (Eq. 8.32) accuracy.

At next step, we applied the algorithm to the randomly generated four series. Results are also summarized in Table 8.8. Note that a significant increase in the temperature amplitude forecast error does not provoke a significant increase in the forecast error of daily solar irradiation. Figure 8.18 proves that predicted values of daily solar irradiation quite accurately follow measured ones even in the most unfavorable forecasting scenario of the daily temperature amplitude ($\sigma_e = 2\,°C$).

Currently, weather services forecast minimum and maximum air temperature in a day with comparable accuracy to that considered in this study. Therefore, to use air temperature-based models in forecasting daily solar irradiance, the priority is to improve the accuracy of the models themselves.

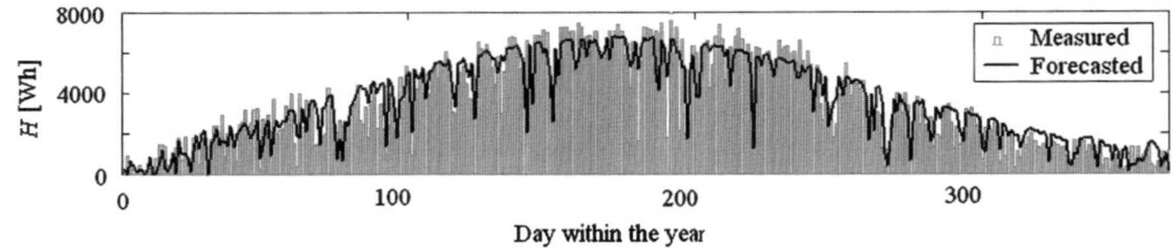

Fig. 8.18 Measured and forecasted daily global solar irradiation during one year. The forecasting procedure described in Fig. (8.1) has been used assuming at the input of Eq. (8.32) the simulated series of air temperature amplitude Eq. (8.31) with $\sigma_e = 2\,^\circ\mathrm{C}$ for white noise δt

8.7 Summary and Discussion

This chapter proposes an algorithm for forecasting solar radiation based on two observations: air temperature is a common forecast parameter everywhere and air-temperature-based models for estimating solar irradiation achieve an acceptable level of accuracy. Thus, the forecasted air temperature may be used as entry in the mentioned models for nowcasting solar irradiance and forecasting solar irradiation.

From the research results in modeling solar irradiation via air temperature data summarized above it can be concluded that air temperature may be successfully used in a recipe for indirect characterization of the state of the sky. Nevertheless, future efforts should improve the models accuracy and extend the geographical application area by finding new ways for weakening the parental place dependence.

Since 2008, our group started studying ways for constructing solar irradiation models based on fuzzy sets theory. For solar energy estimation there are certified models already reported but forecasting procedures are still an emerging theme currently under investigation. This fuzzy approach propose a way for constructing a new generation of solar irradiation models where fuzzy sets theory is used as an alternative to the binary logic which is otherwise so successful in many applications, like all digital electronics, but may lack the flexibility needed for other applications, where the dependencies exist but are too subtle to be described in a Boolean algorithm, which seems to be the case in solar radiation estimation.

A world grid for predicting air temperature already exists. Therefore, when air temperature-based models for solar irradiation will mature and perform accurately enough, they may be straightforwardly employed for forecasting solar irradiation. An excellent ratio between the usefulness of the information delivered to power grid operators and the required cost for obtaining it is to be expected.

References

Ångström A (1924) Solar and terrestrial radiation. Q J Royal Meteorol Soc 50:121

Badescu V (1997) Verification of some of some very simple clear and cloudy sky models to evaluate global solar irradiance. Sol Energy 61:251–264

Badescu V (2002) A new kind of cloudy sky model to compute instantaneous values of diffuse and global solar irradiation. Theor Appl Climatol 72:127–136

Bristow KL, Campbell GS (1984) On the relationship between incoming solar radiation and daily maximum and minimum temperature. Agric For Meteorol 31:159–166

Bugler JW (1977) The determination of hourly insolation on an inclined plane using a diffuse irradiance model based on hourly measured global horizontal insolation. Sol Energy 19:4–11

Donatelli M, Marletto V (1994) Estimating surface solar radiation by means of air temperature. In: Proceedings of the 3rd European Society for Agronomy Congress, Padova, Italy, pp 352–353

Donatelli M, Campbell GS (1998) A simple model to estimate global solar radiation. In: Proceedings of the fifth Congress of the European Society for Agronomy, Nitra, Slovakia, II, pp 133–134

Donatelli M, Bellocchi G (2001) Estimate of daily global solar radiation: new developments in the software Rad-Est3.00. In Proceedings of second International Symposium Modelling Cropping Systems, Florence, pp 213–214

Donatelli M, Bellocchi G, Fontana F (2003) RadEst 3.00: software to estimate daily radiation data from commonly available meteorological variables. Eur J Agron 18:363–367

El Metwally M (2003) Simple new methods to estimate global solar radiation based on meteorological data in Egypt. Atmos Res 69:217–239

Gueymard CA (1995) SMARTS2-A simple model of the atmospheric radiative transfer of sunshine: algorithms and performance assessment. Florida Solar Energy Center Rep. FSEC-PF-270-95. Available online: http://instesre.org/GCCE/SMARTS2.pdf

Gul MS, Muneer T, Kambezidis HD (1998) Models for obtaining solar radiation from other meteorological data Sol Energy 64:99–108

Hottel HC (1976) A simple model for estimating the transmittance of direct solar radiation through clear atmosphere. Sol Energy 18:129–139

ISCI (2012) Research institute for industrial crops, Bologna, Italy. Software tools. http://agsys.cra-cin.it/tools/

Kasten F, Czeplak G (1980) Solar and terrestrial radiation dependent on the amount and type of cloud. Sol Energy 24:177–189

Leckner B (1978) The spectral distribution of solar radiation at the earth's surface—elements of a model. Sol Energy 20:143–150

MetOffice (2012) The UK's National Weather Service. Temperature forecast performance. http://www.metoffice.gov.uk/about-us/who/accuracy/forecasts

NCDC (2012) The global surface summary of day data, version 7. National Climatic Data Center, Asheville, USA. http://www.ncdc.noaa.gov

Paulescu M, Schlett Z (2003) A simplified but accurate spectral solar irradiance model. Theor Appl Climatol 75:203–212

Paulescu M, Schlett Z (2004) Performance assessment of global solar irradiation models under Romanian climate. Renewable Energy 29:767–777

Paulescu M, Fara L, Tulcan-Paulescu E (2006) Models for obtaining daily global solar irradiation from air temperature data. Atmos Res 79:227–240

Paulescu M, Stefu N, Tulcan-Paulescu E (2011) A temperature-based model for global solar irradiance and its application to estimate daily irradiation values. Int J Eng Res 35(6):520–529

SRMS (2012) Solar Platform of the West University of Timisoara, Timisoara, Romania. http://solar.physics.uvt.ro/srms

Supit I, Van Kappel RR (1998) A simple method to estimate global radiation. Sol Energy 63:147–160

Tulcan-Paulescu E, Paulescu M (2008) Fuzzy modeling of solar irradiation using air temperature data. Theor Appl Climatol 91:181–192

Chapter 9
Outdoor Operation of PV Systems

9.1 Introduction

To forecast the power produced by a photovoltaic power plant in a certain time horizon, two mathematical models are in principle required. The first model aims to forecast the solar irradiance at the location and the second model relates to the operation of the photovoltaic converter. Accurate prediction of solar irradiance is essential to forecast the power output of a PV system. Previous chapters have been devoted to this theme. The presentation was focused on the study of those quantities describing the fluctuating character of solar irradiance, such as the sunshine number as an indicator of direct solar radiation occurrence (see Chaps. 3 and 4) and the clearness index (see Chaps. 6 and 7). This chapter deals with modeling PV converters, particularly grid-connected PV systems. Roughly, a grid-connected PV system consists of two essential components: an array of photovoltaic modules (PV generator) and the inverter. For both components, mathematical models will be presented, but most of the chapter will be focused on the modeling of PV modules. The motivation for this choice is that some functional parameters of the PV modules are linked to weather conditions. It is well known that as the temperature of crystalline solar cells increases, their conversion efficiency decreases. This bears a paradox that applies to crystalline solar cells: in summer (or mid-day) when the solar flux is at maximum, the energy conversion efficiency of the crystalline solar cells is at minimum. The heat balance of a solar cell is determined by the energy incident on the surface (solar irradiance) and the heat exchange between cell and environment, which depends on temperature. As will be shown in Sect. 9.4.2, air temperature can influence a few percent of solar cells' conversion efficiency which is significant when affecting the less than 20 % efficiency of current commercial devices. Air temperature as an input parameter in models that describe the operation of a solar cell must be forecasted. Opportunely, meteorology has made significant progress in terms of air temperature forecast (see Sect. 8.1). Cleanliness of the module surface is another parameter that influences the optical properties of solar cells, this parameter being

M. Paulescu et al., *Weather Modeling and Forecasting of PV Systems Operation*, Green Energy and Technology, DOI: 10.1007/978-1-4471-4649-0_9,

predictable with good accuracy. In Sect. 9.3.5 we show that the effect of even a partial shading of a module is destructive for conversion efficiency. At the same time shading is an effect of clouds circulation which is very difficult to predict. It can be concluded that the output of a PV system is not a linear function of solar irradiance, being significantly affected in a complex manner by air temperature, surface cleanliness, and shading. The radiation intensity and the angle of incidence are two other parameters affecting cell efficiency, however, to a lesser degree than the three previously named factors.

A central issue in PV power forecasting is the estimation of the operating regime of a solar cell (as part of a PV module) in generic environmental conditions. These conditions are specified by parameters such as solar irradiance, solar cell temperature, and degree of surface cleanliness.

In this chapter, the solar cell will be treated as a macroscopic device with two contacts. Illuminating the cell and connecting a load to its terminals will produce an electric current. The description of specific physical processes taking place at microscopic scale, such as light absorption in semiconductor material, carrier generation, charge separation, and collection, is beyond the goal of this chapter. Many textbooks are dedicated to this matter; the reader may consult the following: Green (1982), Nelson (2003), Wurfel (2005), etc. In the following, the presentation is focused only on the mathematical description of the solar cell operating, seen as an optoelectronic device. The standard equivalent electrical circuit of the solar cell and their corresponding equations are explained first. Further, this standard equivalent circuit will be the starting point to derive simplified models that approximate the voltage–current (V–I) characteristic.

There are many possible mathematical relationships of varying complexity which can be used to describe the PV V–I behavior of solar cells. Several factors should be considered to assess whether a V–I model is consistent with a particular application. The first criterion is related to the universality of the model, which must be able to calculate the operating point (voltage and current giving the output power) over the entire operating meteorologic conditions. The second criterion is related to the availability of cell functional parameters needed to run the model. Models with a large number of parameters may be more accurate, but may also require more parameters than those given in the manufacturer data sheet. Their determination would require additional laboratory tests, which are neither simple nor cheap. The third criterion is related to the number of required meteorologic input parameters, which must comply with the parsimony principle. The latter criterion is related to the complexity of calculation. A good model is the outcome of an inherent tradeoff between simplicity and accuracy.

The solar cell converts light into electricity, being the *fundamental* element of the PV system. By itself, it cannot be used directly, being of small power (watts) at a voltage too low for most applications and not weather protected. By contrast, a PV module is the smallest usable solar–electric converter, monolithic by construction, and commercially available. It consists of several identical solar cells connected in series and parallel. The PV module is a device for higher power (tens, hundreds of watts) and is weather protected. Thus, we can say that the PV module is the *primary* element of a PV system.

Solar cells and, consequently, PV modules produce output power that depends on the irradiance level and also the temperature of the device. Thus, a set of test conditions had to be standardized regulating the values of both irradiance and temperature at which PV manufacturers rate the output of their devices. The performance PV standards IEC 61215/2005 (IEC 2005) and IEC 61646/2008 (IEC 2008) set the specific test sequences, conditions, and requirements for the design qualification of a PV module. The design qualification is considered to represent the PV module's performance capability under prolonged exposure to standard climates (defined in IEC 60721-2-1). In addition, there are several other standards (IEC 61730-1, IEC 61730-2 and UL1703) that address the safety qualifications for a module.

The PV modules are delivered accompanied by data sheets listing parameters measured in standard test conditions (STC). STC specifies the global solar irradiance incident on the PV module of $G_{STC} = 1000\ \mathrm{W/m^2}$ with a spectral distribution AM1.5G (see Sect. 2.1) and a cell temperature $T_{STC} = 25\ °\mathrm{C}$.

In most of this chapter, we treat the problem of translating the module V–I characteristic from an operation state to another. Photovoltaic translation equations are derived to assess the performance of PV cells, PV modules, and PV network in real operating conditions. Generally, the operation takes place in two steps. First, the characteristic equation parameters are estimated using catalog values measured in STC conditions, then the real operating conditions characteristic is constructed. Four models differentiated by the number of parameters used to build the V–I curve are presented and illustrated. All these models are discussed in the hypothesis of uniform illumination. In practice, PV arrays may be illuminated nonuniformly. The impact on the PV system output power of partial shadowing and the modeling of inverter efficiency are discussed in brief.

9.2 Computing PV Modules' Performance

At first, in this section, the solar cell V–I characteristic is described mathematically and the equivalent circuit is introduced. The influence on the conversion efficiency of some intrinsic parameters such as the junction ideality factor, saturation current, or series and parallel resistances is assessed. Further, a connection is established between the parameters listed in the PV module manufacturer data sheet and the parameters of the V–I characteristic equation of the cells that are component of the module. In the last part, which is more ample, equations are introduced for translating the V–I characteristic of a module from STC to arbitrary conditions. Different numerical methods for solving them are discussed. The theoretical presentation is illustrated with examples of translating the V–I characteristic for some commercial modules.

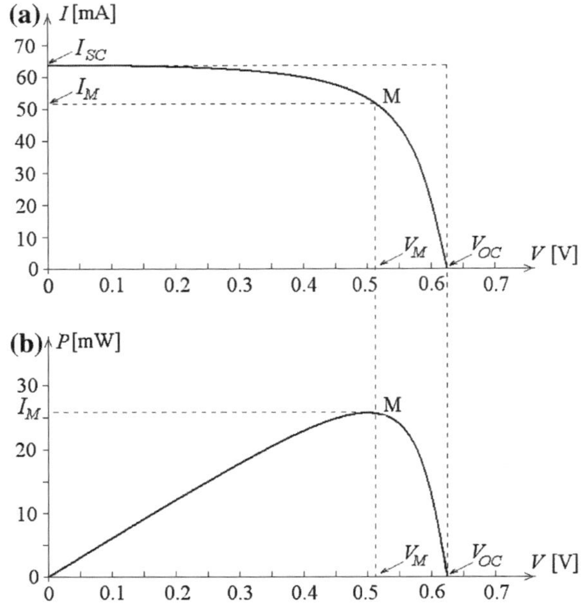

Fig. 9.1 **a** Typical V–I characteristic of a solar cell. V_{OC} and I_{SC} stand for the open-circuit voltage and short-circuit current. At the point M(V_M, I_M), the output power is maximum. **b** The output power curve of the solar cell

9.2.1 Standard V–I Characteristic of a Solar Cell

A solar cell is a solid state current generator whose typical V–I characteristic is depicted in Fig. 9.1a.

At constant temperature and illumination, the working point on the V–I diagram is determined by the load connected to the device terminals. The point of intersection of the characteristic with the voltage axis corresponds to no load and measure the voltage across the cell in open-circuit V_{OC}, while the intersection with the current axis I_{SC} corresponds to the short-circuit current. Each point on the curve delimits in relation to the axis a rectangle whose area is numerically equal to the power output:

$$P = VI \tag{9.1}$$

The power curve (Fig. 9.1b) indicates that there is an optimal point of operation, M(V_M, I_M), in which the power is at maximum.

The maximum power of a solar cell $P_M = V_M I_M$ is correlated with two of the most important parameters that evaluate the quality of solar cells; the fill factor and the conversion efficiency. The fill factor is defined as the ratio of the actual maximum obtainable power in STC to the product of the open-circuit voltage and short-circuit current $V_{OC} I_{SC}$:

$$\mathrm{FF} \equiv \left. \frac{P_M}{V_{OC} I_{SC}} \right|_{\mathrm{STC}} \tag{9.2}$$

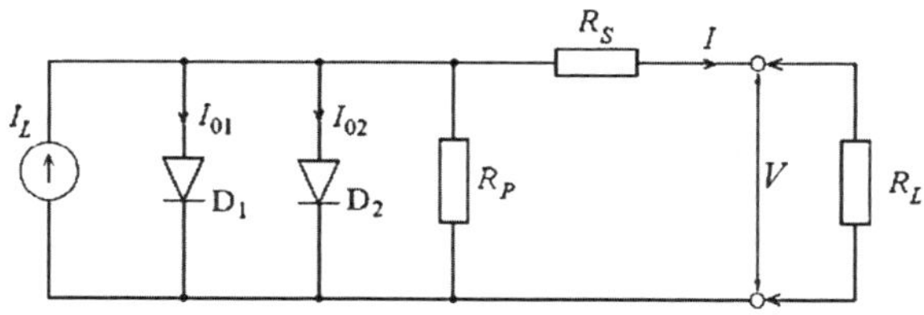

Fig. 9.2 Equivalent electrical circuit of a solar cell. R_L is a load attached to the terminals. Experimentally, the cell V–I characteristic in Fig. 9.1a is raised by sequentially increasing R_L from zero (short-circuit) to infinite (open-circuit) maintaining the same incident solar energy flux and cell temperature

The maximum conversion efficiency of a solar cell is defined as the ratio of maximum power produced in STC at the point M and the solar irradiance G incident on the cell surface of area A_c:

$$\eta_{\mathrm{STC}} \equiv \left.\frac{P_M}{GA_c}\right|_{\mathrm{STC}} \tag{9.3}$$

9.2.1.1 Equivalent Electrical Circuits

The standard equivalent electrical circuit of a solar cell is represented in Fig. 9.2. This circuit emerges from understanding the physical processes underlying solar cell operation. The photocurrent I_L is produced by a current source mainly dependent on solar irradiance. The two diodes mounted in parallel model dark current losses (D_1) and the effect of generation–recombination in the space charge region (D_2). Current loss caused by increased junction conductivity at cell edges is modeled by the shunt resistance R_P. If we denote by V the voltage at the terminals, the effective voltage on the parallel group components is larger, being equal to $V + IR_S$. The series resistance R_S translates resistive losses in the cell, such as contact resistance, neutral regions resistance, and others.

The equivalent electrical circuit of a solar cell allows writing an implicit equation for the current which can only be solved iteratively:

$$I = I_L - I_{01}\left[\exp\frac{e(V+IR_S)}{m_1 k_B T} - 1\right] - I_{02}\left[\exp\frac{e(V+IR_S)}{m_2 k_B T} - 1\right] - \frac{V+IR_S}{R_P} \tag{9.4}$$

where:

- m_1, m_2 are the ideality factors of the diodes D_1 and D_2, respectively. The diode ideality factor is a measure of p–n junction imperfection and is not directly measurable.
- I_{01}, I_{02} [A] are the reverse saturation currents of the diodes D_1 and D_2, respectively. The reverse current described by the sum of the exponential terms

in Eq. (9.4), $I_{01}\left[\exp\frac{e(V+IR_S)}{m_1 k_B T}-1\right]+I_{02}\left[\exp\frac{e(V+IR_S)}{m_2 k_B T}-1\right]$ forms the so-called *dark current*, which flows across the device under an applied voltage, or *bias*, in the dark.

- I_L stands for the photocurrent [A]. In the first approximation, the photocurrent generated by a solar cell under illumination at short circuit is linearly dependent on the incident light level.
- R_S is the series resistance. As the series resistance increases the fill factor of the cell deteriorates. The series resistance is a particular problem at high current densities, for instance when the cell area is large.
- R_P is the shunt resistance. The shunt resistance alters the rectifying characteristic of the device.

Other notations in Eq. (9.4) are: $k_B = 1.38\times 10^{-23}$ J/K is the Boltzmann constant, $e = 1.6\times 10^{-19}$ C is the elementary charge, and T is the cell temperature.

The overall V–I response of the solar cell is approximated by Eq. (9.4) as the superposition of the photocurrent, dark current, and leakage current. The main problem posed by this equation is the following: the characteristic V–I of the cell passes through three given points (with ordinates I_{SC}, I_M, and $I = 0$, see Fig. 9.1); however, Eq. (9.4) does not specify the actual point (V, I) of operation. If one requires this state to be (V_M, I_M), corresponding to the maximum power provided by the cell, another equation must be used, imposing operation of the device at full power:

$$\left.\frac{d(\mathrm{VI})}{dV}\right|_M = 0 \tag{9.5}$$

By solving the system formed by Eqs. (9.4) and (9.5), we can determine optimal values (V_M, I_M) of the cell's current and voltage.

Equation (9.4) represents the standard model of the solar cell. The cell performance is degraded by the presence of series and parallel resistance. For an efficient solar cell, series resistance should be as small as possible and the shunt resistance to be as large as possible. Equation (9.4) can be simplified at different degrees of approximation and idealization, assuming $m_1 = 1$, $m_2 = 2$, $R_S = 0$, or/and $R_P = \infty$.

Figure 9.3 shows how the cell parameters shape the V–I curve given by Eq. (9.4). The curves were raised by a simple procedure. For a given set of parameters the voltage was varied from 0 to 0.6 V with 0.01 V step. For each fixed voltage V_i, Eq. (9.4) was solved numerically to determine the appropriate current I_i. Pairs (V_i, I_i) as determined form the cell characteristic, i.e., a curve in Fig. 9.3. In each of the four graphs that make up Fig. 9.3, there is a default curve calculated with the following set of parameters: $m_1 = 1$, $m_2 = 2$, $I_{01} = I_{02} = 10^{-8}$ A, $R_S = 0.005\ \Omega$ and $R_P = 130\ \Omega$. The other curves are made varying only one of these parameters and keeping all others constant. In all cases, the solar cell was always considered in STC, which means the temperature in Eq. (9.4) is $T = T_{STC} = 25°$. We have assumed typical values for cells equipping commercial

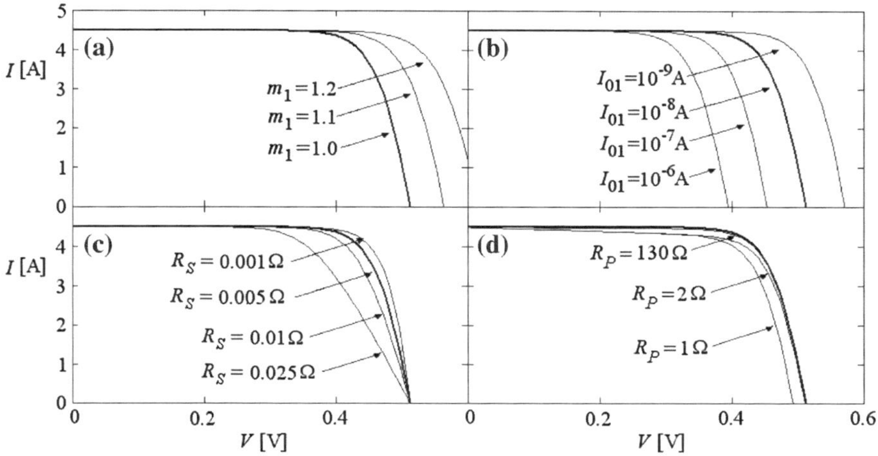

Fig. 9.3 Solar cell V–I curve dependence on the parameters in Eq. (9.4). **a** Diode D_1 ideality factor (m_1). **b** Saturation current of diode D_1 (I_{01}). **c** Series resistance (R_S). **d** Shunt resistance (R_P). The reference curve ($m_1 = 1$, $I_{01} = 10^{-8}$ A, $R_S = 0.005\ \Omega$ and $R_P = 130\ \Omega$) is thickened

PV modules, cell area $A_C = 100\ \text{cm}^2$ and generated photocurrent $I_L = 4.5$ A. Finally, extracting from the V–I characteristic the coordinates of the optimum operating point (V_M, I_M), the conversion efficiency can be calculated using Eq. (9.3). For the reference curve (marked with bold line in each graph), the conversion efficiency is $\eta_0 = 17.7\%$. Figure 9.3a shows that an increased diode ideality factor m_1 leads to an increase in cell efficiency. However, this picture should be interpreted with care. In a cell where m_1 departs from ideality the saturation current is greater. The saturation current tends to increase solar cell output voltage while the nonideality diode factor acts to erode it. The net effect is given by a combination of the increase in voltage shown for increasing m_1 in Fig. 9.3a and the decrease in voltage shown for increasing I_0 in Fig. (9.4b). Typically, I_{01} is a more significant factor and the net result is a reduction in voltage and overall cell efficiency. Assuming $m_1 = 1$, the increase by an order of magnitude of the saturation current from 10^{-8} to 10^{-7} A causes a decrease in efficiency with 3.2 %, from 17.7 to 14.5 %.

As the series resistance increases, the voltage drop on that becomes greater for the same current. The result is that the current-controlled portion of the V–I curve begins to drop toward the origin, producing a significant decrease in the terminal voltage and a slight reduction in the short-circuit current. The effect is shown in Fig. 9.3c. Losses caused by series resistance are in a first approximation given by I^2R_S, rising with the square of the photocurrent. In Fig. 9.3c, doubling the series resistance from 5 to 10 mΩ leads to a decrease of efficiency with 0.9 % from 17.7 to 16.8 %. As shunt resistance decreases (Fig. 9.3d), the current diverted through the shunt resistor increases for a given junction voltage. The result is that the voltage-controlled portion of the current-voltage (I–V) curve begins to drop toward the origin, producing a decrease in the terminal current I and a slight

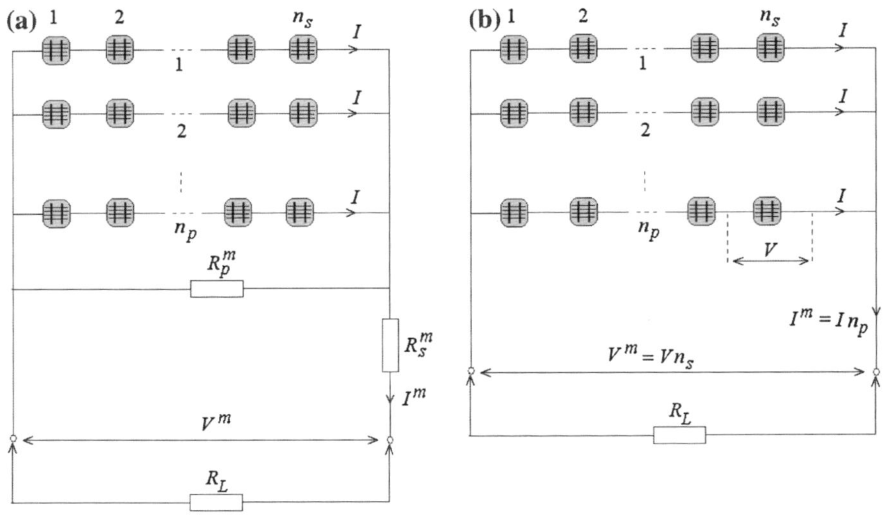

Fig. 9.4 Models of a PV module. **a** The PV module contains n_p rows of solar cells connected in parallel, each string consisting of n_S cells connected in series. R_s^m models, the resistance of the electrical lines between cells and of the contacts while R_p^m models, the loss in the bypass diodes. **b** Simplified scheme of a PV module ($R_s^m = 0$ and $R_p^m = \infty$)

reduction in V_{OC}. Only very low values of R_P will produce a significant reduction in V_{OC}. Actual solar cells are produced with R_P greater than 100 Ω, thus enough to consider this close to ideal. Even large variations of R_P around these values do preserve the solar cells V–I characteristic as shown in Fig. 9.3d.

Several comments have been made about the parameters in V–I Eq. (9.4). The following sections address how to determine the parameters and create workable V–I equations by setting up and solving systems of simultaneous non-linear equations in several unknowns. Once values for each parameter are calculated (or in some cases, chosen), the resultant V–I equation gives a continuous analytical expression of current as a function of voltage, at a reference irradiance and cell temperature. At other irradiances and cell temperatures, some of the parameters vary, and auxiliary equations are needed to calculate updated values at each set of conditions. The updated parameters yield a new V–I equation valid under the new conditions.

9.2.2 PV Modules

The key to modeling a PV system, even a power plant, is the photovoltaic module. A PV module consists of a network of solar cells, having by assembly unitary character and environmental protection. A module must satisfy requirements such as: solar cells protection against environmental actions (wind, rain, hail, snow, wet air, mechanical stress induced by different dilatation) and user safety (electrocution). They should have a minimum 20 years lifespan and have minimal acquisition cost.

A PV module of n_p parallel strings, each of them consisting of n_s identical solar cells is shown in Fig. 9.4a. The series PV module resistance is usually small, depending on technology (e.g. R_s^m is of order of tens of Ohm for crystalline silicon modules (Polverini et al. 2012)). Also, the shunt PV module resistance is usually large (e.g., R_p^m is of order of 100 Ohms for crystalline silicon modules (Celik and Acikgoz 2007)). Therefore, a simplified scheme of the module's electric circuit is adopted here (see Fig. 9.4).

At the module's terminals, the voltage and the intensity of electric current are denoted V^m and I^m, respectively. One supposes that all cells are identical and they work under identical conditions: same values of the incident solar radiation intensity and heat dissipation (ambient temperature and wind speed). In this case, electric currents and voltage across each cell have the same values, still denoted by V and I. Using the model in Fig. 9.4b, it can be written:

$$I = I^m/n_p\,,\ V = V^m/n_s \qquad (9.6\text{a, b})$$

Equation (9.6a, b) express the approximate relationship between voltage and current through a cell and through the whole module, respectively.

9.2.2.1 Product Parameters

Manufacturer's data sheet of PV modules lists the values of some parameters of solar cells, measured in STC. Also, data sheets specify values characterizing the entire module as a cell assembly. Most manufacturers give specifications for the following parameters:

- Cell surface: A_C
- Module surface: A^m
- The number of cells connected in series: n_s
- The number of cell strings connected in parallel: n_p
- The open-circuit voltage across the module: $V^m_{\text{OC,STC}}$, at $I^m = 0$ in STC
- The short-circuit current of the module: $I^m_{\text{SC,STC}}$, at $V^m = 0$ in STC
- The voltage across the module and the intensity of current supplied by the module, at maximum power: $V^m_{M,\text{STC}}$ and $I^m_{M,\text{STC}}$
- Nominal operating cell temperature (NOCT)
- Coefficient of variation with temperature of the cell open-circuit voltage, under STC:

$$\alpha_V = \left.\frac{\mathrm{d}V_{\text{OC}}}{\mathrm{d}T}\right|_{T=T_{\text{STC}}} \qquad (9.7)$$

If the module data sheet does not report values for the coefficient α_V, it can be considered $\alpha_V = -2.3$ mV/°C for crystalline silicon cells. Some manufacturers provide information about α_V in %/°C. Thus, the value given in the catalog is $\alpha'_V = \frac{1}{V_{OC}} \frac{\mathrm{d}V_{OC}}{\mathrm{d}T}$ and:

$$\alpha_V = \frac{\mathrm{d}V_{OC}}{\mathrm{d}T} = \alpha'_V V_{OC} \tag{9.8}$$

- Temperature coefficient of the photocurrent in STC:

$$\alpha_I = \left. \frac{\mathrm{d}I_L}{\mathrm{d}T} \right|_{T=T_{STC}} \tag{9.9}$$

The temperature variation coefficient of the photocurrent is very small and positive ($\sim$0.03–0.04 % of the short-circuit current per Kelvin). Like the temperature coefficient of open-circuit voltage, one can also define for the photocurrent a normalized coefficient $\alpha'_I = \frac{1}{I_L} \frac{\mathrm{d}I_L}{\mathrm{d}T}$ and:

$$\alpha_I = \frac{\mathrm{d}I_L}{\mathrm{d}T} = \alpha'_I I_L \tag{9.10}$$

9.2.2.2 Estimated Parameters of PV Module in Operation

Real world operating conditions differ from STC. Incident global solar irradiance G and ambient temperature T are the external factors that influence most the module's performance in real operating conditions. The operating temperature of the cells is always higher than the ambient temperature and is roughly proportional to the effective flux density of incident solar energy:

$$T = T_a + C_t G_{\mathrm{eff}} \tag{9.11}$$

where C_t takes the value

$$C_t = \frac{\mathrm{NOCT}(^\circ\mathrm{C}) - 20\,^\circ C}{800\ W/m^2} \tag{9.12}$$

Nominal operating cell temperature (NOCT) is the cell temperature when irradiance is 800 W/m^2, ambient temperature is 20 °C, and wind speed is 1 m/s at a module tilt angle of 45°.

The influence of both parameters irradiance level and temperature on the V–I curve of a PV module is illustrated in Fig. 9.5. The V–I curves in Fig. 9.5a indicate that in a first approximation the module short-circuit current I^m_{SC} and consequently the photocurrent I^m_L varies linearly with the incident irradiance level. Figure 9.5b shows that cell temperature affects the V–I characteristic equation in two ways. First, the module open-circuit voltage decreases linearly with increasing

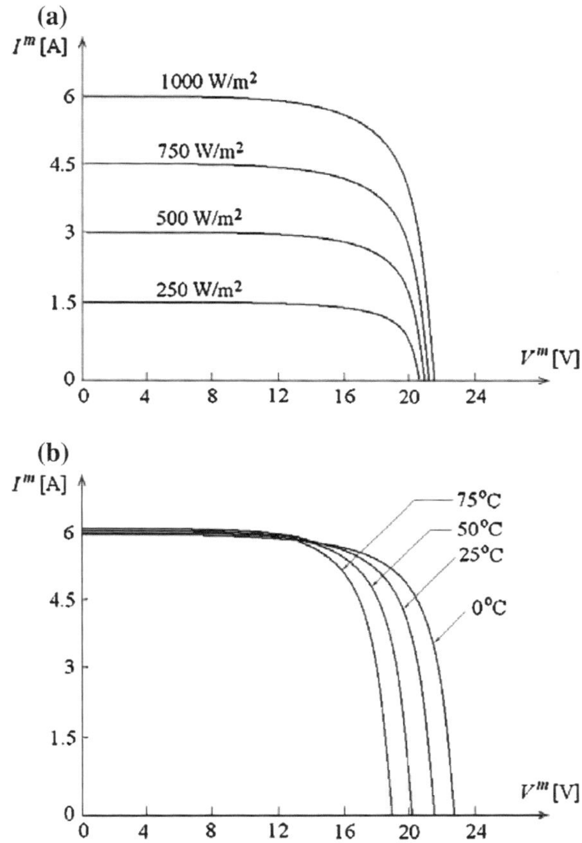

Fig. 9.5 Dependence of a PV module V–I characteristic on **a** level of incident solar irradiance and **b** cell temperature

cell temperature. It is a net effect from the following two facts. First, while increasing T reduces the magnitude of the exponents in Eq. (9.4), the value of the saturation current increases exponentially with T. Second, the amount of photo-generated current in the cells rises slightly with temperature because of an increase in the number of thermally generated carriers in the cell. The overall effect of cell temperature on the module output power is, thus, more complex and will be assessed in detail the next section.

In addition, efficiency calculation needs to take into account the cleanliness degree of the PV module and losses at large angles of incidence of solar flux on the surface of the module. These losses are determined by the dependence of optical materials reflectance and transmittance on the angle of incidence. Most of the time, the angle of incidence of solar radiation is considerably different from the normal incidence assumed at STC. As a result, reflection losses alter the module output power mainly al large incidence angles. The quantification of power losses by reflection on surface can be done using theoretical models from optics, based on Fresnel equations (see e.g. the comprehensive presentation in Hecht (1997). Simplified equations are presented in Ref. Martin and Ruiz (2001), which are

Table 9.1 Values of parameters needed to calculate losses at large angles of incidence due to the dust (Martin and Ruiz 2001)

Cleanliness	$\tau_d(0)/\tau_c(0)$	a_r	c
Perfectly clean	1	0.17	−0.069
Good	0.98	0.20	−0.054
Medium	0.97	0.21	−0.049
Low	0.92	0.27	−0.023

summarized below. These equations take into account both influences; the module cleanliness and incidence angle. Let G_{ef} denote the solar irradiance corrected for effects of incidence angle and cleanliness of the module's cover:

$$G_{\text{ef}}(\beta) = \frac{\tau_d(0)}{\tau_c(0)}\left[G_b(0)\tau_b(\theta)\cos\theta + G_d(0)F_d(\beta)\tau_d(\beta) + G_r(\beta)\right] \tag{9.13}$$

In Eq. (9.13), the cleanliness is defined by the ratio of transmittance in normal use and transmission for perfectly clean surface $\tau_d(0)/\tau_c(0)$, both measured at normal incidence. $F_d(\beta)$ is the conversion factor of diffuse radiation (see Sect. 5.6.1). The relative transmittance of the beam irradiance $\tau_b(\theta)$ is a function of incident angle and a parameter a_r correlated with the cleanliness of the module's cover:

$$\tau_b(\theta) = 1 - \frac{\exp(-\cos\theta/a_r) - \exp(-1/a_r)}{1 - \exp(-1/a_r)} \tag{9.13a}$$

The parameter a_r is given in Table 9.1 for different degrees of cleanliness of the module.

The relative transmittance of the diffuse irradiance $\tau_d(\beta)$ is a function of the module tilt angle:

$$\tau_d(\beta) = 1 - \exp\left[-\frac{1}{a_r}\left(\frac{4}{3\pi}f(\beta) + cf(\beta)^2\right)\right] \tag{9.13b}$$

where $f(\beta) = \sin\beta + (\pi - \beta - \sin\beta)/(1 + \cos\beta)$. The parameter c is also listed in Table 9.1.

9.3 Modeling PV Module Operating Outdoor

The first equations for translating the V–I characteristic of a PV module from an operating point to another have been reported in Sandström (1967). This study was conducted in conjunction with the preparation in the 1960s of space missions to Mars. These equations deserve mentioning, because even while undergoing changes and additions, their underlying principles remain the same. Sandström equation is written:

$$I_2 = I_1 + \Delta I_{SC} \tag{9.14a}$$

$$\Delta I_{SC} = I_{SC1}\left(\frac{G_2}{G_1} - 1\right) + \alpha(T_2 - T_1) \tag{9.14b}$$

$$V_2 = V_1 - \beta(T_2 - T_1) - \Delta I_{SC}R_S - \gamma(T_2 - T_1)I_2 \tag{9.14c}$$

$$P_2 = I_2 V_2 \tag{9.14d}$$

where G is the solar irradiance in W/m^2, T in °C is the cell temperature, and P is the cell output power. α in A/°C is the short-circuit current coefficient, β in V/°C is the voltage coefficient, and γ in Ω/°C is the curve correction factor. The coefficients α and β are measured in the same units as α_I and α_V, respectively, but are basically different. The subscripts 1 and 2 define specific conditions of solar irradiance and temperature. In particular, either of the two states can be identified with STC. Equation (9.14) are difficult to use as they require input of parameters not listed in the data sheet of today's PV devices. From the publication of Sandström equations and to this day many studies have been devoted to the subject, e.g., Hovinen (1994), Martin and Ruiz (2001), Ortiz-Conde et al. (2006), De Soto et al. (2006), Celik and Acikgoz (2007), Chenni et al. (2007), Karatepe et al. (2007), Nguyen and Lehman (2008), Marion (2008), Lo Brano et al. (2010).

In what follows, four models for the V–I characteristic of PV modules will be presented. The first three models run in two steps: (a) calculate the values of certain parameters of the cell in STC and (b) determine the same parameter values in real operating conditions. The fourth model allows the performance parameters determined at one operating conditions to be translated to any different operating conditions.

The models are largely based on the relations written for a solar cell. Since the experimental data reported by manufacturers refer to the entire module it is sometimes necessary to use relationships of the type Eq. (9.6a, b) by which the current and voltage across the cell are related to the current and voltage across the module.

The basic approach to solving for the unknown quantities in V–I equations is to consider what information is normally published by PV manufacturers and how that information can be used to help predict V–I behavior under varying conditions. The most important information provided by the manufacturer are the module short-circuit current, open-circuit voltage, and maximum power point (MPP) current and voltage, all measured at the same reference irradiance and cell temperature. This information fixes three V–I points, all of which must lie on the same V–I curve, and therefore satisfy the same V–I equation (9.4). The three data points on the V–I curve permit three independent versions of the V–I equation to be written. The result is a nonlinear system of three equations, which can be solved for three unknowns. However, Eq. (9.4) includes more than three parameters ($I_L, I_{01}, I_{02}, m_1, m_2, R_S, R_P$). Depending on the simplification being used, the equation system can in some instances be solved explicitly by simple substitution, or in general, it can be solved numerically. The V–I models studied further differ in essence by the assumptions made to reduce the number of unknowns in Eq. (9.4).

9.3.1 Five-Parameter Model

According to the Shockley theory (see for example Sze and Kwok 2007), recombination in the charge zone of a p–n junction of a crystalline solar cell can be neglected, and therefore the diode D_2 from Fig. 9.2 can be omitted. Thus, the two diode solar cell model (Eq. 9.4) can be reduced to a single diode model in which the characteristic V–I is given by the equation:

$$I = I_L - I_0\left[\exp\left(\frac{e(V+\mathrm{IR}_S)}{\mathrm{mk}_B T}\right) - 1\right] - \frac{V+\mathrm{IR}_S}{R_p} \tag{9.15}$$

The temperature dependence of the saturation current of the cell in Eq. (9.15) is nonlinear:

$$I_0 = C_0 T^3 \exp\left(-\frac{E_g}{k_B T}\right) \tag{9.16}$$

where E_g is the bandgap energy of the semiconductor material from which the cell is made and C_0 is a constant depending on material parameters. For equations required to infer Eq. (9.16), see e.g,. Sah (1991). Using Eqs. (9.15) and (9.16) follows:

$$I = I_L - C_0 T^3 \exp\left(-\frac{E_g}{k_B T}\right)\left[\exp\left(\frac{e(V+\mathrm{IR}_S)}{\mathrm{mk}_B T}\right) - 1\right] - \frac{V+\mathrm{IR}_S}{R_p} \tag{9.17}$$

The solar cell model Eq. (9.17) is characterized by the following five parameters: I_L, C_0, R_S, m and R_p. Sometimes, the model described by Eq. (9.17) is called the *five-parameter model.*

9.3.1.1 Calculation of Parameters I_L, C_0, R_S, m, and R_p at STC

The values of the five parameters from the title can be determined by comparing the solar cell V–I characteristic (Eq. 9.17) with the characteristic obtained experimentally by entire module testing under standard conditions. Therefore, whenever reference is made to these experimental points on the characteristic, they must be accompanied by restrictions $T = T_{\mathrm{STC}}$ and $G = G_{\mathrm{STC}}$. Let us note $I_{L,\mathrm{STC}}$, $C_{0,\mathrm{STC}}$, $R_{S,\mathrm{STC}}$, m_{STC} and $R_{p,\mathrm{STC}}$ the solar cell model parameter values (Eq. 9.17) corresponding to STC.

In case of crystalline photovoltaic modules, the calculation of parameters can be done iteratively, assuming an initial infinite parallel resistance. After calculating the other parameters a more realistic value of the parallel resistance can be determined.

Thus, assuming $R_P \to \infty$, Eq. (9.17) for the solar cell under STC is written:

$$I = I_{L,\mathrm{STC}} - C_{0,\mathrm{STC}} T_{\mathrm{STC}}^3 \exp\left(-\frac{E_g}{k_B T_{\mathrm{STC}}}\right)\left[\exp\left(\frac{e(V + IR_{S,\mathrm{STC}})}{m_{\mathrm{STC}} k_B T_{\mathrm{STC}}}\right) - 1\right] \tag{9.18}$$

The algorithm for calculating the parameters is as follows:

1. *Parameter* $I_{L,\mathrm{STC}}$

The short-circuit current of the module in STC $I_{\mathrm{SC,STC}}^m$ is obtained from the technical sheet of the module. With very good approximation the photocurrent $I_{L,\mathrm{STC}}$ of the cell in STC is equal to the STC short-circuit current: $I_{L,\mathrm{STC}} \approx I_{\mathrm{SC,STC}}$. Therefore, $I_{L,\mathrm{STC}}$ can be calculated with Eq. (9.6a):

$$I_{L,\mathrm{STC}} = I_{\mathrm{SC,STC}}^m \Big/ n_p \tag{9.19}$$

2. *Parameter* $C_{0,\ \mathrm{STC}}$ *(depending on* m_{STC}*)*

First, the open-circuit voltage of the module under STC $V_{\mathrm{OC,STC}}^m$ is taken from the catalog. It corresponds to the STC temperature, $T = T_{\mathrm{STC}}$. The open-circuit voltage of the solar cell under STC $V_{\mathrm{OC,STC}}$ is obtained using Eq. (9.6b):

$$V_{\mathrm{OC,STC}} = V_{\mathrm{OC,STC}}^m \Big/ n_s \tag{9.20}$$

The saturation current of the *p–n* junction at STC can be calculated as will be shown below. In open circuit, the cell current is null and from Eqs. (9.18) to (9.20) one obtains:

$$0 = I|_{V=V_{\mathrm{OC,STC}}} \approx I_{L,\mathrm{STC}} - I_{0,\mathrm{STC}} \exp\left(\frac{eV_{\mathrm{OC,STC}}^m / n_s}{m_{\mathrm{STC}}\, k_B T_{\mathrm{STC}}}\right) \tag{9.21}$$

where we took into account that at high voltage like $V_{\mathrm{OC,STC}}^m \Big/ n_s$ the exponential term in Eq. (9.18) is much greater than unity. It follows:

$$I_{0,\mathrm{STC}} = I_{L,\mathrm{STC}} \exp\left(-\frac{eV_{\mathrm{OC,STC}}^m / n_S}{k_B m_{\mathrm{STC}} T_{\mathrm{STC}}}\right) \tag{9.22}$$

The coefficient $C_{0,\mathrm{STC}}$ is calculated from Eq. (9.16) applied in testing conditions:

$$C_{0,\mathrm{STC}} = \frac{I_{0,\mathrm{STC}}(T_{\mathrm{STC}})}{T_{\mathrm{STC}}^3 \exp(-E_g / k_B T_{\mathrm{STC}})} \tag{9.23}$$

Note that the diode ideality factor m_{STC} entering Eq. (9.23) via $I_{0,\mathrm{STC}}(T_{\mathrm{STC}})$ is still undetermined.

3. *Parameter* $R_{S,\mathrm{STC}}$ *(depending on* m_{STC}*)*

The optimal operating point coordinates of the cell are known, corresponding to the module under test conditions ($I_{M,\mathrm{STC}}$, $V_{M,\mathrm{STC}}$). In addition, $I_{0,\mathrm{STC}}$ from Eq. (9.22) is also known. $V^m_{M,\mathrm{STC}}$ and $I^m_{M,\mathrm{STC}}$ are given in the module's data sheet. From Eq. (9.6a, b), the optimal STC cell current and voltage are:

$$I_{M,\mathrm{STC}} = I^m_{M,\mathrm{STC}}/n_p,\ V_{M,\mathrm{STC}} = V^m_{M,\mathrm{STC}}/n_s \tag{9.24a, b}$$

Near the MPP, the exponential value from Eq. (9.18) is high, which allows writing the following approximate relation:

$$I_{M,\mathrm{STC}} \approx I_{L,\mathrm{STC}} - I_{0,\mathrm{STC}} \exp\left(\frac{e(V_{M,\mathrm{STC}} + I_{M,\mathrm{STC}} R_{S,\mathrm{STC}})}{m_{\mathrm{STC}} k_B T_{\mathrm{STC}}}\right) \tag{9.25}$$

In these conditions, the cell series resistance $R_{S,\mathrm{STC}}$ can be determined as follows:

$$R_{S,\mathrm{STC}} = \frac{(m_{\mathrm{STC}} k_B T_{\mathrm{STC}}/e) \ln\left(1 - \frac{I_{M,\mathrm{STC}}}{I_{L,\mathrm{STC}}}\right) + \left(V^m_{\mathrm{OC,STC}} - V^m_{M,\mathrm{STC}}\right)/n_S}{I^m_{M,\mathrm{STC}}/n_p} \tag{9.26}$$

The diode ideality factor m_{STC} which is part of Eq. (9.26) will be determined next.

4. *Parameter* m_{STC}

Cell temperature coefficients for voltage α_V and current α_I are listed in the module's technical sheet, under test conditions. These coefficients are used to determine the value of the diode ideality factor m_{STC}. Solving Eq. (9.15) for $I = 0$ under the assumptions $R_P \to \infty$ and the exponential term much greater than unity, gives the open-circuit voltage: $V_{\mathrm{OC}} = (m k_B T/e) \ln(I_L/I_0)$. Thus, the temperature derivative of the solar cell open-circuit voltage V_{OC} at $T = T_{\mathrm{STC}}$ is:

$$\begin{aligned}\left.\frac{\mathrm{d}V_{\mathrm{OC}}}{\mathrm{d}T}\right|_{T=T_{\mathrm{STC}}} \equiv \alpha_V = {} & \frac{m_{\mathrm{STC}} k_B}{e} \ln\left(\frac{I_L(T_{\mathrm{STC}})}{I_0(T_{\mathrm{STC}})}\right) \\ & + \frac{m_{\mathrm{STC}} k_B T_{\mathrm{STC}}}{e}\left(\frac{1}{I_L(T_{\mathrm{STC}})}\frac{\mathrm{d}}{\mathrm{d}T} I_L(T)\bigg|_{T=T_{\mathrm{STC}}} - \frac{1}{I_0(T_{\mathrm{STC}})}\frac{\mathrm{d}}{\mathrm{d}T} I_0(T)\bigg|_{T=T_{\mathrm{STC}}}\right)\end{aligned} \tag{9.27}$$

To calculate the derivative in Eq. (9.27), it was considered that parameter m is constant, i.e., does not depend on temperature, in the vicinity of V_{OC}. The temperature derivative of the photocurrent at STC $\mathrm{d}I_L(T)/\mathrm{d}T|_{T=T_{\mathrm{STC}}}$ is exactly α_I. The temperature derivative of the diode saturation current is deduced from Eq. (9.23):

$$\left.\frac{\mathrm{d}I_0(T)}{\mathrm{d}T}\right|_{T=T_{\mathrm{STC}}} = C_{0,\mathrm{STC}} T_{\mathrm{STC}}\left(3T_{\mathrm{STC}} + \frac{E_g}{k_B}\right)\exp\left(-\frac{E_g}{k_B T_{\mathrm{STC}}}\right) \tag{9.28}$$

Using Eqs. (9.23) and (9.28), it follows:

$$\left.\frac{1}{I_0}\frac{\mathrm{d}I_0(T)}{\mathrm{d}T}\right|_{T=T_{\mathrm{STC}}} = \frac{3}{T_{\mathrm{STC}}} + \frac{E_g}{k_B T_{\mathrm{TSC}}^2} \tag{9.29}$$

From Eqs. (9.27) and (9.29), one obtains:

$$\alpha_V = \frac{m_{\mathrm{STC}} k_B}{e} \ln\frac{I_L(T_{\mathrm{STC}})}{I_0(T_{\mathrm{STC}})} + \frac{m_{\mathrm{STC}} T_{\mathrm{STC}} k_B}{e} \frac{\alpha_I}{I_L(T_{\mathrm{STC}})} - \left(3 + \frac{E_g}{k_B T_{\mathrm{STC}}}\right) \frac{m_{\mathrm{STC}} k_B}{e} \tag{9.30}$$

from where the parameter m_{STC} is determined:

$$m_{\mathrm{STC}} = \frac{\alpha_V - \frac{V_{\mathrm{OC,STC}}^m / n_s}{T_{\mathrm{STC}}}}{\frac{k_B T_{\mathrm{STC}}}{e}\frac{\alpha_I}{I_L(T_{\mathrm{STC}})} - \frac{k_B}{e}\left(3 + \frac{E_g}{k_B T_{\mathrm{STC}}}\right)} \tag{9.31}$$

The value of m_{STC} obtained with Eq. (9.31*)* can be used in Eqs. (9.22) and (9.23) for calculating $C_{0,\mathrm{STC}}$ and in Eq. (9.26) for $R_{S,\mathrm{STC}}$.

5. *Parameter* $R_{p,\mathrm{STC}}$

The above calculus was constructed assuming $R_{p,\mathrm{STC}} = \infty$. Now, taking into account that all other parameters $I_{L,\mathrm{STC}}$, $C_{0,\mathrm{STC}}$, $R_{S,\mathrm{STC}}$ and m_{STC} have been determined, a more realistic value of the resistance $R_{p,\mathrm{STC}}$ may be calculated with Eq. (9.15), using the replacements $V = V_{M,\mathrm{STC}}$ and $I = I_{M,\mathrm{STC}}$ from Eq. (9.24a, b):

$$R_{p,\mathrm{STC}} = \frac{V_{M,\mathrm{STC}} + I_{M,\mathrm{STC}} R_{S,\mathrm{STC}}}{I_{\mathrm{SC,STC}} - C_{0,\mathrm{STC}} T_{\mathrm{STC}}^3 \exp\left(-\frac{E_g}{k_B T_{\mathrm{STC}}}\right) \exp\left(\frac{e(V_{M,\mathrm{STC}} + I_{M,\mathrm{STC}} R_S)}{m_{\mathrm{STC}} k_B T_{\mathrm{STC}}}\right) - I_{M,\mathrm{STC}}} \tag{9.32}$$

Further on, the expression on the cell series resistance under test conditions may be recalculated including the value of parallel resistance Eq. (9.32). This is leading to an iterative computation for the resistances $R_{S,\mathrm{STC}}$ and $R_{P,\mathrm{STC}}$ until the desired accuracy is reached.

9.3.1.2 Calculating the Parameters I_L, C_0, R_S, m, and R_p in Outdoor Operating Conditions

PV module manufacturers report experimental values for current and voltage temperature coefficients, which are valid around the standard test temperature T_{STC}. Using these coefficients, an approximate value for the photocurrent I_L can be determined:

$$I_L = I_{L,\mathrm{STC}} \frac{G_{\mathrm{ef}}}{G_{\mathrm{STC}}} \left[1 + \alpha_I'(T - T_{\mathrm{STC}})\right] \tag{9.33}$$

and the open-circuit voltage reads:

$$V_{OC}(T) = V_{OC,STC} + (T - T_{STC})\frac{dV_{OC}}{dT} \tag{9.34}$$

It is important to remember that we have experimental information under STC about three states on the V–I diagram: short-circuit, open-circuit, and maximum power state. This information allowed the calculation of all parameters $I_{L,STC}$, $C_{0,STC}$, $R_{S,STC}$, m_{STC} and $R_{p,STC}$. Outside the STC we have information about one single state, the open circuit, for which the voltage value V_{OC} is given by Eq. (9.34). Knowing this value allows for the calculation of a single parameter under real conditions, which is R_p, as will be demonstrated below. I_L under working conditions is given by Eq. (9.33). We assume that three of the cell's parameters have the same values in- and outside of STC, respectively:

$$C_0 = C_{0,STC}, \quad R_S = R_{S,STC}, \quad m = m_{STC} \tag{9.35a, b, c}$$

The cell's V–I Eq. (9.15) for the open circuit state is:

$$0 = I|_{V=V_{OC}} = I_L - I_0\left[\exp\left(\frac{eV_{OC}}{\mathrm{m}k_B T}\right) - 1\right] - \frac{V_{OC}}{R_p} \tag{9.36}$$

From Eq. (9.36), one obtains:

$$R_p = \frac{V_{OC}}{I_L - I_0\left[\exp\left(\frac{eV_{OC}}{\mathrm{m}k_B T} - 1\right)\right]} \tag{9.37}$$

where I_0 is calculated using C_0 from Eq. (9.16).

The five-parameter model is illustrated in Sect. 9.4.3.

Another formulation of the five-parameter model is reported in Ref. Lo Brano et al. (2010) and should be acknowledged here. As authors state, the model is capable to describe analytically the V–I characteristic of a PV module for each generic operating condition. The five parameters I_L, I_0, m, R_S and R_P, are obtained by imposing the following conditions on both the calculated V–I characteristics and those issued by manufacturers: equality of the short-circuit current, equality of the open-circuit voltage, correspondence of the MPP and equal values of the curve derivative in the points of short circuit and open circuit for nominal conditions.

9.3.2 Four-Parameter Model

A PV cell model with five parameters (I_L, I_0, R_S, m and R_p) was studied in the previous section. Below, a simplifying assumption is adopted, namely that the shunt resistance is infinite ($R_P = \infty$). This assumption is realistic, given the influence of R_P on the V–I characteristic of a crystalline solar cell (see Fig. 9.3d). The new model contains a total of four parameters (i.e. I_L, I_0, R_S, m), hence the name "four-parameter model". The equivalent circuit of the solar cell is derived from the standard model (Fig. 9.2)

Fig. 9.6 Equivalent circuit of the four parameters solar cell model

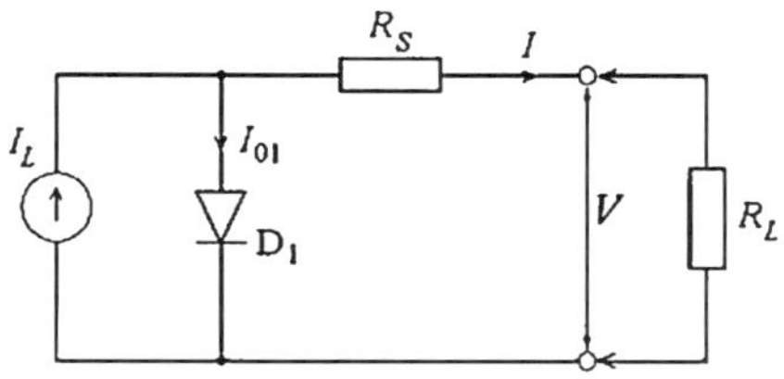

and is drawn in Fig. 9.6. The V–I characteristic of the cell in the four parameters model is given by:

$$I = I_L - I_0\left[\exp\left(\frac{e(V + \mathrm{IR}_S)}{\mathrm{mk}_B T}\right) - 1\right] \tag{9.38}$$

Equation (9.38) and the equivalent circuit (Fig. 9.6) are probably the most employed in computing the V–I characteristic of a crystalline solar cell, offering a good balance between accuracy and simplicity.

The four specific parameters (I_L, I_0, m, R_S) of Eq. (9.38) can be determined from manufacturer data as follows. As already stated above, the data sheet usually gives three points on the V–I characteristic: (1) Short-circuit current ($V = 0, I = I_{\mathrm{SC,STC}}$), (2) Open-circuit voltage ($V = V_{\mathrm{OC,STC}}$, $I = 0$), and (3) MPP ($V = V_{M,\mathrm{STC}}$, $I = I_{M,\mathrm{STC}}$). Replacing in Eq. (9.38) results the following three equations:

$$-I_{\mathrm{SC,STC}} + I_L - I_0\left[\exp\left(\frac{eR_S I_{\mathrm{SC,STC}}}{\mathrm{mk}_B T_{\mathrm{STC}}}\right) - 1\right] = 0 \tag{9.39a}$$

$$I_L - I_0\left[\exp\left(\frac{eV_{\mathrm{OC,STC}}}{\mathrm{mk}_B T_{\mathrm{STC}}}\right) - 1\right] = 0 \tag{9.39b}$$

$$-I_{M,\mathrm{STC}} + I_L - I_0\left[\exp\left(\frac{e\left(V_{M,\mathrm{STC}} + R_S I_{M,\mathrm{STC}}\right)}{\mathrm{mk}_B T_{\mathrm{STC}}}\right) - 1\right] = 0 \tag{9.39c}$$

The fourth equation needed to form a system of four equations with four unknowns is obtained by imposing the condition (Eq. 9.5) which forces the device to function at the point of maximum power. Thus, expressing power as$P(V) = I(V) \cdot V$, Eq. (9.5) is written:

$$\frac{\mathrm{d}P(V)}{\mathrm{d}V} = \frac{\partial I(V)}{\partial V}V + I(V) = 0 \tag{9.40}$$

Differentiating $I(V)$ with respect to V, we obtain:

$$\frac{\partial I(V)}{\partial V} = \frac{-I_0 \dfrac{e}{\mathrm{mk}_B T}\exp\left(\dfrac{e(V + \mathrm{IR}_S)}{\mathrm{mk}_B T}\right)}{1 + R_S I_0 \dfrac{e}{\mathrm{mk}_B T}\exp\left(\dfrac{e(V + \mathrm{IR}_S)}{\mathrm{mk}_B T}\right)} \tag{9.41}$$

Replacing Eqs. (9.41) and (9.38) in Eq. (9.40) and taking into account $I = I_{M,\mathrm{STC}}$ and $V = V_{M,\mathrm{STC}}$, after a few computations we obtain:

$$I_L + \frac{-I_0 \frac{e}{\mathrm{mk}_B T} \exp\left(\frac{e\left(V_{M,\mathrm{STC}} + R_S I_{M,\mathrm{STC}}\right)}{\mathrm{mk}_B T}\right)}{1 + R_S I_0 \frac{e}{\mathrm{mk}_B T} \exp\left(\frac{e\left(V_{M,\mathrm{STC}} + I_{M,\mathrm{STC}} R_S\right)}{\mathrm{mk}_B T}\right)} V_{M,\mathrm{STC}} - I_0\left[\exp\left(\frac{e\left(V_{M,\mathrm{STC}} + R_S I_{M,\mathrm{STC}}\right)}{\mathrm{mk}_B T}\right) - 1\right] = 0 \tag{9.42}$$

Equations (9.39a, b, c) and (9.42) are independent and sufficient to solve for the four parameters I_L, I_0, m, R_S. The equation system can be solved numerically using, for example, the iterative Newton–Raphson method for solving nonlinear equations. A short introduction to the Newton–Raphson method follows, after which it is particularized to the equation system (9.39a, b, c) and (9.42).

9.3.2.1 Newton–Raphson Method

The Newton–Raphson method (for example see Epperson 2007) is used to solve systems of nonlinear equations. It find roots of a nonlinear function $f{:}\mathbb{R}^k \to \mathbb{R}^k$, by computing the Jacobian matrix of first partial function around an initial guess point $x^{(0)} = \left(x_1^{(0)}, x_2^{(0)}, \ldots, x_k^{(0)}\right)$ and using iterations to move closer to the nearest zero. We will use a parenthesized superscript for the iteration counter and a subscript denoting the element of the vector. Thus, $x^{(n)}$ will refer to the vector at iteration n and $x_j^{(n)}$ will refer to the jth component of the vector.

A set of nonlinear equations in matrix form is given by:

$$f(x) = \begin{pmatrix} f_1(x_1, x_2, \ldots, x_k) \\ f_2(x_1, x_2, \ldots, x_k) \\ \vdots \\ f_k(x_1, x_2, \ldots, x_k) \end{pmatrix} = 0 \tag{9.43}$$

where 0 is understood to be the zero vector in $\mathbb{R}^k$. We want to find the vectors $x \in \mathbb{R}^k$ such that $f(x) = 0$. We might have:

$$\begin{aligned} 0 &= f_1(x_1, x_2, \ldots, x_k) \\ 0 &= f_2(x_1, x_2, \ldots, x_k) \\ &\ldots \\ 0 &= f_k(x_1, x_2, \ldots, x_k) \end{aligned} \tag{9.44}$$

We take $x^{(0)} = \left(x_1^{(0)}, x_2^{(0)}, \ldots, x_k^{(0)}\right)$ an initial approximation to the solution and expand all component functions in a Taylor's series around that point:

$$f(x) = f\left(x^{(0)}\right) + J_f\left(x^{(0)}\right)\left(x - x^{(0)}\right) + R \tag{9.45}$$

Here, J_f is the Jacobian matrix of first partial derivatives function f:

$$J_f = \begin{vmatrix} \frac{\partial f_1}{\partial x_1} & \frac{\partial f_1}{\partial x_2} & \cdots & \frac{\partial f_1}{\partial x_k} \\ \frac{\partial f_2}{\partial x_1} & \frac{\partial f_2}{\partial x_2} & \cdots & \frac{\partial f_2}{\partial x_k} \\ \vdots & \vdots & & \vdots \\ \frac{\partial f_k}{\partial x_1} & \frac{\partial f_k}{\partial x_2} & \cdots & \frac{\partial f_k}{\partial x_k} \end{vmatrix} \tag{9.46}$$

Now, set $f(x) = 0$ and drop the remainders:

$$0 = f\left(x^{(0)}\right) + J_f\left(x^{(0)}\right)\left(x - x^{(0)}\right) \tag{9.47}$$

And solving for x,

$$x = x^{(0)} - J_f^{-1}\left(x^{(0)}\right)f\left(x^{(0)}\right) \tag{9.48}$$

The value $x^{(0)}$ is replaced by $x^{(n)}$ and x by the new value$x^{(n+1)}$, so

$$x^{(n+1)} = x^{(n)} - J_f^{-1}\left(x^{(n)}\right)f\left(x^{(n)}\right) \tag{9.49}$$

Now $x^{(n)} = \left(x_1^{(n)}, x_2^{(n)}, \ldots, x_k^{(n)}\right)$ is the nth approximate solution vector.

9.3.2.2 Solving the Four Parameters Equations System Using the Newton–Raphson Method

The system that we propose to solve consists of Eq. (9.39a, b, c) and (9.42) with the unknowns I_L, I_0, m, R_S. Therefore, in the Newton–Raphson algorithm described above $k = 4$. First, for each of the four unknowns should be given a guess value. Then, a first-order Taylor's series expansion (Eq. 9.45) is used to establish the new set of equations in matrix representation. This requires computing the Jacobian matrix (Eq. 9.46). Solving the new set of equations provides a better guess for each unknown, and the process is repeated until the solution reaches the demanded tolerance.

Let $f_i(I_L, I_0, m, R_S), i = 1, 2, 3, 4$ be the function from Eq. (9.43) with notations: $x_1 \equiv I_L$, $x_2 \equiv I_0$, $x_3 \equiv m$ and $x_4 \equiv R_S$. Every function $f_i(I_L, I_0, m, R_S)$ represents the left sides of Eqs. (9.39a, b, c) and (9.42), in that order. The Taylor's series expansion of the first function $f_1(I_L, I_0, m, R_S)$ reads:

$$
\begin{aligned}
f_1\left(I_L^{(n)},I_0^{(n)},m^{(n)},R_s^{(n)}\right) \simeq{} & f_1\left(I_L^{(n-1)},I_0^{(n-1)},m^{(n-1)},R_s^{(n-1)}\right) \\
& + \left.\frac{\partial f_1(I_L,I_0,m,R_s)}{\partial I_L}\right|_n \left(I_L^{(n)}-I_L^{(n-1)}\right) + \left.\frac{\partial f_1(I_L,I_0,m,R_s)}{\partial I_0}\right|_n \left(I_0^{(n)}-I_0^{(n-1)}\right) \\
& + \left.\frac{\partial f_1(I_L,I_0,m,R_s)}{\partial m}\right|_n \left(m^{(n)}-m^{(n-1)}\right) + \left.\frac{\partial f_1(I_L,I_0,m,R_s)}{\partial R_s}\right|_n \left(R_s^{(n)}-R_s^{(n-1)}\right)
\end{aligned}
\tag{9.50}
$$

where the superscript n counts the iteration. The development for the other equations will have the same form. Thus, the Jacobian matrix is given for each iteration as:

$$
J_f^{(n)} = \begin{bmatrix}
\frac{\partial f_1}{\partial I_L} & \frac{\partial f_1}{\partial I_0} & \frac{\partial f_1}{\partial m} & \frac{\partial f_1}{\partial R_s} \\
\frac{\partial f_2}{\partial I_L} & \frac{\partial f_2}{\partial I_0} & \frac{\partial f_2}{\partial m} & \frac{\partial f_2}{\partial R_s} \\
\frac{\partial f_2}{\partial I_L} & \frac{\partial f_3}{\partial I_0} & \frac{\partial f_3}{\partial m} & \frac{\partial f_3}{\partial R_s} \\
\frac{\partial f_2}{\partial I_L} & \frac{\partial f_3}{\partial I_0} & \frac{\partial f_4}{\partial m} & \frac{\partial f_4}{\partial R_s}
\end{bmatrix}
\tag{9.51}
$$

The Jacobian matrix elements are (superscript n is omitted):

$$
\frac{\partial f_1}{\partial I_L} = 1,\ \frac{\partial f_1}{\partial I_0} = 1 - \exp\left(\frac{eR_S I_{\mathrm{SC,STC}}}{\mathrm{m}k_B T_{\mathrm{STC}}}\right)
\tag{9.52a, b}
$$

$$
\frac{\partial f_1}{\partial m} = I_0 \frac{eR_S I_{\mathrm{SC,STC}}}{m^2 k_B T_{\mathrm{STC}}} \exp\left(\frac{eR_S I_{\mathrm{SC,STC}}}{\mathrm{m}k_B T_{\mathrm{STC}}}\right)
\tag{9.52c}
$$

$$
\frac{\partial f_1}{\partial R_S} = -I_0 \frac{eI_{\mathrm{SC,STC}}}{\mathrm{m}k_B T_{\mathrm{STC}}} \exp\left(\frac{eR_S I_{\mathrm{SC,STC}}}{\mathrm{m}k_B T_{\mathrm{STC}}}\right)
\tag{9.52d}
$$

$$
\frac{\partial f_2}{\partial I_L} = 1,\ \frac{\partial f_2}{\partial I_0} = 1 - \exp\left(\frac{eV_{\mathrm{OC,STC}}}{\mathrm{m}k_B T_{\mathrm{STC}}}\right)
\tag{9.53a, b}
$$

$$
\frac{\partial f_2}{\partial m} = I_0 \frac{eV_{\mathrm{OC,STC}}}{m^2 k_B T_{\mathrm{STC}}} \exp\left(\frac{eV_{\mathrm{OC,STC}}}{\mathrm{m}k_B T_{\mathrm{STC}}}\right),\ \frac{\partial f_2}{\partial R_s} = 0
\tag{9.53c, d}
$$

$$
\frac{\partial f_3}{\partial I_L} = 1,\ \frac{\partial f_3}{\partial I_0} = 1 - \exp\left(\frac{e\left(V_{M,\mathrm{STC}} + R_S I_{M,\mathrm{STC}}\right)}{\mathrm{m}k_B T_{\mathrm{STC}}}\right)
\tag{9.54a, b}
$$

$$
\frac{\partial f_3}{\partial m} = I_0 \frac{e\left(V_{M,\mathrm{STC}} + R_S I_{M,\mathrm{STC}}\right)}{m^2 k_B T_{\mathrm{STC}}} \exp\left(\frac{e\left(V_{M,\mathrm{STC}} + R_S I_{M,\mathrm{STC}}\right)}{\mathrm{m}k_B T_{\mathrm{STC}}}\right)
\tag{9.54c}
$$

$$
\frac{\partial f_3}{\partial R_s} = -I_0 \frac{eI_{M,\mathrm{STC}}}{\mathrm{m}k_B T_{\mathrm{STC}}} \exp\left(\frac{e\left(V_{M,\mathrm{STC}} + R_S I_{M,\mathrm{STC}}\right)}{\mathrm{m}k_B T_{\mathrm{STC}}}\right)
\tag{9.54d}
$$

$$\frac{\partial f_4}{\partial I_L} = 1, \ \frac{\partial f_4}{\partial I_0} = 1 - A\exp\left(\frac{e\left(V_{M,\mathrm{STC}} + R_S I_{M,\mathrm{STC}}\right)}{\mathrm{m}k_B T_{\mathrm{STC}}}\right) \tag{9.55a, b}$$

$$\begin{aligned}\frac{\partial f_4}{\partial m} =& \frac{1}{R_0}\frac{V_{M,\mathrm{STC}} + R_S I_{M,\mathrm{STC}}}{m}\left(1 + \frac{eV_{M,\mathrm{STC}}}{\mathrm{m}k_B T_{\mathrm{STC}}(1 + R_S/R_0)}\right) \\ &+ \frac{V_{M,\mathrm{STC}}}{R_0}\left[\frac{1}{m(1 + R_S/R_0)} - \frac{R_S/R_0}{m(1 + R_S/R_0)^2}\left(1 + \frac{e\left(V_{M,\mathrm{STC}} + I_{M,\mathrm{STC}}R_S\right)}{\mathrm{m}k_B T_{\mathrm{STC}}}\right)\right]\end{aligned} \tag{9.55c}$$

$$\frac{\partial f_4}{\partial R_S} = -\frac{I_{M,\mathrm{STC}}}{R_0}\left(1 + \frac{eV_{M,\mathrm{STC}}}{\mathrm{m}k_B T_{\mathrm{STC}}(1 + R_S/R_0)}\right) + \frac{V_{M,\mathrm{STC}}}{R_0^2}\frac{1 + \frac{eR_S I_M}{\mathrm{m}k_B T_{\mathrm{STC}}}}{(1 + R_S/R_0)^2} \tag{9.55d}$$

where

$$\frac{1}{R_0} = I_0\frac{e}{\mathrm{m}k_B T_{\mathrm{STC}}}\exp\left(\frac{e\left(V_{M,\mathrm{STC}} + R_S I_{M,\mathrm{STC}}\right)}{\mathrm{m}k_B T_{\mathrm{STC}}}\right) \tag{9.56a}$$

$$A = 1 + \frac{eV_{M,\mathrm{STC}}}{\mathrm{m}k_B T_{\mathrm{STC}}}\frac{1}{1 + R_S/R_0} - \frac{eV_{M,\mathrm{STC}}R_S/R_0}{\mathrm{m}k_B T_{\mathrm{STC}}(1 + R_S/R_0)^2} \tag{9.56b}$$

Solving the system of equations requires determination of *guess values*. In order to find the functions $f_i(I_L, I_0, m, R_S), \quad i = 1, \ldots, 4$, defined by the left side of Eqs. (9.39a, b, c) and (9.42) can be simplified by making adequate assumptions, based on the physics of the solar cell. After that, the resulted system of equations can be solved for the guess values.

The notation $V_T = k_B T/e$ is used in the following.

In Eq. (9.39b) at open circuit voltage, the exponential term is much greater than one which can be neglected:

$$f_2(I_L, I_0, m, R_S) \cong I_L - I_0\exp\left(\frac{V_{\mathrm{OC,STC}}}{mV_T}\right) \tag{9.57}$$

In Eq. (9.39c), at the MPP, the exponential term is also much greater than one which can be neglected:

$$f_3(I_L, I_0, m, R_S) \cong -I_{M,\mathrm{STC}} + I_L - I_0\exp\left(\frac{V_{M,\mathrm{STC}} + R_S I_{M,\mathrm{STC}}}{mV_T}\right) \tag{9.58}$$

In Eq. (9.39a) at short circuit there is no reason to neglect the unity because the magnitude of the exponential term may be in the same range. However, the entire term $I_0[\exp(R_S I_{\mathrm{SC}}/mV_T) - 1]$ representing the dark current at short circuit is very small and can be neglected. So,

$$f_1(I_L, I_0, m, R_S) \cong -I_{\mathrm{SC}} + I_L \tag{9.59}$$

In order to obtain $f_4(I_L, I_0, m, R_S)$, the equation for V–I characteristic has been simplified, assuming that generally $\exp[(V + \mathrm{IR}_S)/mV_T] \gg 1$:

$$I \cong I_L - I_0 \exp\left(\frac{V + \mathrm{IR}_S}{mV_T}\right) \tag{9.60}$$

In order to calculate the maximum power, we have to derive the current (Eq. 9.60) in respect to the voltage:

$$\frac{\partial I}{\partial V} = \frac{-\frac{I_0}{mV_T}\exp\left(\frac{V+\mathrm{IR}_S}{mV_T}\right)}{1 + \frac{R_S I_0}{mV_T}\exp\left(\frac{V+\mathrm{IR}_S}{mV_T}\right)}$$

Next, we apply Eq. (9.5):

$$\frac{\mathrm{d}P}{\mathrm{d}V} = \frac{\partial I}{\partial V}V + I = \frac{-\frac{I_0}{mV_T}\exp\left(\frac{V+\mathrm{IR}_S}{mV_T}\right)}{1 + \frac{R_S I_0}{mV_T}\exp\left(\frac{V+\mathrm{IR}_S}{mV_T}\right)}V + I_L - I_0\exp\left(\frac{V + \mathrm{IR}_S}{mV_T}\right)$$

Finally, $f_4(I_L, I_0, m, R_S)$ results as:

$$f_4(I_L, I_0, m, R_S) \cong \frac{-\frac{I_0}{mV_T}\exp\left(\frac{V+\mathrm{IR}_S}{mV_T}\right)}{1 + \frac{R_S I_0}{mV_T}\exp\left(\frac{V+\mathrm{IR}_S}{mV_T}\right)}V + I_L - I_0\exp\left(\frac{V + \mathrm{IR}_S}{mV_T}\right) \tag{9.61}$$

The guess values of the Newton–Raphson algorithm are the solutions of the system:

$$f_i(I_L, I_0, m, R_S) = 0, \quad i = 1, \ldots, 4 \tag{9.62}$$

After some algebra it results:

$$I_L^{(0)} = I_{\mathrm{SC,STC}} \tag{9.63a}$$

$$I_0^{(0)} = I_{\mathrm{SC,STC}} \exp\left(-\frac{V_{\mathrm{OC,STC}}}{mV_T}\right) \tag{9.63b}$$

$$m^{(0)} = \frac{1}{V_T} \cdot \frac{\left(2V_{M\mathrm{,STC}} - V_{\mathrm{OC,STC}}\right)\left(I_{\mathrm{SC,STC}} - I_{M\mathrm{,STC}}\right)}{I_{M\mathrm{,STC}} + \left(I_{\mathrm{SC,STC}} - I_{M\mathrm{,STC}}\right)\ln\left(1 - \frac{I_{M\mathrm{,STC}}}{I_{\mathrm{SC,STC}}}\right)} \tag{9.63c}$$

$$R_S^{(0)} = \frac{V_M\left(\frac{I_{\mathrm{SC,STC}}}{I_{M\mathrm{,STC}}} - 1\right) + \frac{V_{\mathrm{OC,STC}} - V_{M\mathrm{,STC}}}{\ln\left(1 - \frac{I_{M\mathrm{,STC}}}{I_{\mathrm{SC,STC}}}\right)}}{I_{\mathrm{SC,STC}} - I_{M\mathrm{,STC}} + \frac{I_M}{\ln\left(1 - \frac{I_{M\mathrm{,STC}}}{I_{\mathrm{SC,STC}}}\right)}} \tag{9.63d}$$

Using the guess values as given by Eq. (9.63), the Newton–Raphson algorithm can be executed.

9.3.3 Three-Parameter Model

This method starts from Eq. (9.38), resulted from the standard model which neglected the effects of generation–recombination in the space charge region and the leakage current ($R_p \rightarrow \infty$). In addition, three other simplifying assumptions are made, generally valid for crystalline silicon PV cells:

1. Photogenerated and short-circuit current are equal: $I_L = I_{\text{SC}}$.
2. $\exp(e(V + IR_S)/(k_B T)) \gg 1$ in any working conditions.
3. The *p–n* junction is ideal: $m = 1$.

These assumptions allow rewriting the V–I characteristic given by Eq. (9.38) in a simplified form:

$$I = I_{\text{SC}} - I_0 \exp\left[\frac{e(V + \text{IR}_S)}{k_B T}\right] \tag{9.64}$$

For the state of open circuit, $I = 0$ we obtain the following expression for the parameter I_0:

$$I_0 = I_{\text{SC}} \exp\left(-\frac{eV_{\text{OC}}}{k_B T}\right) \tag{9.64a}$$

Replacing Eq. (9.64a) in Eq. (9.64), the V–I characteristic of the cell is obtained, expressed in terms of the parameters I_{SC} and V_{OC}:

$$I(V) = I_{\text{SC}}\left[1 - \exp\left(\frac{e(V - V_{\text{OC}} + \text{IR}_S)}{k_B T}\right)\right] \tag{9.64b}$$

This is a simpler model, having only three parameters, I_{SC}, V_{OC} and R_S, hence its name of "three-parameter model".

9.3.3.1 Determination of the Parameters I_{SC}, V_{OC} and R_S in Testing Conditions

In STC, the values of the cell parameters $I_{\text{SC,STC}}$ and $V_{\text{OC,STC}}$ can be obtained using the catalog STC parameters of the module, $I^m_{\text{SC,STC}}$ and $V^m_{\text{OC,STC}}$ in Eq. (9.6a, b):

$$I_{\text{SC,STC}} = I^m_{\text{SC,STC}} \Big/ n_p,\ V_{\text{OC,STC}} = V^m_{\text{OC,STC}} \Big/ n_s \tag{9.65a, b}$$

Series resistance $R_{S,\text{STC}}$ can be calculated from the MPP coordinates ($V_{M,\text{STC}}$, $I_{M,\text{STC}}$) with Eq. (9.64b). Otherwise, the determination of parameter $R_{S,\text{STC}}$ can be made using an approximate approach. For a solar cell, the voltage and current $V_{M,\text{STC}}$ and $I_{M,\text{STC}}$ corresponding to the MPP in STC, are given by Eq. (9.24a, b).

Supposing that $I_{M,\text{STC}}$ and $V_{M,\text{STC}}$ obey the following empirical relationships (Green 1982):

$$FF_{\mathrm{STC}} = \frac{V_{M,\mathrm{STC}} I_{M,\mathrm{STC}}}{V_{\mathrm{OC,STC}} I_{\mathrm{SC,STC}}} = F_{f0,\mathrm{STC}}(1 - r_{S,\mathrm{STC}}) \tag{9.66}$$

where

$$F_{f0,\mathrm{STC}} = \frac{v_{\mathrm{OC,STC}} - \ln(v_{\mathrm{OC,STC}} + 0.72)}{v_{\mathrm{OC,STC}} + 1} \tag{9.67}$$

and

$$v_{\mathrm{OC,STC}} = eV_{\mathrm{OC,STC}}/(k_B T_{\mathrm{STC}}),\ r_{S,\mathrm{STC}} = R_{S,\mathrm{STC}} I_{\mathrm{SC,STC}}/V_{\mathrm{OC,STC}} \tag{9.68a, b}$$

which are the normalized open-circuit voltage and the normalized series resistance, respectively. Equations (9.66) and (9.67) are valid for $v_{\mathrm{OC}} > 15$ and $r_S < 0.4$, having accuracy better than 1 %. They can be applied directly to calculate the energy produced by a photovoltaic module, given that all cells within it are identical and the voltage drop on the electrical wires that connect the cells is negligible, which usually happens in practice. Applying Eqs. (9.66) and (9.67) and using notations Eq. (9.68a, b) is found:

$$\frac{V_{M,\mathrm{STC}} I_{M,\mathrm{STC}}}{V_{\mathrm{OC,STC}} I_{\mathrm{SC,STC}}} = \frac{v_{\mathrm{OC,STC}} - \ln(v_{\mathrm{OC,STC}} + 0.72)}{v_{\mathrm{OC,STC}} + 1}(1 - r_{S,\mathrm{STC}}) \tag{9.69}$$

Because $v_{\mathrm{OC,STC}}$ can be determined from Eq. (9.68a), relationship (9.69) contains only one unknown, namely $r_{S,\mathrm{STC}}$. After solving Eq. (9.69) and finding $r_{S,\mathrm{STC}}$, the series resistance at STC can be calculated with Eq. (9.68b):

$$R_{S,\mathrm{STC}} = r_{S,\mathrm{STC}} \frac{V_{\mathrm{OC,STC}}}{I_{\mathrm{SC,STC}}} \tag{9.70}$$

This concludes the determination of parameters $I_{\mathrm{SC,STC}}$, $V_{\mathrm{OC,STC}}$ and $R_{S,\mathrm{STC}}$ in standard testing conditions.

9.3.3.2 Determination of Parameters I_{SC}, V_{OC}, and R_S in Operating Conditions

The values of the three parameters can be found starting from the respective STC values as follows.

The short-circuit current of solar cell operating in outdoor conditions is expressed as:

$$I_{\mathrm{SC}}(G_{\mathrm{ef}}) = \frac{I_{\mathrm{SC,STC}}}{G_{\mathrm{STC}}} G_{\mathrm{ef}} \tag{9.71}$$

where G_{ef} is given by Eq. (9.13). Open-circuit voltage of the cell operating in outdoor conditions is calculated with:

$$V_{OC}(T) = V_{OC,STC} + (T - T_{STC})\frac{dV_{OC}}{dT} \tag{9.72}$$

Equation (9.72) can use the approximation:

$$\left.\frac{dV_{OC}}{dT}\right|_{T=T_{STC}} \cong \frac{dV_{OC,STC}}{dT_{STC}} = \alpha_V \tag{9.73}$$

The calculation of the R_S parameter for real conditions can be done in two ways. In the first variant, the hypothesis is that the series resistance is not depending on operating conditions. Therefore:

$$R_S = R_{S,STC} \tag{9.74}$$

In the second, it is considered that the cell series resistance depends on operating conditions, i.e. $R_S \neq R_{S,STC}$. It is assumed that values I_M and V_M for the solar cell under real working conditions are respecting Eqs. (9.66) and (9.67), with notations (9.68a, b). Note that Eq. (9.68a) can be used to determine v_{OC}, as V_{OC} is given by Eq. (9.72) and T by Eq. (9.11). Replacing Eq. (9.67) in Eq. (9.66) it follows:

$$\frac{V_M I_M}{V_{OC} I_{SC}} = \frac{v_{OC} - \ln(v_{OC} + 0.72)}{v_{OC} + 1}(1 - r_S) \tag{9.75}$$

Also, it is assumed that V_M and I_M obey the relationships (Arajuo and Sanchez 1982):

$$\frac{V_M}{V_{OC}} = 1 - \frac{b}{v_{OC}}\ln a - r_S\left(1 - a^{-b}\right) \tag{9.76}$$

$$I_M/I_{SC} = 1 - a^{-b} \tag{9.77}$$

where a and b are expressed as:

$$a = v_{OC} + 1 - 2v_{OC}r_S,\ b = \frac{a}{1 + a} \tag{9.78a, b}$$

The set Eqs. (9.75), (9.76), (9.77) and (9.78a, b) is a system of five equations with five unknowns: V_M, I_M, r_S, a, and b. Solving the system gives the value of r_S, among others. Using this value and Eq. (9.68b) results in R_S from the V–I equation, under real working conditions:

$$R_S = r_S\frac{V_{OC}}{I_{SC}} \tag{9.79}$$

With that, all three parameters are determined in different operating conditions than STC.

An example of using the three-parameter model is given in Sect. 9.4.3.

9.3.4 Translation Equations

Unlike previous models, the following set of equations will allow direct translation of the operating state determined by temperature and irradiance into another state, with different temperature and irradiance. These equations are given by Anderson (1996).

In principle, equations were determined to meet the following conditions:

- Equations must accurately translate short-circuit current I_{SC} and open-circuit voltage V_{OC}.
- Equations should translate simple and accurate the V–I characteristic of devices. The MPP translation has to be performed without involving the translation of values I_{SC}, V_{OC}.

In the following the deduction of a translation set of equations is sketched, based on the Anderson (1996) approach. The subscript indices 1 and 2 designate quantities that characterize the solar cell in state 1 and 2, respectively.

Approximating derivatives with finite variations, Eqs. (9.7) and (9.9) becomes:

$$\alpha_V = \frac{V_{OC2} - V_{OC1}}{T_2 - T_1}, \; \alpha_I = \frac{I_{SC2} - I_{SC1}}{T_2 - T_1} \tag{9.80a, b}$$

from where the values V_{OC2} and I_{SC2} are obtained:

$$V_{OC2} = V_{OC1} + \alpha_V(T_2 - T_1) \tag{9.81}$$

$$I_{SC2} = I_{SC1} + \alpha_I(T_2 - T_1) \tag{9.82}$$

The coefficients α_V and α_I by the definitions Eqs. (9.7) and (9.9) have dimensions of V/°C and A/°C, respectively. They can be normalized as in Eqs. (9.8) and (9.10), respectively (with dimension 1/°C). Approximating derivatives with finite variations Eqs. (9.8) and (9.10) becomes:

$$\alpha'_V = \frac{V_{OC2} - V_{OC1}}{V_{OC2}(T_2 - T_1)}, \; \alpha'_I = \frac{I_{SC2} - I_{SC1}}{I_{SC2}(T_2 - T_1)} \tag{9.83a, b}$$

Equation (9.83a, b) lead to the relations:

$$I_{SC1} = I_{SC2}\left[1 + \alpha'_I(T_1 - T_2)\right] \tag{9.84a}$$

$$V_{OC1} = V_{OC2}\left[1 + \alpha'_V(T_1 - T_2)\right] \tag{9.84b}$$

Equation (9.84a, b) correlate short-circuit current and open-circuit voltage in the two states, 1 and 2, differentiated by cell temperature. If the solar irradiance values are also different in the two states, Eq. (9.84a, b) have to be adapted accordingly. Given the direct proportionality relationship between irradiance and photocurrent I_L, and considering that $I_{SC} \approx I_L$, Eq. (9.84a) transforms into:

$$I_{SC1} = I_{SC2}\left[1 + \alpha'_I(T_1 - T_2)\right]\frac{G_1}{G_2} \tag{9.85}$$

A correction with solar irradiance can be introduced for V_{OC}:

$$V_{OC1} = V_{OC2}\left[1 + \alpha'_V(T_1 - T_2)\right]\left[1 + \delta \ln\frac{G_1}{G_2}\right] \tag{9.86}$$

The logarithmic term is justified in the general equation of a solar cell Eq. (9.62) where at $V = V_{OC}$ we have $I = 0$ and an exponential relation between photocurrent and open-circuit voltage, $I_L \cong I_0 \exp(eV_{OC}/k_B T)$. As the photocurrent is directly proportional to solar irradiance, choosing a logarithmic correction term is justified. Equation (9.86) introduces an additional factorδ, which usually is not listed in the data sheet of solar modules and has to be determined.

Equations (9.85) and (9.86) translate only the extreme points of the characteristic from a functioning state of the cell to another. The next step is to extend the procedure to translate *any* point on the V–I characteristic from a state to another. In Reference Anderson (1996) a very simple set of equations is proposed. Each point on the characteristic in state 1 is translated into state 2 by multiplying the current with the ratio of short-circuit currents and the voltage with the ratio of open-circuit voltages:

$$I_2 = I_1\frac{I_{SC2}}{I_{SC1}},\ V_2 = V_1\frac{V_{OC2}}{V_{OC1}} \tag{9.87a, b}$$

Because each of the products I_1V_1 and I_2V_2 represents a point on the power curve in the given state, if we multiply Eq. (9.87a) and (b) we obtain a very simple equation for the translation of power (including the MPP):

$$P_2 = P_1\frac{I_{SC2}}{I_{SC1}}\frac{V_{OC2}}{V_{OC1}} \tag{9.88}$$

Substituting Eqs. (9.85) and (9.86) in Eq. (9.88) gives a translation equation for the power having as input parameters the cell temperature and the solar irradiance:

$$P_2 = \frac{P_1G_2/G_1}{[1 + \alpha'_I(T_1 - T_2)][1 + \alpha'_V(T_1 - T_2)][1 + \delta \ln(G_1/G_2)]} \tag{9.89}$$

If $T_1 = T_2$, Eq. (9.89) becomes:

$$P_2 = \frac{P_1G_2/G_1}{1 + \delta \ln(G_1/G_2)} \tag{9.90}$$

Equation (9.90) serves for a new interpretation of the coefficient δ and the term $\delta \ln(G_1/G_2)$. From Eq. (9.90) one observes that if $G_1 \approx G_2$, $\ln(G_1/G_2) \to 0$ and $P_2 \approx P_1G_2/G_1$. This means that at small changes in solar irradiance, regardless of its value, the power is in good approximation proportional to the irradiance. At higher irradiance changes the logarithmic term is different from zero,

$\ln(G_1/G_2) \neq 0$. Thus, the coefficient δ appears as a measure of deviation from the linear relation between power and irradiance.

The deduction of Eq. (9.90) is assuming $T_1 = T_2$. The effect of temperature on power can be quantified considering a new temperature coefficient γ defined as (Anderson 1996):

$$\gamma = \frac{P_{M2} - P_{M1}}{P_{M2}(T_2 - T_1)} \tag{9.91}$$

Definition Eq. (9.91) together with Eq. (9.90) lead to the final equation for translating maximum power:

$$P_{M2} = \frac{P_{M1} G_2/G_1}{[1 + \gamma(T_1 - T_2)][1 + \delta \ln(G_1/G_2)]} \tag{9.92}$$

Knowing the translation equation for maximum power Eq. (9.92), the fill factor could also be translated using the definition Eq. (9.2):

$$\mathrm{FF}_2 = \mathrm{FF}_1 \frac{P_{m2}}{P_{m1}} \frac{I_{SC1} V_{OC1}}{I_{SC2} V_{OC2}} \tag{9.93}$$

Knowledge of the coefficients α_I, α_V, γ and δ in relation with Eqs. (9.85–9.87a, b, 9.92) allows translating the solar cell operating point from a state 1 characterized by environmental parameters solar irradiance G_1 and temperature T_1 to a state 2 characterized by G_2 and T_2. These equations are experimentally validated (Anderson 1996).

To translate the operating point of a module from one state to another, one can directly use relationships linking the cell operating point to that of the module. For a module consisting of n_p parallel rows each containing n_s cells connected in series, Eq. (9.6a, b) makes this connection. If the translation of data is from STC to outdoor operation, in the above relations state 1 identifies with STC ($I_1 \to I_{1,\mathrm{STC}}$, $V_1 \to V_{1,\mathrm{STC}}$, $T_1 = 25$ °C and $G_1 = 1000$ W/m^2) and state 2 with the current state of operation ($I_2 \to I$, $V_2 \to V$, $T_2 \to T$ and $G_2 \to G$). In this case, coefficients α_V and α_I defined in Eqs. (9.7) and (9.9) coincide with the standard coefficients listed in the manufacturer's data sheet.

Therefore, the set of equations for finding the quantities describing the module operating point under environmental conditions is as follows:

$$I_{SC}^m = \frac{I_{SC,STC}^m}{1 + \alpha_I'(25 - T)} \frac{G}{1000} \tag{9.94a}$$

$$V_{OC}^m = \frac{V_{OC,STC}^m}{[1 + \alpha_V'(25 - T)] \cdot [1 + \delta \ln(1000/G)]} \tag{9.94b}$$

$$V^m = V_{STC}^m \left(\frac{V_{OC}^m}{V_{OC,STC}^m} \right), I^m = I_{STC}^m \left(\frac{I_{SC}^m}{I_{SC,STC}^m} \right) \tag{9.94c, d}$$

Table 9.2 The coefficients δ and γ used in Eq. (9.94) for different PV silicon technologies (Anderson 1996)

Coefficient	Monocrystalline	Multicrystalline	Thin film
δ	0.085	0.110	0.063
γ [1/°C]	−0.0033	−0.0047	−0.0020

$$P^m = I^m V^m \tag{9.94e}$$

$$P_M^m = \frac{P_{M,\mathrm{STC}}^m G/1000}{[1+\gamma(25-T)][1+\delta\ln(1000/G)]} \tag{9.94f}$$

Equation (9.94c, d) correlate the operating point of the module (V^m, I^m) in outdoor conditions with the operating point in standard condition (V_{STC}^m, I_{STC}^m) and the other way around. Practical importance of this set of equations is related to determining the coefficients δ and γ experimentally for each type of module of interest. If the user does not have the necessary experimental conditions (does not have a solar simulator), he can use for the coefficients δ and γ the values listed in Table 9.2.

Given that the subject of this book is about forecasting the PV plant output power, the translation equations Eqs. (9.85–9.87a, b, 9.92) are of real interest. Effectively, the operating point of the module is measured (V_i^m, I_i^m) at the time i under meteorological conditions G_i and T_i. Then, the irradiance G_{i+1} and temperature T_{i+1} are forecasted for the time instant t_{i+1}. The values ($V_{i+1}^m I_{i+1}^m$) are determined using the following set of equations derived from Eqs. (9.85–9.87a, b, 9.92):

$$V_{i+1}^m = V_i^m \frac{1}{[1+\alpha_V'(T_i - T_{i+1})][1+\delta\ln(G_i/G_{i+1})]} \tag{9.95}$$

$$I_{i+1}^m = I_i^m \frac{1}{1+\alpha_I'(T_i - T_{i+1})} \frac{G_{i+1}}{G_i} \tag{9.96}$$

Equations (9.95, 9.96) above are a compromise between simplicity, accuracy and convenience. At time $t+1$ a new measurement of the operating point (V_{i+1}, I_{i+1}) and environmental conditions (G_{i+1}, T_{i+1}) is made and the algorithm is repeated. In principle a single measurement could be made of an operating point (outdoor or STC), to be used as reference for any translation. Because accurate translation of the operating point is better at small variations of environmental parameters, at least for nowcasting it is preferable to perform the measurement in situ.

9.3.5 PV Shading

The mismatch losses of a PV system can be defined as the difference between the summed maximum power per each array module and the maximum power of the entire plant (Picault et al. 2010). Mainly two causes determine PV module mismatch: (1) dispersion of electrical properties due to manufacture's tolerance and degradation in time (either way there are no perfect identical modules) and (2) nonuniform illumination. In this section only the second issue is addressed.

PV plants can be subject to partial shading occasioned by passing clouds. In solar tracking plants, shadows of one tracker may appear over modules at early morning or late evening. Poles and power lines may bring shadows across a part of the plant during the day. The consequence of partially shading is a substantial decrease of the PV array output power. The power loss is worse than proportional to the shaded area; it increases nonlinearly (Nguyen and Lehman 2008). In recent years, the impact of partial shadowing on the energy yield of PV systems and the solutions to overcome the power losses have been widely studied (Quaschning and Hanitsch 1996; Woyte et al. 2003; Kaushika and Gautam 2003; Karatepe et al. 2007; Nguyen and Lehman 2008; Picault et al. 2010).

As shown in Fig. 9.4, in the simplest representation, a PV module contains n_p arrays connected in parallel, each containing n_s cells connected in series. Shadowing a single solar cell in a series array leads to a reverse bias of this. The photocurrent generated by the shadowed cell will be smaller than that of the illuminated cells. Because of the serial connection, the current flowing in the string is strictly the same; its value will be limited to the current generated by the shadowed cell. On the other hand, the high photocurrent generated by the cell not shadowed will force the shadowed cell to work in the reverse bias diode regime, dissipating the power generated by the other cells. The situation when a single cell from the series is shadowed is the worst. Depending on the number of series connected cells, the reverse bias of the shadowed unit or group may increase as much to reach breakdown voltage and cause irreversible damage. The breakdown voltage V_B marks the backward voltage of a diode at which, instead of the extremely low saturation current, an exponential current surge arises (avalanche breakdown).

In order to protect the shadowed solar cells and reduce the power loss it is customary to insert bypass diodes (Fig. 9.7). For a partly shaded module, it would be ideal to have one diode for each cell. Contrary, if partial shadow is not expected it would be desirable to omit bypass diodes. In practice, one bypass diode is applied per 18 cells in series (Woyte et al. 2003). Thus, a standard crystalline PV module with 36 series connected cells is regularly equipped with two bypass diodes.

Considerable effort has also been made in simulating the electrical behavior of shadowed PV arrays. The PV module V–I characteristic is generically illustrated in Fig. 9.8. Essentially, modeling of a partial shadowed PV system starts with the model of the basic element, the solar cell discussed in detail in Sect. 9.3. For

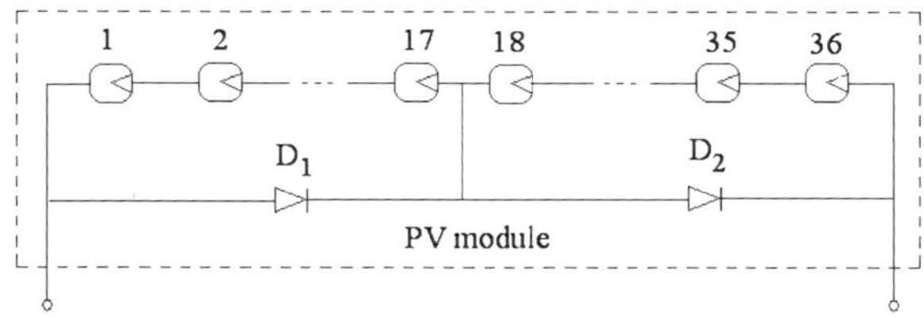

Fig. 9.7 Schematics of a standard PV module equipped with 36 cells and 2 bypass diodes

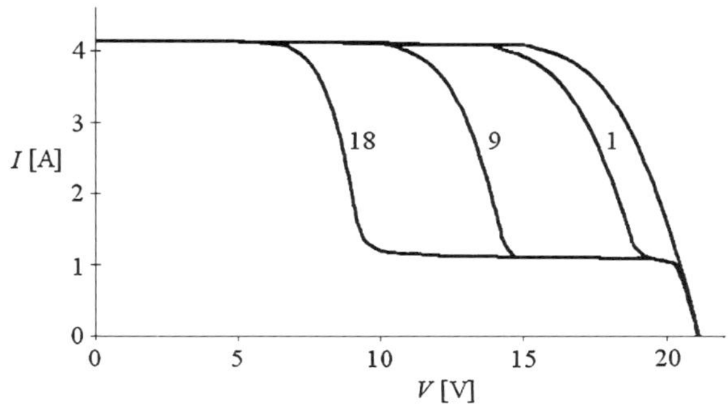

Fig. 9.8 Generic V–I characteristic of a PV module at STC with 36 crystalline soar cells when a number of cells are shaded. The curves are plotted assuming bypass diodes connected to 1, 9 and 18 cells

example, Quaschning and Hanitsch (1996) describe a model for V–I characteristics of PV generator with shaded solar cells. This is based on the two diode model of the cell (Fig. 9.2). An additional term $-a(V + \mathrm{IR}_s)\left(1 - \frac{V+\mathrm{IR}_s}{V_B}\right)^{-m}$ is included in the right side of Eq. (9.4) in order to model the avalanche breakdown at negative voltage. V_B stands for the breakdown voltage. This model is used for describing solar cells in interconnection with bypass diodes, cables and other elements constituting the PV module. The Kirchhoff laws are applied to provide an equation system relating all currents and voltages in the network. The authors argue by comparing simulation results with measured data that the model is able to trace accurately the specificity of module V–I curves with shaded cells (Quaschning and Hanitsch 1996).

A remarkable study on the effects of nonuniform solar irradiation distribution on the energy output for different interconnected configurations in PV arrays is reported in Karatepe et al. (2007). The model can take into consideration the effects of bypass diodes and the variation of the equivalent circuit parameters with respect to operating conditions. A distinctive feature is that all model parameters are estimated by using ANN. The authors claim that the model can provide a similar degree of precision like a solar cell-based analysis in assessing large-scale PV arrays.

Based on the results of Picault et al. (2010), an example of forecasting PV power output taking into account mismatch losses is given in Chap. 10.

9.4 PV Modules Operating in Outdoor Conditions

The aim of this chapter is to calculate the PV system efficiency in operation conditions, in order to convert the forecasted solar energy flux into forecasted values of output power. The change of PV module efficiency with changing weather has a major impact on the overall system efficiency. First, seasonal change of the PV modules efficiency due to solar irradiance and air temperature variation during a year is discussed. Then, the methods to calculate the V–I characteristic of a PV module following equations from Sect. 9.3 are illustrated. The examples are based on measurements made on the Solar Platform of the West University of Timisoara, Romania.

9.4.1 Experimental Setup

The Solar Platform (SRMS 2012) is operating at the West University of Timisoara since November 2008. The platform geographic coordinates are: 45°44′49.57"N, 21°13′50.32"E and 87 m altitude. The Solar Platform includes:

- Solar Radiation Monitoring Station equipped with seven DeltaOHM First Class pyranometers (DeltaOhm 2012) in accordance with ISO 9060 standard, measuring global, diffuse, reflected and total (on four directions) solar irradiance and a standard meteorologic station measuring temperature, air pressure, relative humidity, and wind speed;
- Three experimental setups for testing PV systems. The first stand is for testing PV modules, with fixed orientation, directly connected an active load, emulating grid connection. The active load can be replaced by a fixed resistive load to get a device suitable for verifying forecasting models of the energy provided by a PV module operating outdoor. The second setup monitors a stand-alone PV system. The third setup also monitors a stand-alone PV system but with the modules mounted on a sun-tracker system in polar axis configuration. Voltage, currents, and modules temperature are measured.

All sensors are integrated into a National Instruments PXI acquisition data system NI (2011). Measurements of all parameters (electrical, meteorologic, and actinometric) are simultaneously performed at time intervals of 15 s, 24 h/7 day. Hourly solar irradiation on all channels along with hourly mean of meteorologic parameters is available online at http://solar.physics.uvt.ro/srms.

To illustrate the above methods an experimental bench is used, in which the module is directly connected to a resistive load $R_L = 4.15\ \Omega$ at 25 °C. The measuring circuit is represented in Fig. 9.9. The schematics include both electrical and ambient parameters measuring instruments. Figure 9.10 shows a picture of the PV system experimental setup.

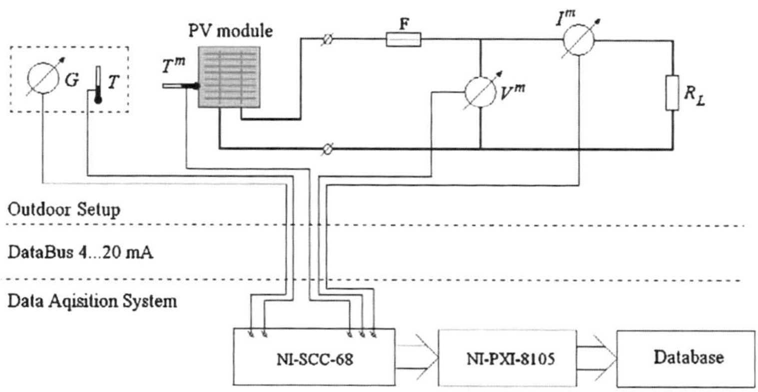

Fig. 9.9 Schematics of the experimental bench for testing solar modules in outdoor operating conditions. *G* and *T* are a pyranometer measuring solar irradiance on the module surface normal and a standard ventilated and shadowed thermal sensor measuring the environmental temperature. The sensor T_m monitors the temperature on the backside of the module

In Fig. 9.10a, two silicon crystalline PV modules are mounted on the testing bench. In the next two sections, data collected from the module FVG 90M (right side in Fig. 9.10a) are used for running examples of outdoor PV systems operating. The main catalog data are presented in Table 9.3.

9.4.2 Variation of Modules Efficiency

This section illustrates the change of a PV module efficiency during 1 year as a result of temperature and irradiance variations. The simulation is done for the monthly mean of hourly values under the following hypotheses:

(1) Optical and electrical parameters of the PV module FVG 90M given in Table 9.3 are assumed. The module is mounted horizontally.
(2) Monthly means of daily solar irradiation given in Table 9.4 are used at input. These values, taken from Paulescu et al. (2010), have been measured in Timisoara.
(3) Also, monthly means of daily air temperature given in Table 9.4 are used as input. The source of data is NASA Surface Meteorology and Solar Energy (SMSE 2012).

Note that we chose to initiate the simulation starting from monthly averages of daily solar irradiation and temperature. The task of calculating the conversion efficiency for an arbitrary set of operating conditions is thus extended by inserting a preliminary stage at which the mean monthly values of the two quantities of interest, solar irradiation and air temperature, have been broken down into hourly tranches.

Fig. 9.10 Experimental setup for PV systems testing on the Solar Platform at West University of Timisoara. **a** Bench for outdoor testing PV modules efficiency. **b** PV module mounted on a solar sun-tracker in polar configuration

Table 9.3 The main characteristics of the FVG 90M module (FVG 2010)

$V^m_{\mathrm{OC,STC}}$	22.3 V
$I^m_{\mathrm{SC,STC}}$	5.37 A
$V^m_{M,\mathrm{STC}}$	18.5 V
$I^m_{M,\mathrm{STC}}$	4.86 A
$\alpha'_V \equiv \frac{1}{V_{\mathrm{OC}}} \frac{\mathrm{d}V_{\mathrm{OC}}}{\mathrm{d}T}\Big\vert_{T=T_{\mathrm{STC}}}$	-0.0034 °C^{-1}
$\alpha'_I \equiv \frac{1}{I_{\mathrm{SC}}} \frac{\mathrm{d}I_{\mathrm{SC}}}{\mathrm{d}T}\Big\vert_{T=T_{\mathrm{STC}}}$	0.0005 °C^{-1}
NOCT	45 ± 2 °C
Module surface, A^m	1.17×0.51 m $= 0.5967$ m^2

We selected this presentation of the calculation, because monthly mean of daily solar irradiation and daily air temperature (mean, minimum, and maximum) are available elsewhere or can be easily estimated. The algorithm is general in that it can be applied in any location and for any PV module with known catalog specifications. The procedure quickly calculates the variation of conversion efficiency of a PV module during a year in specific location conditions. This information is necessary not only in forecasting the output power of a PV system but also for sizing of either grid-connected or stand-alone PV systems.

As stated above, the simulation is carried out in two steps. At the first step the mean monthly solar irradiation and temperature are broken down into hourly tranches. In the second step, the average module converting efficiency is calculated.

Monthly collectable energy values were distributed in hourly tranches according to the following algorithm (Lorenzo 2003). In terms of extraterrestrial

Table 9.4 Monthly mean of estimated daily solar irradiation H and of measured air temperatures mean T_m, minimum $T_{\min}$ and maximum $T_{\max}$ in Timisoara

Month	Jan	Feb	Mar	Apr	May	Jun	Jul	Aug	Sep	Oct	Nov	Dec
H [kWh/m^2]	0.92	1.63	2.72	5.01	5.43	5.35	6.13	5.07	3.82	2.01	1.19	0.64
T_m [°C]	−0.46	0.57	5.01	10.9	16.5	19.9	22.4	22.5	17.7	12.2	5.54	0.66
$T_{\min}$ [°C]	−2.8	−2.2	1.48	6.36	11.20	14.0	16.5	17.4	13.4	8.7	2.9	−1.6
$T_{\max}$ [°C]	1.9	3.6	8.7	15.4	21.5	24.8	27.3	27.3	22.0	16.0	8.4	2.9

solar radiation, the ratio between solar irradiance G_{ext} and daily irradiation H_{ext} can be determined using Eqs. (5.2), (5.12) and (5.13) with $C = S/2\pi$ as:

$$\frac{G_{0,\text{ext}}}{H_{0,\text{ext}}} = \frac{\pi}{S} \frac{\cos\omega - \cos\omega_s}{\omega_s \cos\omega_s - \sin\omega_s} \tag{9.97}$$

where ω_s is the hour angle at sunrise (in radians) and S is the daylight (in hours).

By examining meteorologic data from different stations it was observed that the measured ratio of diffuse components $r_d = G_d/H_d$ is approximately equal with the theoretical expression Eq. (9.97), if one considers long time mean values (Liu and Jordan 1960). At the same time, the correspondence between the measured ratio of global components $r_g = G/H$ and the theoretical expression Eq. (9.97) is acceptable, although imperfect. To fit experimental data in the second case, a correction is needed. Thus, the relationship Eq. (9.97) computed at extraterrestrial level is applied at ground as follows (Collares-Pereira and Rabl 1979):

$$r_d = \frac{G_d}{H_d} = \frac{G_{\text{ext}}}{H_{\text{ext}}} \tag{9.98a}$$

and

$$r_g = \frac{G}{H} = \frac{G_{\text{ext}}}{H_{\text{ext}}}(a + b\cos\omega) \tag{9.98b}$$

where the parameters a and b have the following empirical expressions: $a = 0.409 + 0.5016\sin(\omega_S + \pi/3)$ and $b = 0.6609 + 0.4767\sin(\omega_S + \pi/3)$. The ratios r_d and r_g can be applied to calculate solar radiation for short periods centered on the instantaneous value of the hourly angle considered.

The above procedure has been applied to monthly mean values of daily global solar irradiation listed in Table 9.4. The hour angle in Eq. (9.97) and (9.98b) was computed in the following Julian days: 17, 46, 75, 105, 135, 161, 198, 231, 261, 289, 319, and 345, each belonging to a month from January to December. The results, monthly mean values of hourly solar irradiation, are graphical presented in Fig. 9.11. The graph is the result of interpolation the resulted 15 × 12 values, 15 hourly values for each of the 12 month of a year. It is useful to note that solar irradiation during 1 h (in Wh/m^2) can be considered numerically equal to the

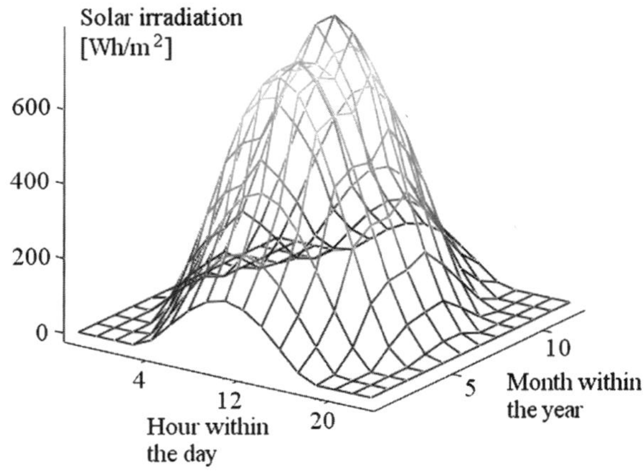

Fig. 9.11 Monthly average of hourly solar irradiation

average solar irradiance (in W/m^2) during the same hour. Thus, the values in Fig. 9.11 can be also identified with hourly mean solar irradiance in W/m^2.

The next step is the decomposition of ambient temperature into hourly samples. In order to do this the following equation has been used:

$$t_a(j,\omega) = \begin{cases} t_1(j,\omega) & if \quad -180^\circ \leq \omega < \omega_0 \\ t_2(j,\omega) & if \quad \omega_0 \leq \omega \leq 30^\circ \\ t_3(j,\omega) & if \quad 30^\circ < \omega \leq 180^\circ \end{cases} \tag{9.99}$$

where t_a is the environmental temperature in the Julian day j at the time measured by the hour angle ω. The equations $t_i(j,\omega)$, $i = 1,2,3$ are:

$$t_1(j,\omega) = t_{\max}(j-1) - \frac{t_{\max}(j-1) - t_{\min}(j)}{2}[1 + \cos(a_1\omega + b_1)] \tag{9.100a}$$

$$t_2(j,\omega) = t_{\min}(j) + \frac{t_{\max}(j) - t_{\min}(j+1)}{2}[1 + \cos(a_2\omega + b_2)] \tag{9.100b}$$

$$t_3(j,\omega) = t_{\max}(j) - \frac{t_{\max}(j) - t_{\min}(j+1)}{2}[1 + \cos(a_3\omega + b_3)] \tag{9.100c}$$

where $a_1 = -180/(\omega_s + 330)$, $b_1 = a\omega_s$, $a_2 = 180/(\omega_s + 30)$, $b_2 = -30a_2$ and $a_3 = 180/(\omega_s + 330)$, $b_3 = -(30a_3 + 180)$. The model is using the observation that the temperature evolves in a similar manner to global radiation with a delay of about 2 h: the minimum temperature $t_{\min}$ is at sunrise ($\omega = \omega_s$), the maximum $t_{\max}$ 2 h after midday ($\omega = 30°$). Between these two moments, the ambient temperature follows two cosine half-cycles: one from dawn until 14:00 (solar time) and the other between 14:00 and the sunrise of the next day.

To apply these equations, in addition to the current day maximum and minimum temperatures it is necessary to know the previous day's maximum $t_{\max}(j-1)$ and minimum temperature $t_{\min}(j+1)$. As the calculation of temperature is for an

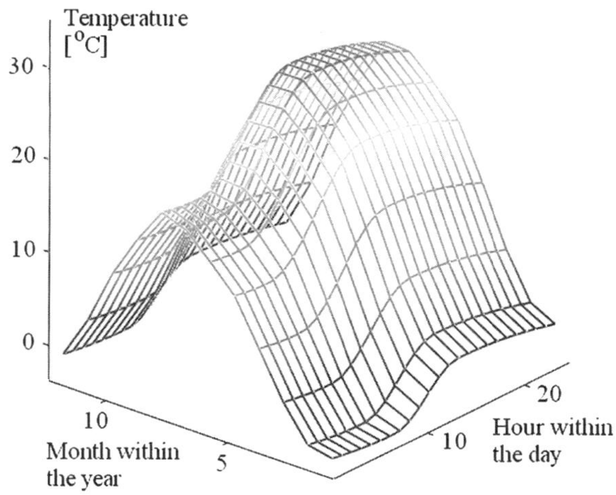

Fig. 9.12 Monthly average of hourly air temperature

average day of the respective month, it was assumed that they are equal to those of the day in question. The matrix with monthly averages of air temperature calculated for every hour in Timisoara resulting from applying the procedure to the values from Table 9.4 is represented in Fig. 9.12.

The two procedures presented above are preliminary, for fixing the weather conditions (monthly average of hourly solar irradiation and air temperature) in location. In order to calculate monthly average of module efficiency in each hour of the day, the V–I characteristic of PV module has been translated from STC (datasheet) to the actual meteorologic conditions using the three-parameter model (Sect. 9.3.3). The module efficiency has been calculated with Eq. (9.3) in which specific values of maximum power P_M and solar irradiance G as calculated above, are assumed.

Figure 9.13 displays the FVG 90M PV module conversion efficiency during a year in Timisoara. At first sight, a continue variation of the conversion efficiency can be seen, both over a year from day to day and throughout a day from hour to hour. In winter days and in spring and autumn mornings, efficiency exceeds 15 %, the calculated value for STC. In summer days the efficiency falls below STC, reaching the minimum at midday, down even under 13 %. This calculation demonstrates the importance of translation of V–I characteristics of a PV module from STC in real operating conditions in both stages of development of a photovoltaic project; sizing the system and forecasting the power output.

9.4.3 Numerical Examples

This section illustrates five- and three-parameter models functioning. Electrical characteristics of FVG 90M measured at STC given in Table 9.3 have been assumed as entry in models. Data measured on the Solar Platform of the West

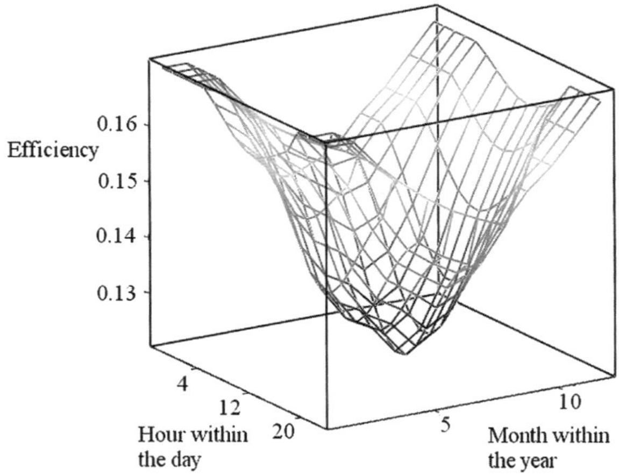

Fig. 9.13 Conversion efficiency of the FVG 90M PV module during a year, in specific meteorological condition of Timisoara. The plotted surface is the result of interpolating the monthly mean of hourly values

University of Timisoara in two days of 2010, July 20 and September 23, are used for comparison. Figure 9.14 displays the total solar irradiance measured on South direction on a surface tilted at 45° and the sunshine number (see Sect. 9.3.2 for definition) for these two days. It can be seen that the day July 20 has been characterized by a substantially unstable radiative regime (sunshine stability number $\zeta = 0.0113$) while the day September 23 was a fully stable sunny day ($\zeta = 0.0018$), excepting two episodes of several minutes after noon.

The five-parameter model runs in outdoor conditions as follows. Parameters are first calculated at STC: (1) The photocurrent $I_{L,\mathrm{STC}}$ with Eq. (9.19); (2) Diode ideality factor m_{STC} with Eq. (9.31) where $\mathrm{V}_{\mathrm{OC,STC}}$ is given by Eq. (9.20); (3) Diode saturation current $I_{0,\mathrm{STC}}$ with Eq. (9.22); (4) Serial resistance $R_{S,\mathrm{STC}}$ with Eq. (9.26) and (5) Parallel resistance $R_{S,\mathrm{STC}}$ with Eq. (9.32). Further on, the serial and parallel resistance may be subject of an iterative calculation in order to reach the desired precision. In the particular case of this example after three steps the difference between two consecutive values has become less than 1 %. In outdoor operating first the cells temperature is calculated with Eq. (9.11). Three of the V–I parameters calculated at STC are preserved: m, C_0, and R_S. Optionally, R_p may be subject of a new refinement using Eq. (9.37). The most important adjustment is for the photocurrent given by Eq. (9.33).

An illustration of the way in which solar irradiance and air temperature levels shape the module V–I curve is given in Fig. 9.15. The graphs resemble the theoretical features discussed in Sect. 9.2. Notable is the strong influence of the environmental temperature on the maximum output power (Fig. 9.16). A variation of air temperature with 40 °C, from −10 to +30 °C leads to a variation of solar cells temperature with 50 °C, from +20 to +70 °C. In the same irradiance conditions, this causes a decreasing of the module output power with 14.2 W representing 12.2 %, from 89.6 to 75.4 W.

Figure 9.17 assesses the module output power calculated with the five-parameter model against data measured in the first half of the two test days.

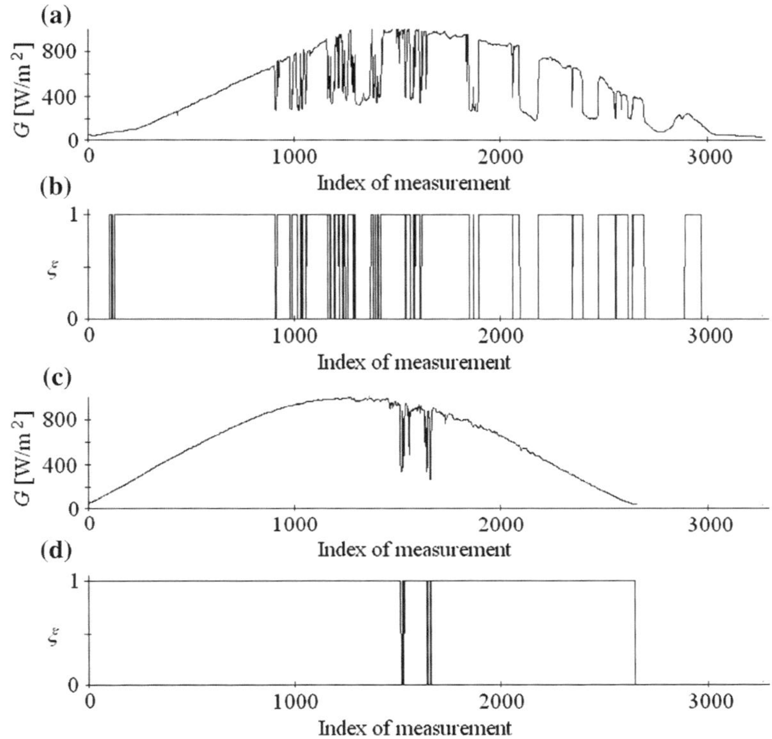

Fig. 9.14 Global solar irradiance G and sunshine number ξ in the days July 20 (**a**, **b**) and September 23 (**c**, **d**) measured in Timisoara

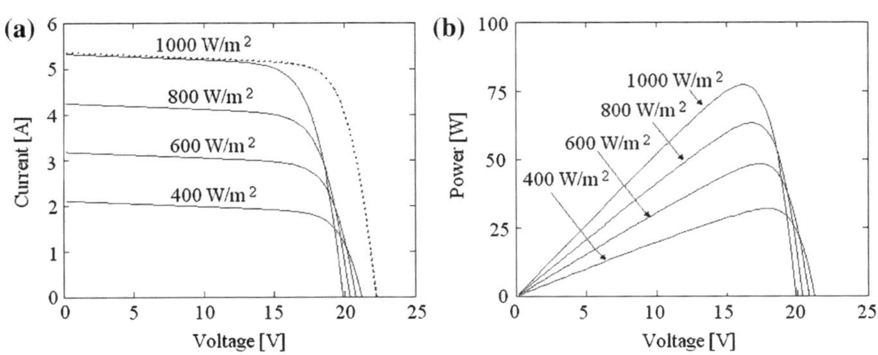

Fig. 9.15 **a** V–I curves of the PV module operating outdoor calculated with Eq. (9.17). The dotted curve corresponds to the ideal PV module ($R_s = 0$, $R_p = \infty$, m = 1). **b** Output power. The curve parameter is the incoming solar irradiance. Cells temperature is $T = T_{STC} = 25$ °C

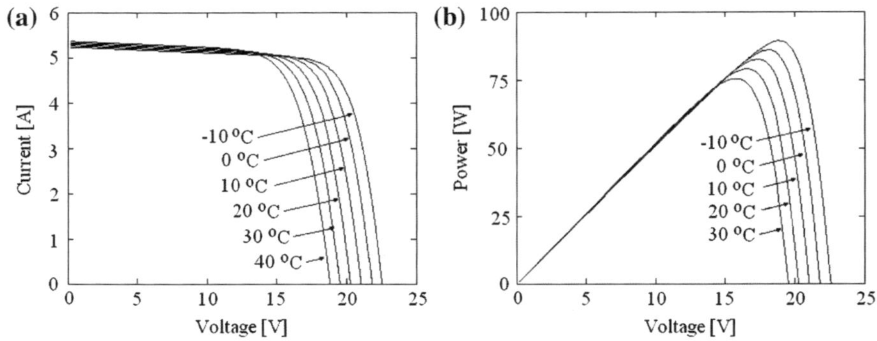

Fig. 9.16 **a** V–I curves of the PV module operating outdoor calculated with Eq. (9.17). **b** Output power. The curve parameter is the cell temperature. Incoming solar irradiance is $G = G_{STC} = 1000\ \mathrm{W/m^2}$

A unitary cleanness coefficient has been assumed for day July 20 since in the morning of this day the module has been cleaned. The event is visible in Fig. 9.17a, where close to the index measurement 400 the module output power was down for a short time. For the day September 23, the cleanness coefficient has been assumed 0.98. Visual inspection shows a good agreement between the measured and calculated. The model traces the measurements even in the periods of radiative regime instability. An important conclusion can be drawn for this example: if the model is proper calibrated, its accuracy in estimating the output power is high enough not to negatively influence the overall forecasting accuracy.

The radiative solar regime may be a source of errors in estimating the output power of a PV system. To illustrate this, the estimation results for the first period of instability in September 23 are detailed in Fig. 9.18.

Let us look at the first moment when the sun was obstructed by clouds. It is visible that significant errors occur. But these errors are not due to the PV model. The cause is the different response times of the pyranometer and the PV device. The pyranometer has a response time of tens of seconds while the PV module response is almost instantaneous. Solar irradiance measured by the pyranometer is used as entry in the PV output power model causing this delay in response. These type of errors occur always only in transitory regime. Their influence on the forecasting accuracy in days with stable radiative regimes is negligible.

Figure 9.18a displays the results obtained with the five-parameter model while Fig. 9.18b displays the results obtained with the three-parameter model. These graphs show no notable difference between these two methods, confirming that even simple models can offer satisfactory results in calculating PV modules outdoor operation.

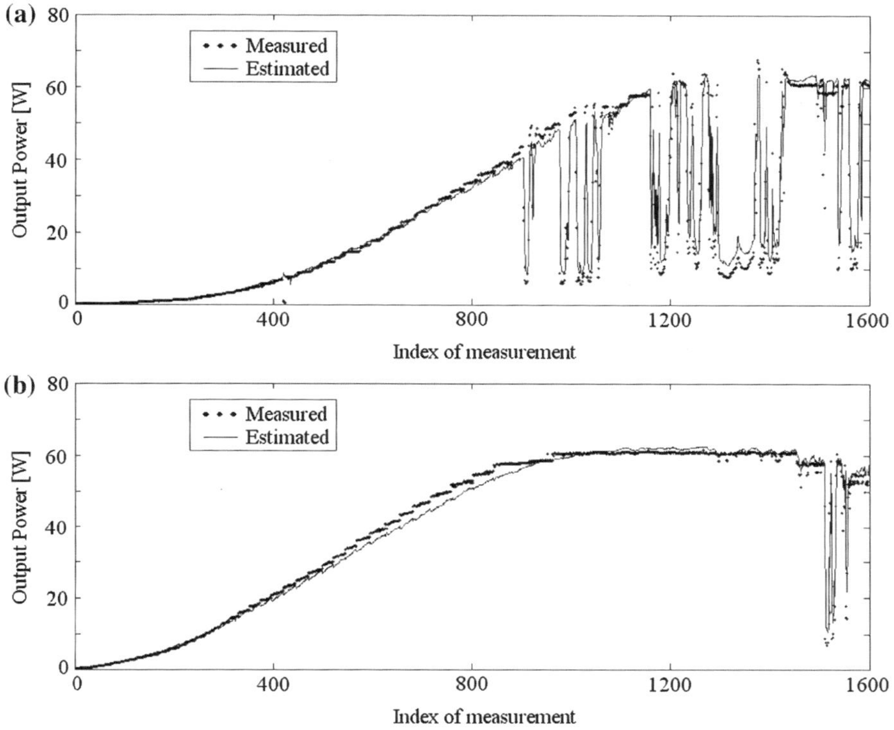

Fig. 9.17 Measured and calculated module output power in the first part of the days **a** July 20 and **b** September 23

9.5 Inverters

The inverter converts direct current produced by a PV system into alternating current with the frequency and voltage of the electrical grid.

In the simplest configuration, an inverter works as follows: the current generated by a PV system is switched during a half period on a transformer line and during the second half on the other line. This results in a rectangular alternating voltage. The rectangular waveform harmonic content is so high that it may disturb or destroy electronic devices. To obtain a small AC voltage harmonic content, the direct current is switched at a high frequency; the pulse width is varied so as to obtain a smooth sine wave. Inverter operating by this principle is known as pulse width modulation inverter and is one of the most used nowadays. The electronics of inverter is far-off from the subject of this book. Interested readers may consult specific books, e.g., Teodorescu et al. (2011), Rashid (2011). In this section, we discuss only the inverter performance as component of a PV system.

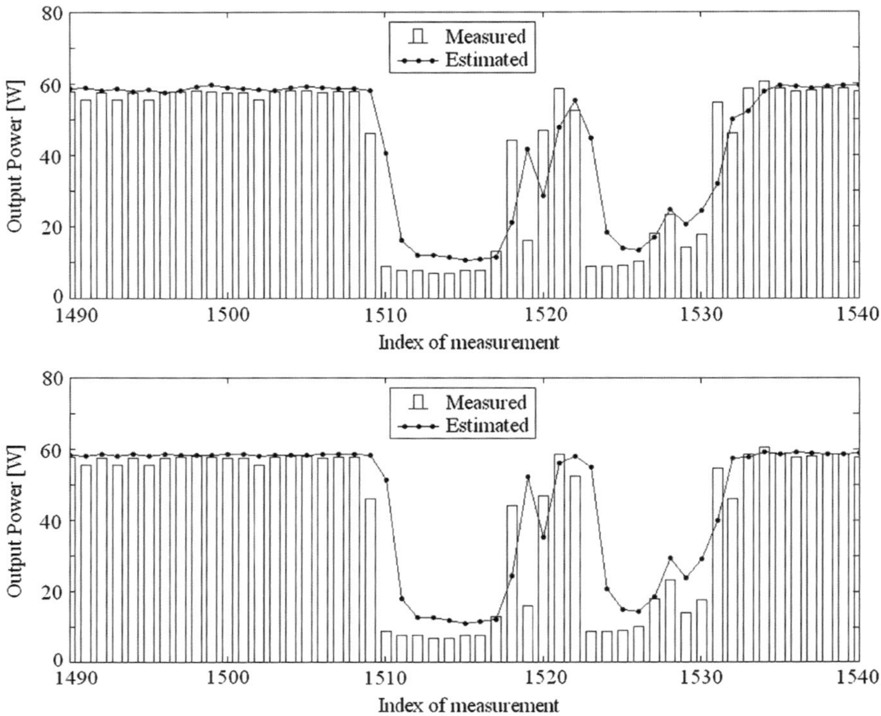

Fig. 9.18 50 samples of measured and calculated of output power in September 23. The following models are used **a** five-parameter and **b** Three-parameter

9.5.1 Inverter Parameters

The parameters that characterize an inverter, usually specified by manufacturers in the data sheet, are summarized below.

At input:

- Maximum DC power, which represents the absolute allowable maximum power under any operating conditions.
- Maximum DC voltage, which represents the absolute allowable maximum voltage under any operating conditions. When sizing a PV array, it should be compared to highest voltage that occur at high irradiance (typical $G = 1000$ W/m^2) and low temperature (e.g. -15 °C for Eastern Europe).
- Maximum DC current, which represents the absolute maximal current admissible at the input of the inverter. When sizing a PV array, it is assumed to be equal to the short-circuit current at STC.
- Power threshold, i.e., the minimum input power needed to operate. It is admitted to be the internal inverter operating consumption.

- Nominal voltage at MPP. When sizing a PV array this is an indication for the optimal number of modules connected in series.
- MPP voltage range, which define the voltage window in which the inverter is able to search for the MPP.

At output:

- Nominal grid frequency, usually 50 or 60 Hz for the US zone. In operation, the output frequency of the inverter follows the grid frequency at any time.
- Nominal grid voltage. During operation, the output voltage of the inverter follows the grid voltage at any time.
- Nominal AC power, which represents the power which the inverter can feed continuously.
- Nominal AC current, defined as the current under nominal power and nominal grid voltage
- Maximum AC power
- Maximum AC current

In addition to these parameters, another important parameter to be considered is the distortion factor (or total harmonic distortion—THD), describing the quality of the alternative voltage produced. The distortion factor is defined by the ratio of the sum of the actual AC voltage V_i harmonics and the same sum of harmonics plus the actual value V_1 of the fundamental oscillation:

$$\mathrm{DF} = \frac{\sqrt{\sum_{i=2}^{\infty} V_i^2}}{V_{\mathrm{tot}}} \tag{9.101}$$

Another quality parameter of an inverter is the output power factor Q, which is defined as the ratio of active power and apparent power:

$$Q = \frac{P_{\mathrm{active}}}{P_{\mathrm{apparent}}} \tag{9.102}$$

For proper operation this ratio should be close to 1.

9.5.2 Inverter Efficiency

The most important characteristic of an inverter is the efficiency η. It is defined as the ratio of AC power P_{AC} output to DC power P_{DC} input:

$$\eta = \frac{P_{\mathrm{AC}}}{P_{\mathrm{DC}}} = 1 - \frac{P_L}{P_{\mathrm{DC}}} \tag{9.103a, b}$$

In the above equation, AC output power is substituted with the difference between the DC power input P_{DC} and the power loss P_L. A very simple method to determine the dependence of inverter efficiency as function of input DC power is presented in Eicker (2003). This is briefly summarized next.

The efficiency is affected at low input power by internal consumption of the inverter and at high power mainly due to resistive losses. In Eq. (9.103a, b) P_L includes three independent power losses in the conversion process: (1) Inverter power electronic circuits supply, P_i; (2) The losses of the power semiconductor switching devices, which are linearly dependent on the inverter output power vP_{AC}; (3) Ohmic losses, increasing with the square of the AC power, rP_{AC}^2. Thus, with good approximation P_L can be represented by a second-order polynomial:

$$p_l = \frac{P_L}{P_N} = \frac{P_i}{P_N} + v\frac{P_{\mathrm{AC}}}{P_N} + r\left(\frac{P_{\mathrm{AC}}}{P_N}\right)^2 = p_i + vp_{\mathrm{ac}} + rp_{\mathrm{ac}}^2 \tag{9.104}$$

In order to obtain dimensionless expressions of the coefficients, all measured powers are divided by the nominal power P_N of the inverter.

From Eqs. (9.103b) and (9.104) it follows:

$$\eta = \frac{(P_{\mathrm{DC}} - P_L)/P_N}{P_{\mathrm{DC}}/P_N} = \frac{p_{\mathrm{dc}} - \left(p_i + \mathrm{v}p_{\mathrm{ac}} + \mathrm{r}p_{\mathrm{ac}}^2\right)}{p_{\mathrm{dc}}} \tag{9.105}$$

In Eq. (9.105), both input and output power appear. Our goal is to calculate the efficiency of the inverter as a function of the input power supplied by the PV generator. Replacing p_{ac} with ηp_{dc}, we obtain an equation in η:

$$\eta = 1 - \frac{p_i}{p_{\mathrm{dc}}} - v\eta - r\eta^2 p_{\mathrm{dc}} \tag{9.106}$$

having the solution:

$$\eta = -\frac{1+v}{2\mathrm{r}p_{\mathrm{dc}}} \pm \sqrt{\frac{(1+v)^2}{(2\mathrm{r}p_{\mathrm{dc}})^2} + \frac{p_{\mathrm{dc}} - p_i}{\mathrm{r}p_{\mathrm{dc}}^2}} \tag{9.107}$$

Only the positive solution of Eq. (9.107) leads to an expression of physical significance. The three loss coefficients p_i, v, and r can be calculated from three values of efficiency η_1, η_2 and η_3 provided by most manufacturers in the data sheet (commonly at $p_1 = 0.1$, $p_2 = 0.5$ and $p_3 = 1$. After solving the algebraic system of three equations with three unknowns is obtained:

$$p_i = \frac{p_1p_2p_3\left(\eta_1^2 p_1(\eta_2 - \eta_3) + \eta_1\left(\eta_3^2 p_3 - \eta_2^2 p_2\right) + \eta_2\eta_3(\eta_2 p_2 - \eta_3 p_3)\right)}{\eta_1^2 p_1^2 - \eta_1 p_1(\eta_2 p_2 + \eta_3 p_3) + \eta_2\eta_3 p_2 p_3(\eta_2 p_2 - \eta_3 p_3)} \tag{9.108a}$$

$$\begin{aligned} v = {} & \frac{\eta_1^2 p_1^2(\eta_2 p_2 - \eta_3 p_3 - p_2 - p_3) + \eta_1 p_1\left(\eta_3^2 p_3^2 - \eta_2^2 p_2^2\right)}{(\eta_1 p_1 - \eta_2 p_2)(\eta_1 p_1 - \eta_3 p_3)(\eta_3 p_3 - \eta_2 p_2)} \\ & + \frac{\eta_2^2 p_2^2(\eta_3 p_3 + p_1 - p_3) - \eta_2\eta_3^2 p_2 p_3^2 + \eta_3^2 p_3^2(p_2 - p_1)}{(\eta_1 p_1 - \eta_2 p_2)(\eta_1 p_1 - \eta_3 p_3)(\eta_3 p_3 - \eta_2 p_2)} \end{aligned} \tag{9.108b}$$

$$r = \frac{\eta_1 p_1(p_2 - p_3) + \eta_2 p_2(p_3 - p_1) + \eta_3 p_3(p_1 - p_2)}{\eta_1^2 p_1^2 - \eta_1 p_1(\eta_2 p_2 + \eta_3 p_3) + \eta_2 \eta_3 p_2 p_3(\eta_2 p_2 - \eta_3 p_3)} \tag{9.108c}$$

Figure 9.19 illustrates the variation of efficiency given by Eq. (9.107) with solutions (9.108a, b, c) for different generic inverters defined by the three points $p_1 = 0.1$, $p_2 = 0.5$, and $p_3 = 1$. It can be seen that at small input power ($p_{dc} < 0.1$) the inverter efficiency is small but increase sharply with p_{dc}. It means that at small solar irradiance the inverter efficiency may substantially affect the PV system overall efficiency. At $p_{dc} > 0.1$ the inverter efficiency increases gradually to a maximum and then decreases also gradually due to Ohmic losses.

In order to standardize comparison between inverters the so-called *Euro efficiency* has been introduced which mediates weighted efficiencies according to the average solar radiation collected in Central Europe:

$$\begin{aligned} \eta = {} & 0.03\eta_{p_{dc}=5\,\%} + 0.06\eta_{p_{dc}=10\,\%} \\ & + 0.13\eta_{p_{dc}=20\,\%} + 0.1\eta_{p_{dc}=30\,\%} + 0.48\eta_{p_{dc}=50\,\%} + 0.2\eta_{p_{dc}=100\,\%} \end{aligned} \tag{9.109}$$

Today, the maximum efficiency and Euro efficiency of inverters reach very high values, beyond 98 % (Siemens 2009).

9.5.3 Inverter Sizing

As we have seen in Fig. 9.19, the inverter efficiency is strongly dependent on the input power which, in turn, depends on the actual value of solar irradiance. In principle Eq. (9.107) with parameters p_i, v, and r given by Eq. (9.108a, b, c) is enough to forecast the output power of a PV system.

On the other hand, an inverter must be sized according to the local solar radiative conditions. For example, in Romania, at 45°N latitude, solar irradiance in the summer at noon is usually ranges between 700 and 800 W/m^2. Designing the inverter in STC may lead to oversizing. Choosing the proper inverter for a PV system should take into account the inverter efficiency over a typical meteorologic year.

To determine the annual inverter efficiency, we must know the temporal distribution of energy production of the PV modules, i.e., the frequency of occurrence for the various p_{dc} values. Partial inverter efficiencies for different input powers p_{dc} multiplied by the weights of p_{dc} will determine the annual inverter efficiency.

The following simple procedure can be employed when sizing an inverter required for grid connection of a PV system (Eicker 2003).

(1) The frequency distribution of average hourly solar irradiance is determined. First, hourly values are distributed into N classes, indexed $i = 1–N$. Each class has the same width ΔG. Absolute annual frequency is determined for each class in number of hours per year n_i, (number of hours in which the hourly

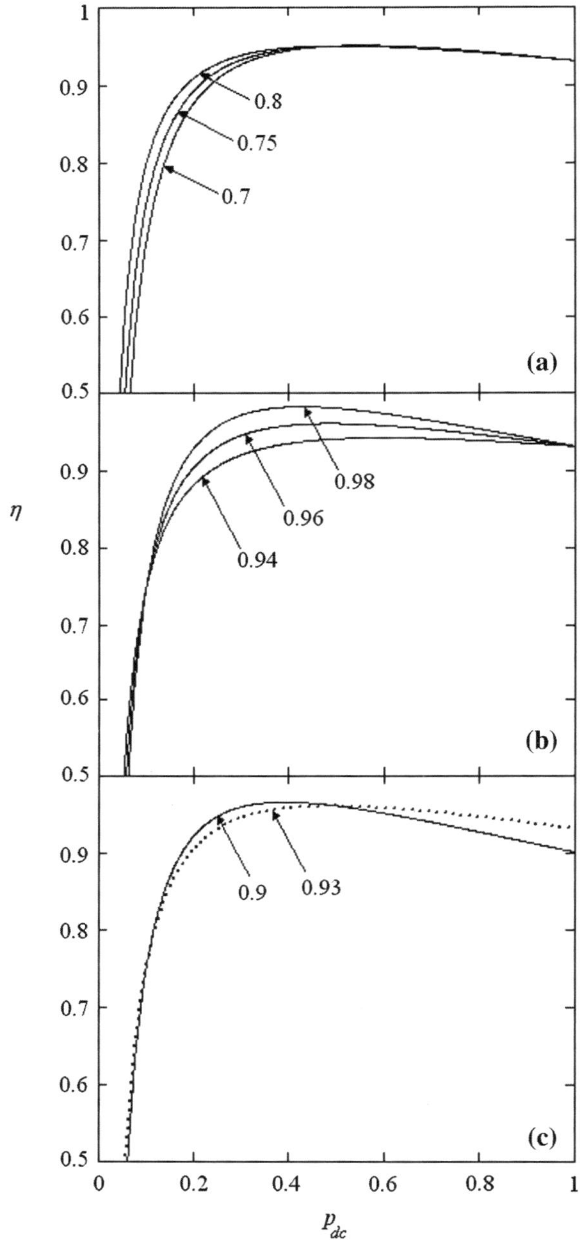

Fig. 9.19 Conversion efficiency of an inverter in respect to the input power normalized to the nominal power. The curves parameter is the inverter efficiency at **a** $p_{dc} = p_1 = 0.1$, **b** $p_{dc} = p_2 = 0.5$, and **c** $p_{dc} = p_3 = 1$

average irradiance falls into class i). Second, the frequency determined for each class is multiplied by the average class irradiance G_i, $i = 1 \ldots N$. Thus, for each interval ΔG_i one obtains the annual collectable energy per unit area,

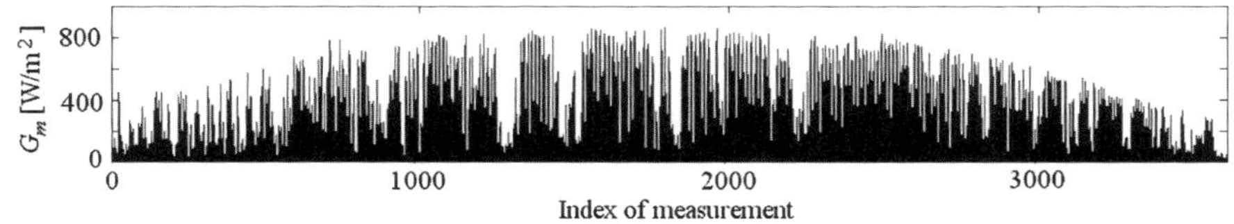

Fig. 9.20 Hourly average of global solar irradiance measured on a horizontal surface in Timisoara during 2010

usually expressed in Wh/m^2. In general, it is sufficient to divide irradiances in the range 0–1000 W/m^2 in 10 classes of 100 W/m^2 width.

(2) The energy collected in each interval $G_i{\cdot}n_i$ is converted into electricity$P_{\text{dc}} \cdot n_i$, based on PV module efficiency η_m, and modules surface A_m:

$$P_{\text{dc},i}n_i = \eta_m G_i n_i A_m \tag{9.110}$$

For each class of irradiance, the normalized input power is $p_{\text{dc},i} = P_{\text{dc},i}/P_n$ and the inverter efficiency $\eta(p_{\text{dc},i})$ is calculated using equation (9.107). P_n is the nominal power of the inverter. Therefore, AC power is expressed with the formula:

$$P_{\text{ac},i}n_i = P_{\text{dc},i}n_i\eta(p_{\text{dc},i}) \tag{9.111}$$

Average annual efficiency of the inverter is obtained by dividing the total AC energy to total DC energy:

$$\eta_y = \frac{\sum_{i=1}^{N} P_{\text{ac},i}n_i}{\sum_{i=1}^{N} P_{\text{dc},i}n_i} = \frac{\sum_{i=1}^{N} \eta_i P_{\text{dc},i}n_i}{\sum_{i=1}^{N} P_{\text{dc},i}n_i} \tag{9.112}$$

In the following, we present a numerical example of calculating the efficiency of an inverter with Eq. (9.112) by following the above algorithm. Illustration is done using hourly mean global solar irradiance measured on the Solar Platform of the West University of Timisoara during 2010. A total of 3,609 values were considered corresponding to integer hourly slots. In other words fractions of hourly measurements after sunrise and before sunset have been neglected. The data are represented in Fig. 9.20. The first value corresponds to the interval 9:00–10:00 on January 1, 2010 and the last time interval is 15:00–16:00 on December 31, 2010.

At first step, the values of hourly global solar irradiance were distributed in $N = 10$ classes of equal width $\Delta G = 100$ W/m^2, resulting in the histogram shown in Fig. 9.21 with a unimodal distribution of frequencies. The irradiance class having most hours (836) is the first ($G < 100$ W/m^2). In 1,551 h (about 43 %) the irradiance falls in the first two classes ($G < 200$ W/m^2). This means that a large part of operating time during the year, the inverter is forced to work at low input power. We denote by n_i the number of hours corresponding to each class.

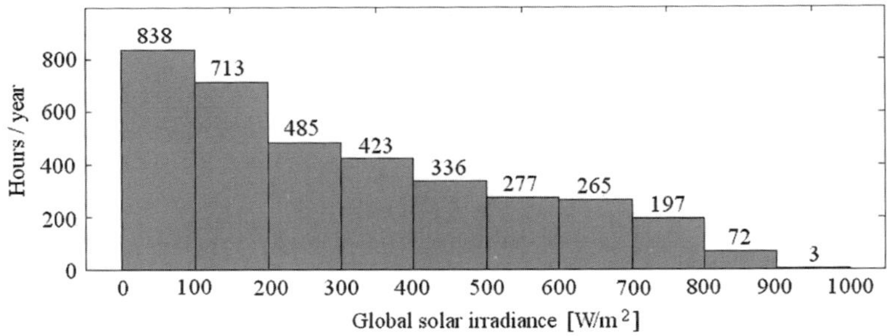

Fig. 9.21 Frequency distribution (hours/year) of monthly mean solar irradiance of data from Fig. 9.20

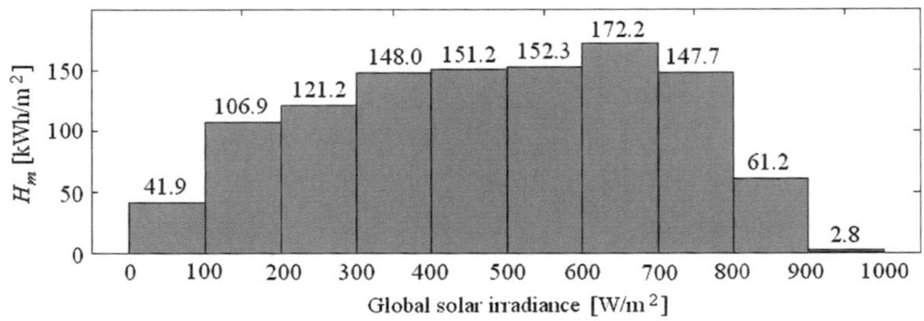

Fig. 9.22 Solar energy available on surface unit H_m for different classes of irradiance

The collectable solar energy for each class is obtained by multiplying the number of hours n_i with the average class irradiance G_i. The result, i.e., the solar energy per unit area in each class of irradiation is presented in Fig. 9.21. The distribution is also unimodal. The most energy is collected in Class 7 of irradiance, $600\ \mathrm{W/m^2} < G < 700\ \mathrm{W/m^2}$. Ideally, an inverter should work with maximum efficiency in this class of irradiance (Fig. 9.22).

In the second step, the power produced by photovoltaic modules shall be calculated. We assume a system equipped with $N_m = 100$ modules of type FVG 90 M as in Sect. 9.4.2. Taking into account Fig. 9.13 and considering that small irradiance values are met early morning and late evening and all day in winter while high values in the middle of summer days, in the first approximation it is simple and reasonable to assume that the conversion efficiency linearly depends on average hourly irradiance. Thus, we attribute the maximum conversion efficiency 16.6 % for the first class of irradiance ($i = 1$) and the minimum for the last ($i = 10$). The energy provided by photovoltaic modules $E_{\mathrm{PV},i}$ for every class is calculated with Eq. (9.110). DC power for each class of irradiation is obtained simply by dividing $E_{\mathrm{PV},I}$ to the number of hours at which the irradiance is in that class:

$$P_{\mathrm{cc},i} = E_{\mathrm{PV},i} \big/ n_i \tag{9.113}$$

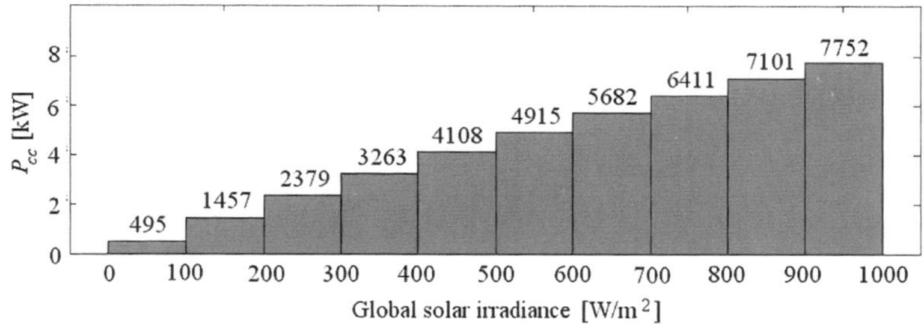

Fig. 9.23 Direct power at the inverter input for each class of irradiance

Equation (9.113) gives us the DC power at the inverter input for the 10 classes of irradiation. This is represented in Fig. 9.23. P_{dc} is virtually the global solar irradiance in the middle of each class multiplied by the conversion efficiency of the class, the area of a module and the total number of modules.

In step 3, we calculate the annual inverter efficiency η_y with Eq. (9.112). At this step, we choose the inverter. In the present case, we assume an inverter with nominal power $P_n = 8$ kW. We suppose that the inverter efficiency depending on the input power is known in three points: $\eta_1(p_1 = 0.1) = 0.75$, $\eta_2(p_2 = 0.5) = 0.98$ and $\eta_1(p_1 = 1) = 0.93$. Using these values in Eq. (9.108), one can calculate the inverter efficiency dependence of the normalized input power $p_{dc} = P_{dc}/P_n$ and further on the efficiency on every class of irradiance. Finally, applying Eq. (9.112) the annual inverter efficiency is calculated, in our case $\eta_y = 95.1\,\%$.

9.5.3.1 Maximum Power Point Tracking And Efficiency Optimization

Besides the function to produce alternative current, the inverter has to manage the PV modules arrays operating point: the effective input impedance is continuously adapted depending on solar irradiance and modules temperature so that the DC power to be at maximum. Setting the optimal operating point is achieved by varying the input voltage, so that the output power reaches a maximum value.

Many MPP tracking (MPPT) methods for PV systems are well represented in the literature. A detailed review of them is beyond the scope of this chapter so here is just an enumeration, as done in Esram et al. (2006). The most commonly known are hill-climbing, fractional open-circuit voltage control, perturb and observe (P & O), and incremental conductance (IncCond). There are lesser known methods such as maximizing load current or voltage, fractional short-circuit current control, array reconfiguration, linear current control, fuzzy control, neural network, dc link capacitor droop control, pilot cells, current sweep, limit-cycle control, and several others. In addition to this, Esram et al. (2006) describe the Ripple Correlation Control (RCC) method, which has distinctive advantages for professional inverters, such as: it converges asymptotically to the MPP; uses array current and voltage ripple, which must already be present if a switching converter is used, to determine gradient information; no artificial perturbation is required; achieves convergence at a rate limited by switching period and the controller gain; does not rely on assumptions or characterization of the array or an individual cell; can compensate for array capacitive effects; has several straightforward circuit implementations and has a well-developed theoretical basis.

9.6 Summary and Conclusion

All algorithms for forecasting output power of a PV system include the task of modeling the conversion of solar energy into electricity. The conversion efficiency depends on environmental parameters, mainly air temperature, but these parameters do not change suddenly. The operation of a solar cell can be described by analytical functions. In comparison with the accuracy (or lack thereof) of forecasting solar irradiance, the modeling of the solar converter is precise enough not to adversely affect the output power forecasting. However, there are specific conditions (e.g. partial shading of the module) that may radically change the prior statement. There are models to evaluate the generator performance in such conditions. In fact, the shadow is not an intrinsic property of the converter so the main problem is to forecast the shadow occurrence, which is an issue related to the state of the sky. Measures for the state of the sky are introduced in Chaps. 3 and 4 of this book.

The primary element of a solar generator is the PV module. The manufacturer provides a data sheet in STC, which are never met outdoor. In this chapter, four various models of different degree of complexity for the V–I curve of a PV module in operating condition have been presented. These models allow the determination of parameters for the V–I characteristic in environmental conditions specified by incoming solar irradiance, air temperature, and modules surface cleanness. The operating point is positioned on the curve by the load. A high-performance inverter usually contains a controller forcing the device to work at the MPP. Examples given in this chapter illustrate the relative impact of environmental parameters on the PV module output and assess the models quality. The change in commercial module efficiency during a year with monthly mean of hourly samples is an edifying example. This highlights the major difference between real versus STC. Also, it reveals the dynamics of outdoor module operation.

The inverter is also an element that influences the output power forecasting. Sizing of the inverter is an important task for achieving high efficiency. Low irradiance, i.e., low levels of input power, may dramatically alter the inverter performance. There are known models to evaluate the inverter efficiency; a very simple model was illustrated.

To conclude, it takes at least two energy transformations, accompanied by unavoidable losses, from solar irradiance to electric power fed into the grid. Modeling of electronic devices (modules, inverters) allows accurate estimation of energy losses. The major problem in forecasting PV output power remains weather forecasting, including solar irradiance, to specify the operating conditions of devices.

References

Anderson AJ (1996) Photovoltaic translation equations: a new approach. NREL Final Subcontract Report, January 1996. NREL/TP-411-20279. http://www.osti.gov/bridge/product.biblio.jsp?osti_id=177401

Arajuo GL, Sanchez E (1982) Analytical expressions for the determination of the maximum power point and the fill factor of a solar cell. Solar Cells 5:377–386

Celik AN, Acikgoz N (2007) Modelling and experimental verification of the operating current of mono-crystalline photovoltaic modules using four-and five-parameter models. Appl Energy 84:1–15

Chenni R, Makhlouf M, Kerbache T, Bouzid A (2007) A detailed modeling method for photovoltaic cells. Energy 32(3):1724–1730

Collares-Pereira M, Rabl A (1979) The average distribution of solar radiation-correlations between diffuse and hemispherical and between daily and hourly insolation values. Sol Energy 22:155–164

DeltaOHM (2012) Pyranometers, Albedometers, Net Irradiance Meter. Manual. http://www.Deltaohm.Com/ Ver2008/Uk/Pyra02.Htm

De Soto W, Klein SA, Beckman WA (2006) Improvement and validation of a model for photovoltaic array performance. Sol Energy 80:78–88

Eicker U (2003) Solar technologies for buildings. Wiley, chichester

Epperson JF (2007) An introduction to numerical methods and analysis, Wiley—Interscience, Hoboken

Esram T, Kimball JW, Krein PT, Chapman PL, Midya P (2006) Dynamic maximum power point tracking of photovoltaic arrays using ripple correlation control. IEEE Trans Power Electron 21:1282–1291

FVG (2010) FVG90M module datasheet. http://www.fvgenergy.com/

Green MA (1982) Solar cells: operating principles, technology, and system applications. Englewood Cliffs, NJ, Prentice-Hall

Hecht E (1997) Optics, 4th edn. Addison Wesley, San-Francisco

Hovinen A (1994) Fitting of the solar cell IV-curve to the two diode model. Phys Scr T54:175–176

IEC (2005) Crystalline silicon terrestrial photovoltaic (PV) modules—Design qualification and type approval. IEC 61215. http://webstore.iec.ch/preview/info_iec61215%7Bed2.0%7Den_d.pdf

IEC (2008) Thin-film terrestrial photovoltaic (PV) modules—Design qualification and type approval. IEC 61646. http://webstore.iec.ch/preview/info_iec61646%7Bed2.0%7Db.pdf

Karatepe E, Boztepe M, Colak M (2007) Development of a suitable model for characterizing photovoltaic arrays with shaded solar cells. Sol Energy 81:977–992

Kaushika ND, Gautam NK (2003) Energy yield simulations of interconnected solar PV arrays. IEEE Trans Energy Convers 18:127–134

Liu B, Jordan R (1960) The interrelationship and characteristic distribution of direct, diffuse and total solar radiation. Sol Energy 4:1–19

Lo Brano V, Orioli A, Ciulla G, Di Gangi A (2010) An improved five-parameter model for photovoltaic module. Sol Energy Mater Sol Cells 94:1358–1370

Lorenzo E (2003) Energy collected and delivered by PV modules. In: Luque A, Hegedus S (eds) Handbook of photovoltaic science and engineering. Wiley, Chichester, p 950

Marion B (2008) Comparison of predictive models for photovoltaic module performance. In Proceedings of 33th IEEE Photovoltaic Specialists Conference, San-Diego, California, May 11–16, 2008

Martin N, Ruiz JM (2001) Calculation of the PV modules angular losses under field conditions by means of an analytical model. Sol Energy Mater Sol Cells 70(1):25–38

Nelson J (2003) The physics of solar cells. Imperial College Press, London

NI (2011) Data acquisition with PXI and PXI express, National Instruments Corporation. http://www.ni.com/
Nguyen D, Lehman B (2008) An adaptive solar photovoltaic array using model-based reconfiguration algorithm. IEEE Trans Ind Electron 55(7):2644–2654
Ortiz-Conde O, Garcia Sanchez FJ, Muci J (2006) New method to extract the model parameters of solar cells from the explicit analytic solutions of their illuminated I–V characteristics. Sol Energy Mater Sol Cells 90:352–361
Paulescu M, Dughir C, Tulcan-Paulescu E, Lascu M, Gravila P, Jurca T (2010) Solar radiation modeling and measurements in Timisoara, Romania: Data and model quality. Environ Eng Manag J 9:1089–1095
Picault D, Raison B, Bacha S, De La Casa J, Aguilera J (2010) Forecasting photovoltaic array power production subject to mismatch losses. Sol Energy 84:1301–1309
Polverini D, Tzamalis G, Müllejans H (2012) A validation study of photovoltaic module series resistance determination under various operating conditions according to IEC 60891. Prog Photovoltaics. doi:10.1002/pip.1200
Quaschning V, Hanitsch R (1996) Numerical simulation of current–voltage characteristics of photovoltaic systems with shaded solar cells. Sol Energy 56:513–520
Rashid M (2011) Power electronics handbook: devices, circuits, and applications, 3rd edn. Elsevier, Amsterdam
Sah CT (1991) Fundamentals of solid-state electronics. World Scientific, Singapore, pp 430–436
Sandstrom JD (1967) A method for predicting solar cell current- voltage curve characteristics as a function of incident solar intensity and cell temperature. Jet Propulsion Laboratory Technical Report 32-1142 July 15, 1967. http://ntrs.nasa.gov/archive/nasa/casi.ntrs.nasa.gov/19670021539_1967021539.pdf
Siemens (2009) New transformerless solar inverters achieve 98 percent efficiency. Press release. http://www.siemens.com/press/en/pressrelease/?press=/en/pressrelease/2009/industry_automation/iia2009051934.htm
SMSE (2012) NASA surface meteorology and solar energy. http://eosweb.larc.nasa.gov/sse/
Sze SM, Kwok KNg (2007) Physics of semiconductors devices. Wiley-Interscience, Hoboken
SRMS (2012) Solar Platform of the West University of Timisoara, Romania. http://solar.physics.uvt.ro/srms
Teodorescu R, Liserre M, Rodríguez P (2011) Grid converters for photovoltaic and wind power systems. Wiley, London—IEEE
Woyte A, Nijs J, Belmans R (2003) Partial shadowing of photovoltaic arrays with different system configurations: literature review and field test results. Sol Energy 74:217–233
Wurfel P (2005) Physics of solar cells: From principles to new concepts. Wiley-VCH, Weinheim

Chapter 10
Forecasting the Power Output of PV Systems

This book started from the paradigm that increasing the renewable energy weight in the energy mix is feasible and desirable in the long term. There are short term limitations mainly generated by the inherent variability of the resource, calling for tools for forecasting the power output produced by solar and wind generators. These tools should operate as accurate as possible on various horizons of time, from several minutes to hours and up to one or two days ahead.

We have seen that the accuracy of forecasting the output power of a PV system is intimately related to the accuracy of forecasting of solar irradiance. More clearly, Bacher et al. (2009) state that forecasting PV system power output is basically the same problem of forecasting solar irradiance. Along this book a variety of approaches to forecast solar irradiance were commented and evaluated against data. Also, the weather effect on PV systems conversion efficiency was assessed.

In this chapter some models for forecasting PV system output power are discussed. This study is based on data reported in recent papers. The chapter is structured in two parts. The first deals with distinct forecasting group of methods, statistic and artificial intelligence, the accuracy of different models being compared. The second part deals with the smoothing effect on the solar power output variability when more spatially distributed solar systems are connected in the same power grid.

Before discussing the results of forecasting output power, it is useful to review some aspects related to the calculation of the solar irradiance and the energy provided by a solar generator in operating conditions. We have chosen to do this in the viewpoint of two recently published studies: Martin et al. (2010) which emphasized the importance of post-processing data in the procedures of forecasting solar irradiance and Almonacid et al. (2011) which compared results of different approaches for calculating the energy produced by a PV system and emphasized the importance of meteorological inputs for the estimation accuracy. The two models are briefly summarized next.

M. Paulescu et al., *Weather Modeling and Forecasting of PV Systems Operation*, Green Energy and Technology, DOI: 10.1007/978-1-4471-4649-0_10,

Martin et al. (2010) compared three different models based on time series applied to predict half day values of global solar irradiance with a temporal horizon of 3 days.

The dataset used for this consisted of ground solar irradiance measured at four stations of the Spanish National Weather Service: Murcia (S1), Albacete (S2), Madrid (S3), and Lerida (S4). Half day time series are constructed from hourly values of global solar irradiance accumulated from sunrise to solar noon and from noon until sunset for each day. Thus, for each day there are two values encapsulating the dynamic characteristic of global irradiance which are separated by the solar noon. Due to the fact that half day solar irradiance time series is non-stationary, it was necessary to translate it into a stationary one. Thus, the half daily global solar irradiance time series has been transformed into two new series: clearness index (see. Sect. 4.1.3) and lost component. Lost component is defined as the difference between solar irradiance values at extraterrestrial and ground levels. These series were used as input of the predictive models.

The models tested by Martin et al. (2010) are the following. The first is an autoregressive model of order p, AR(p), given by the following equation:

$$G_t = c + \sum_{i=1}^{p} a_i G_{t-i} + \varepsilon_i \tag{10.1}$$

where c is a constant, in this case equal to mean values of time series and ε_i is a white noise series. The second is an Artificial Neural Networks model (ANN), a category that was briefly described in Sect. 7.1.1 of this book. The third is an Adaptive Network-based Fuzzy Inference System (ANFIS) model. ANFIS defines a class of neural networks which are functionally equivalent to fuzzy logic inference systems (Jang 1993). The model incorporates a five-layer network to implement fuzzy rules of Takagi–Sugeno type (see Sect. 7.1.2.4 for an introduction to Takagi–Sugeno fuzzy approach).

In the testing stage the three models have been used to predict half day values of solar irradiance for the next 3 days, i.e., six values. For the best fitted models in each location at first time horizon RMSE was found as follows (Martin et al. 2010): between 20.65 and 26.54 % for AR, 20.58 to 26.37 % for ANN and 20.86 to 27.36 % for ANFIS, showing that the models perform roughly similar. For the last time horizon RMSE was found from 25.71 to 31.06 % for AR, from 24.68 to 30.39 % for ANN, and 25.69 to 30.42 % leading to the same conclusion of comparable performance.

This leads to a conclusion of significance: if models that are in principle very different (in this example autoregressive, neural network or fuzzy) are fine fitted to measured data then the accuracy of the predicted results is roughly the same.

Improvement of the models expressed in terms of RMSE:

$$\text{Improvement} = 100 \times \frac{\text{RMSE}_{\text{reference}} - \text{RMSE}}{\text{RMSE}_{\text{reference}}}\ [\%] \tag{10.2}$$

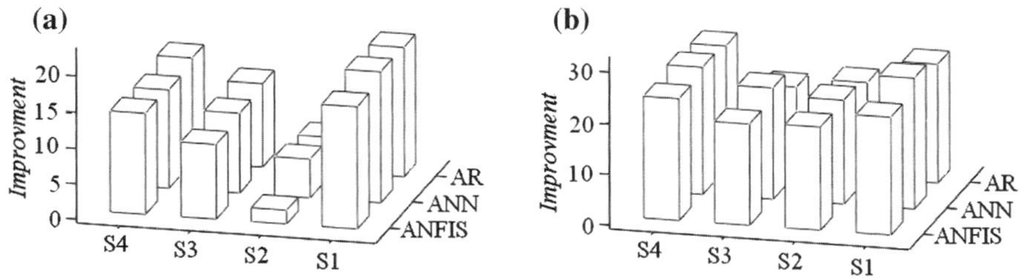

Fig. 10.1 Improvement of the best performance models against persistence in the Spanish locations Murcia (*S1*), Albacete (*S2*), Madrid (*S3*), and Lerida (*S4*) at forecast horizon (half day): **a** first and **b** six. Data from Martin et al. (2010) have been used to build the graphs

was further compared against the simplest predictor, namely persistence. Figure 10.1 shows the results for the first and last horizon times, respectively. This demonstrates that all models improve persistence prediction and the improvement increase with increasing the time horizon.

Another conclusion of Martin et al. (2010) is that the ANN and ANFIS models forecast with better accuracy half daily values of solar irradiance if lost component is used as input. Clearness index time series obtain better results in models of lower order compared to lost component.

These notable results show at least two things. First, selecting a model by its nature is not critical. The current performance of different models for forecasting solar irradiance, i.e., statistical, neural network, or fuzzy, is about the same, as will appear more evident further on in this chapter. Critical is the post processing of data in order to obtain stationary series. The unusual approach of *lost component* seems to be as good a solution as clearness index and in some cases even better.

The teaching that we can draw from here is that we cannot talk about a general recipe for irradiance forecasting based on past records. Characteristics of solar irradiance time series are determined by local weather conditions. This leads to the idea that post-processing data may be site specific and the choosing of the model cannot be dissociated from the task of post-processing data.

The Martin et al. (2010) study is related to the first stage of a procedure for forecasting the output power of a PV plant. Like said before, this stage is decisive in determining the overall accuracy of the forecasting procedure. Still, the accurate modeling of the PV system, of transforming solar energy into electricity, cannot be neglected. In Chap. 9, we showed that the conversion efficiency is influenced by weather. Energy losses in a year influenced by environmental conditions vary between 11 and 45 % (Almonacid et al. 2011), depending on the characteristics of the considered system. In Fig. 10.2 the losses are presented according to their nature. Ambient temperature, shading of modules, low irradiance, dirt, and dust are environmental factors that influence the magnitude of energy losses. It is intuitive that the transformation equations must include these parameters to achieve an acceptable forecasting accuracy of the energy provided by the PV generator. Almonacid et al. (2011) does an analysis in this respect that deserves a closer look.

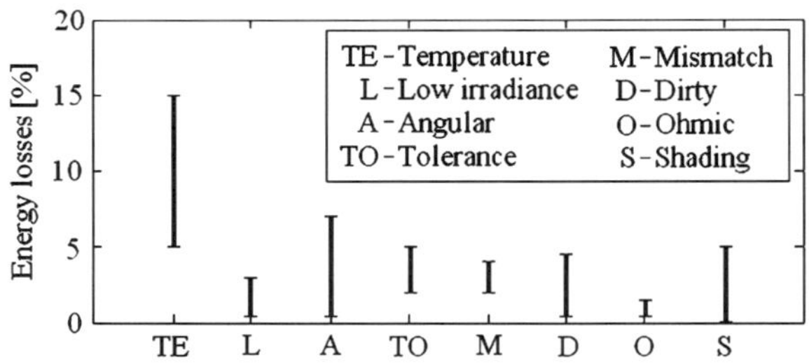

Fig. 10.2 Typical range of annual energy loses of a PV system. Data from Almonacid et al. (2011) have been used to build the graph

Almonacid et al. (2011) compare the results of four different methods for estimating the annual energy produced by a PV generator: three of them are classical methods and the fourth one is based on an artificial neural network. These models are briefly described next.

1. The Osterwald (1986) model which provides one of the simplest equation for the V—I characteristic of a PV cell:

$$P_m = P_{m,\mathrm{STC}} \frac{G\,[\mathrm{W}/m^2]}{1,000\mathrm{W}/m^2}[1 - \gamma(T_c - 25)] \tag{10.3}$$

where P_m and $P_{m,\mathrm{STC}}$ are the cell maximum power in operating conditions and at STC, respectively. γ is the cell maximum power temperature coefficient (see Eq. 9.91).
2. The Arajuo–Green model, which is briefly described in Sect. 9.3.3 of this book.
3. The Green model (Green 1982), which estimates the power provided by a PV system based on the one diode model of a solar cell (Eq. 9.64b):

$$I = N_p^m N_p^c I_{\mathrm{SC}}\left[1 - \exp\left(\frac{V/\left(N_s^m N_s^c\right) - V_{\mathrm{OC}} + IR_S/\left(N_p^m N_p^c\right)}{V_T}\right)\right] \tag{10.4}$$

where N_s^c is the number of cells of PV module connected in series in an array and N_p^c is the number of arrays connected in parallel. N_s^m is the number of modules connected in series in an array of the PV generator and $N_p^m$is the number of arrays connected in parallel. The maximum power is obtained by including Eq. (9.5) and solving the resulted two equations system.
4. An ANN model developed by Almonacid and colleagues. It includes three layers: the input layer with two neurons (global solar irradiance and cell temperature), a hidden layer with three neurons and an output layer. The result is the V–I curve of a PV module for a pair of determined values of solar irradiance and cell temperature. This ANN-based method, apart from the effect of solar irradiance and cells temperature, also takes into account two other second-order effects: low irradiance and angular losses.

The models have been applied to four different PV systems with different settings and types of modules: two identical with nominal power of 68 kWp each, a third

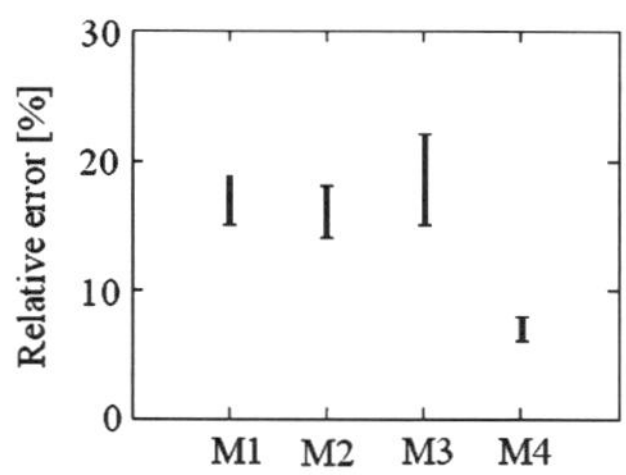

Fig. 10.3 Percentage error range at the output power calculation when the four models (*M1*—Osterwald, *M2*—Arajuo-Green, *M3*—Green and *M4*—ANN) have been applied to operational PV systems. Data from Almonacid et al. (2011) have been used to build the graph

with 20 kWp nominal power, and the fourth with 40 kWp nominal power. These systems are part of the Univer Project (Drif et al. 2007) consisting of the installation of a grid-connected PV system with a total power of 200 kWp, in Jaen University Campus. The relative errors of the above models applied to these systems are summarized in Fig. 10.3 using data reported in Almonacid et al. (2011).

In the study of Almonacid et al. (2011) the method based on artificial neural network provides better results than the simple equations, mainly due to the fact that this method also considers second-order effects, such as low irradiance, angular, and spectral effects. The magnitude of relative errors is another relevant point. This proves that very simple methods for calculating the output power may induce considerable errors (over 15 % in this case). The using of more complex models (ANN in this case) which take into account the environment influence in a more detailed manner may reduce significantly the errors in calculating the output power (three times in this case).

10.1 Forecasting the Output Power: Facts

10.1.1 Statistical-Based Models

Generally, the statistical approach for forecasting the output power runs in two stages: first a post-processing procedure (e.g., data are normalized with a clear sky or extraterrestrial model) is applied aiming to isolate the stochastic component (mainly related to the cloud cover amount) of the forecasted quantity (solar irradiance or PV output power); (2) the forecasting procedure itself (e.g., fitting an ARIMA model). In the following, we discuss the results of two inspiring studies (Bacher et al. 2009; Brabec et al. 2010) which are based on different premises and use different approach to point out remarkable conclusions.

Bascher et al. (2009) propose an approach for online forecasting hourly output power of a PV system up to 36 h ahead. The model is briefly summarized next.

Data used in Bascher's study are records of solar power from 21 PV systems located in a small village in Denmark. The data cover the entire year 2006.

The average values of solar power $p_{i,t}$ over 15 min observed for a PV system at time t are used to form a time series: $\{p_t, t = 1, \ldots, N\}$ where $p_t = (1/21)\sum_{i=1}^{21} p_{i,t}$.

The total number of samples in series is $N = 35040$. Forecasts of global solar irradiance have been provided by the Danish Meteorological Institute using the HIRLAM mesoscale NWP model. The NWPs of global solar irradiance are given in forecasts of average values for every third hour. The forecasts are updated at 00:00 and 12:00 each day. The ith update of the forecasts is the time series $\left\{\widehat{g}_{i,k}, k = 1, \ldots, 12\right\}$ which then covers the horizons up to 36 h ahead.

The global irradiance is estimated as function of cloud transmittance τ_C as:

$$G = G_0 \tau_C \tag{10.5}$$

where G_0 is global clear sky irradiance. Cloud transmittance is modeled as a stochastic process using ARIMA. A similar equation with (10.5) is applied for PV power output P:

$$P = P_{CS}\tau \tag{10.6}$$

where P_{CS} is the clear sky solar power. τ is referred as normalized solar power. This choice is motivated by the fact that τ is almost stationary series. P_{CS} is defined as $P_{\mathrm{CS}} = f_{\max}$ (Julian day; time within the day), which is estimated ($\hat{P}_{\mathrm{CS}}$) as a local maximum by the weighted quantile regression method. This is used to form the clear sky estimated power time series: $\left\{\hat{P}_t^{\mathrm{CS}}, t = 1, \ldots, N\right\}$. Finally, the normalized solar power is calculates as:

$$\tau_t = P_t / \hat{P}_t^{\mathrm{CS}} \tag{10.7}$$

and then to form the normalized solar power time series: $\{\tau_t, t = 1, \ldots, N\}$.

Bacher et al. (2009) tested three models which differ by inputs: (1) Autoregressive model (AR) which has only lagged past observations τ_t as input; (2) A model with only $\hat{\tau}_t^{\mathrm{nwp}}$ as input, referred to as Local Meteorological (LM_{nwp}); (3) An autoregressive with exogenous inputs (ARX), i.e., with both type of inputs.

The AR model is of order two and is expressed as:

$$\tau_{t+k} = m + a_1\tau_t + a_2\tau_{t-s(k)} + e_{t+k} \tag{10.8}$$

where $s(k) = 24 + k \bmod 24$ ensures that the latest observation of the diurnal component is included and $e_{t+k} = p_{t+k} - \hat{p}_{t+k|_t}$ is the k step prediction error.

The LM_{nwp} model is expressed as:

$$\tau_{t+k} = m + b_1\hat{\tau}_{t+k}^{\mathrm{nwp}} + e_{t+k} \tag{10.9}$$

The model using both lagged observations of τ_t and *NWPs* as input is an *ARX* model:

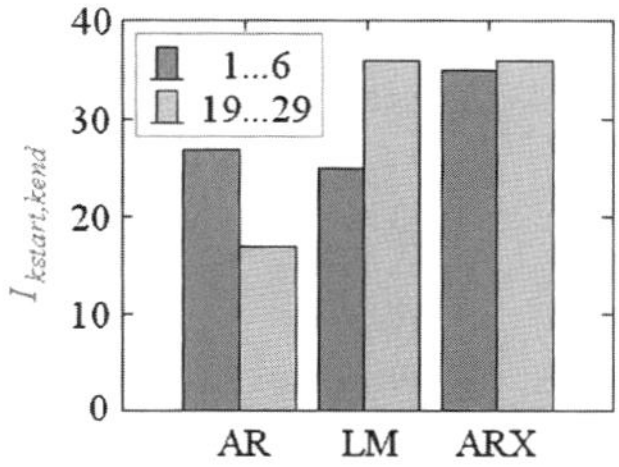

Fig. 10.4 Improvement indicator for short time horizons $k = 1,\ldots,6$ and next day horizon $k = 19\ldots29$. Data from Bacher et al. (2009) are used to build the graph

$$\tau_{t+k} = m + a_1\tau_t + a_2\tau_{t-s(k)} + b_1\hat{\tau}^{\text{nwp}}_{t+k|} + e_{t+k} \tag{10.10}$$

The summary error $\langle \text{RMSE}\rangle_{k_{\text{start}},k_{\text{end}}} = (k_{\text{end}} - k_{\text{start}} + 1)^{-1} \sum_{k=k_{\text{start}}}^{k_{\text{end}}} \text{RMSE}_{\text{k}}$ was used to assess the quality of a model over a range forecasting time horizon. The models performance was compared by means of the improvement indicator given by Eq. (10.2). The reference model used in Bacher's paper is the best performing naive predictor for a given horizon. This is the diurnal persistence $p_{t+k} = p_{t-s(k)} + e_{t+k}$, $k > 2$.

Bacher et al. (2009) reports calculations of the improvement indicator for the four models: $I_{1,6}$ for short-time horizon and $I_{19,29}$ for next day horizon. The barplot presented in Fig. 10.4 is based on these results. It shows that a RMSE improvement of around 35 % over the reference model can be achieved by using the ARX model. Another very important find is that for time horizons below 2 h the solar power is the main input, but for a next day time horizon it is adequate to use NWPs as input.

An outstanding study on energy generated by PV plants reported in Brabec et al. (2010) is summarized next. A sample of 97 PV systems from about 6,000 connected to the CEZ company grid (Czech Republic) in 2010 was used to build the statistical model. For the sample PV systems hourly measurements of produced electric energy have been used. The authors use NWP output solar irradiance as the primary driver of the prediction model. The NWP model runs with 9×9km grid resolution, covering most of Europe. Outputs from the model grid were interpolated in order to find solar irradiance in the geographic locations of each individual system.

The near linear model between solar irradiance G and standardized generated energy E (obtained by subtracting the minimum and dividing the result by the range) might not be the same for all PV systems. The regression coefficients can vary from system to system, motivating the linear mixed effects model. Thus, for the jth E measurement at ith PV system:

$$E_{ij} = (\beta + b_i)G_{ij} + \varepsilon_{ij} \tag{10.11}$$

where $b_i \sim N(0, \sigma_b^2)$ and $\varepsilon_{ij} \sim N(0, \sigma^2)$ are two independent, random, normally distributed terms. It splits the E variability in two easy interpretable components: one is connected with the within-system variability, while the other is connected with the between-systems variability.

The modeling of E using the approaches described above is feasible only for the in-sample PV systems. In order to generalize the model, i.e., to make it usable for all PV systems, including those that did not exist at the time of model training, the authors adopted a different approach. The new regression approach starts from the identity:

$$\mathbf{E}\left[E_{ij}\middle|i\right] = r_i\left(G_{ij}\right)C_i \tag{10.12}$$

where i indexes PV systems, j indexes time, and $\mathbf{E}[\cdot|i]$ is the conditional expected value operator. Installed output capacity C_i is a time-invariant attribute of a particular system. $r_i(G)$ is a function that for a given value of solar irradiance G, for system i, gives the expected ratio of E–C_i. The function is assumed to be smooth, and hence it can be estimated by a non-parametric regression for each system in the sample. For PV systems that are not in the sample, the function $r_i(G)$ cannot be estimated individually. Instead, the following form was proposed:

$$\mathbf{E}\left[E_{ij}\right] = r_C\left(G_{ij}\right)C_i \tag{10.13}$$

where $\mathbf{E}(\cdot)$ is an approximation of the conditional expected value operator $\mathbf{E}[\cdot|i]$ obtained by averaging over the all PV systems. The common function $r_C(G) = \mathbf{E}[r_i(G)]$ can be easily estimated from all sampled systems using non-parametric regression.

With these two models given by Eqs. (10.12) and (10.13) the evaluation for in-sample systems and out-sample systems can be done. Assuming normal (Gaussian) distribution of errors, the above regression models are written as:

$$E_{ij} = r_i\left(G_{ij}\right)C_i + \varepsilon_{ij} \tag{10.14}$$

$$E_{ij} = r_C\left(G_{ij}\right)C_i + \zeta_{ij} \tag{10.15}$$

where $\varepsilon_{ij} \sim N(0, \sigma^2)$, $\zeta_{ij} \sim N(0, \eta^2)$ are the normal distributed errors. These assumptions over errors sometimes may lead to unphysical values of E, due to the inherent positive definition of E and hence of functions r. In order to avoid this problem, the authors modeled the variable r, which takes values in the range $[0, 1]$, using an inflated beta distribution (Brabec et al. 2010; Brabec et al. 2011).

For assessing the effect of the NWP on the quality of standardized energy prediction, the model described above was tested in Brabec et al. (2010). The solar irradiance G used at input was derived either from NWP or from local climatology. The authors concluded that forecasts of E given by Eqs. (10.14) and (10.15) compared with the measurements indicate that using NWP input radically improves the quality of E estimates in comparison to local climatology input. The forecasting accuracy of the two variants is compared in Fig. 10.5 in terms of mean absolute errors MAE. Brabec et al. (2010) also tested the beta regression model with solar irradiance derived from NWP at input. For comparison MAE is also included in Fig. 10.5.

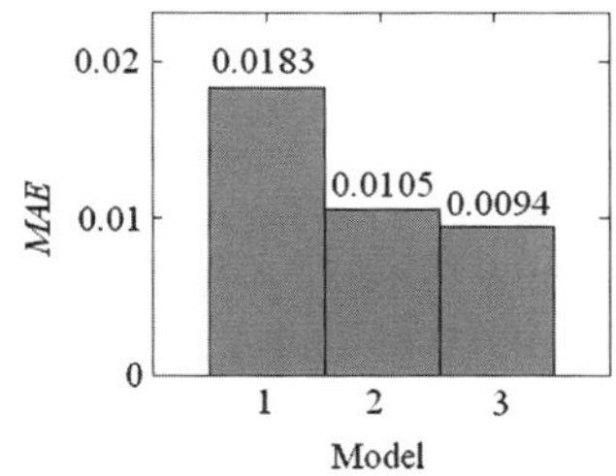

Fig. 10.5 Mean absolute errors of the following models: (*1*) Gaussian (Eqs. 10.14, 10.15) with solar irradiance derived from local climatology; (*2*) Gaussian (Eqs. 10.14, 10.15) with solar irradiance derived from NWP; (*3*) Beta regression with solar irradiance derived from NWP

Figure 10.5 shows that using at input solar irradiance derived from NWP improves prediction in comparison with the case when solar irradiance is derived from local climatology. Beta regression model brings a further small improvement comparing to Gaussian approach.

Joining the conclusion of both papers Bacher et al. (2009) and Brabec et al. (2010) we can draw some important conclusions. The accuracy of forecasting the output power of a PV system using autoregressive models is rather limited. Using models with exogenous data at input may increase considerable the prediction accuracy. We came to a similar conclusion in Chap. 6 of this book but from a different perspective, studying the ARIMA models at forecasting clearness index. We have noted that the forecasting accuracy is strongly related to radiative regime stability. We concluded that models that allow integrating at input parameters related to the state of the sky may increase accuracy. Also we concluded that exogenous inputs are more necessary as the forecasting horizon time increases, which is confirmed by Bacher et al. (2009). Brabec et al. (2010) shows that the use of NWP gives better results than local climatology inputs. In our opinion this is a too radical conclusion drawn in a concrete situation that may not apply otherwise. In general, the forecasting output power accuracy is influenced by the quality of the estimating/forecasting exogenous inputs. If there is well a calibrated local meteorological model, its output may be used successfully for a forecasting model input.

The conclusions above are based on results of traditional statistical models. The next section is dedicated to artificial intelligence approaches, mainly ANN models.

10.1.2 ANN-Based Models

A short introduction to ANN along with two examples of using ANN for forecasting solar radiation were previously briefed in Sect. 7.1.1 of this book. Here we continue the illustration of ANN models focusing on accuracy.

Mellit and Pavan (2010) developed a model based on a Multi-Layer Perceptron for 24 h ahead forecasting solar irradiance. This model and test results were briefly presented in Sect. 7.1.1 to exemplify the use of ANN in the tentative solving of a complicated, nonlinear relationship like Eq. (7.6). Mean daily solar irradiance, mean daily air temperature, and the Julian day are the parameters used to activate the three-neuron input layer. After passing two hidden layers of 11 and 17 neurons, respectively, the output is represented by the 24 values of solar irradiance in the next 24 h.

A comparable approach with some refinements is proposed in Chen et al. (2011). They attempt to directly predict the hourly output of a PV power plant. In their example, the plant has a peak power of around 18 kWp and is situated in the Renewable Energy Research Center of Huazhong University of Science and Technology (HUST), China. Data collected in the PV power system include SCADA registers with average 5 min power delivered by the PV power system to the grid. Numerical weather predictions models were used to supplement the number of inputs with the predicted daily mean of solar irradiance $\bar{G}_{t+1}$, air temperature $\bar{T}_{t+1}$, humidity $\overline{RH}_{t+1}$and wind speed $\bar{v}_{t+1}$. These forecasted values together with the time and current day power output amount to six inputs feed into the ANN. Their equivalent of Eq. (7.6) reads:

$$\left(P_{t+1}^{(1)}, P_{t+1}^{(2)}, \ldots, P_{t+1}^{(24)}\right) = f\left(\bar{G}_{t+1}, \bar{T}_{t+1}, \overline{RH}_{t+1}, \bar{v}_{t+1}, P_t, t\right) \tag{10.16}$$

The use of meteorological inputs has the potential to improve the prediction accuracy, but they have to be available and of reasonable accuracy themselves. The authors received NWPs outputs from the meteorological services of Wuhan and the accuracy of short-term weather forecasting of Wuhan can reach 90 %.

A notable feature is the use of radial basis functions (RBF) as activating functions for the (single) hidden layer. The RBF is similar to the Gaussian function which is defined by a *center* position and a *width* parameter. The Gaussian function gives the highest output when the incoming variables are closest to the center position and decreases monotonically as the distance from the center increases. The width of the RBF unit controls the rate of decrease and is a parameter to be experimentally optimized for best results, together with the number of hidden layer neurons. The authors tested with 5, 10, and 15 hidden layer neurons, the results proving no significant differences. Sunny, cloudy and rainy days being three different regimes, the authors took another refinement: instead of a general training of the same ANN they used three corresponding subsets, each leading to a specific network layout and optimization through the learning process. This way, they obtained excellent results for both sunny and cloudy days (correlation coefficient 98–99 %) and "acceptable" results for rainy days (correlation coefficient 50–80 %). In terms of mean absolute percentage error the model performance is summarized in Fig. 10.6.

Figure 10.6 also emphasizes the difficulty for the models to predict PV energy in weather instability conditions. Comparing *MAPE* in sunny and cloudy days, it

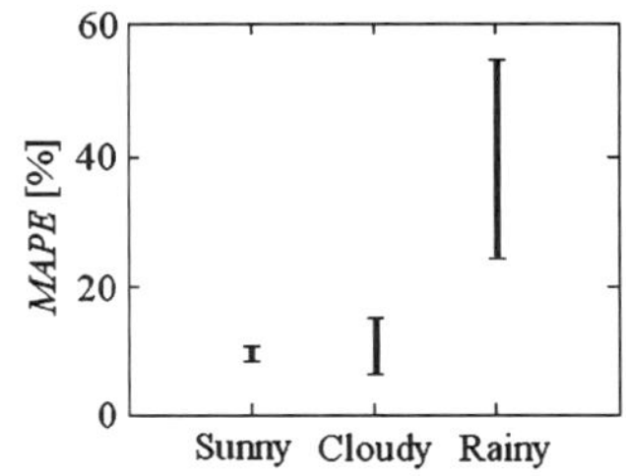

Fig. 10.6 Mean absolute percentage error for prediction distribution of solar irradiance in consecutive clear, cloudy and rainy days. The plot has been build with data from Chen et al. (2011)

can be seen that the range of errors is larger in the cloudy days than in the sunny days. This proves again that the meteorological stability is a key parameter in forecasting PV plant output accuracy. Further improvements of the model performance may be obtained if measures for characterizing the state of the sky and its stability are used as inputs. Similar findings have been discussed in Chap. 6 related to the very different ARIMA approach.

Prior to the above-mentioned study, Zervas et al. (2008) also used a RBF network, consisting of three layers, to predict daily distribution of global solar irradiance. The first input is a forecasted quantity S, the "state", characterizing the presence and type of clouds, using six discrete values: 1—rainfall, 2—heavy clouds, 3—cloudy, 4—partly cloudy, 5—few clouds, 6—clear. A *human expert meteorologist* is asked to perform the classification. The second input is L, being the half of the "number of daylight tenths". L is a function of $abs(\omega_s)$, with ω_s the sunrise (or sunset) angle. A Gaussian-type function is proposed for approximating the daily global solar irradiance distribution:

$$J(x) = M\exp(-a\frac{x^2}{L^2})$$

where M is the maximum value, x the distance from solar noon in daylight tenths, L is half of the number of daylight tenths and a is a tuning parameter. The model described by this equation is called the "SGGSI" (Simple Gaussian Global Solar Irradiance model). Comparing ratios of J/M from the model with measured ones, it follows that SGGSI is not able to fit them too well for the entire interval. The authors propose a correction such that the parameter a will be chosen to obtain the best fit for the region $-0.75 < x/L < 0.75$. This means high accuracy for the center of the Gaussian admitting errors on the tail sections. The corrected model is called the "AGGSI model" (Adjusted Gaussian Global Solar Irradiance model).

A fuzzy partition of the input space was used, produced by defining a number of triangular fuzzy sets in the domain of each input variable. In order to tune the model, a database was utilized which contained global solar irradiation measurements for an entire year (1 January 2004–31 December 2004). The measurements were recorded every 10 min by the ''ITIA'' Meteorological station of the National Technical University of Athens, Zografou Campus. Validation of the model was performed by testing it on the set of data set aside for this purpose. The corresponding coefficient of determination r^2 was 0.985, which is representative of the high correlation between experimental and predicted values using the RBF network methodology.

Paoli et al. (2010) was a second example already presented in Chap. 7 of this book to illustrate the use of ANN. A comparable approach can be considered Izgi et al. (2012), aiming at short-and medium-term prediction of generated electricity by solar cells under climatic conditions of Istanbul, Turkey (mild Mediterranean during summer). Their attempt to minimize prediction errors is also based on the ANN learning ability. The PV system of 750 Wp used in this study was put into operation in February 2009 at the meteorological park of Istanbul Technical University. Parameters of the PV system, ambient temperature, cell temperature global and diffuse solar irradiance are monitored. Data of electricity generation by PVs have been collected in 1 min time horizon. For short-term power prediction, only April and August data are considered in detail. In Istanbul conditions, April is a spring month and during this time uncertainties are very high. In this month, sometimes thunderstorms are observed with synoptic systems and sometimes convective systems are effective. This means that uncertainty is very high in this time interval and high prediction errors can be expected. In April, approximately 23,000 readings with 1 min time interval are used. During ANN application 70 % of data are used for training and the remaining 30 % for test procedures. The same procedure was applied in August, when solar irradiance values increased to the highest level and during this month cloud uncertainty conditions decreased to the lowest level. The ANN is a four-layer (two hidden) feed-forward network using sigmoid activation function and back propagation learning procedure, taking as input four values $P(t)$.. $P(t + 3)$. The output is the predicted value $P(t + 4)$. They concentrated on optimal time horizons for prediction and concluded that "in April between 5 and 35 min time horizons could be used for power prediction of PV modules, unfortunately, less than 5 and greater than 35 min time horizons getting worse prediction situations. During August, stationary solar irradiation conditions are prevalent allowing the ANN to predict accurately the generated electricity from 30 to 300 min ahead. Additionally, it is estimated that in August, between 3 and 40 min time horizons stable data conditions are conserved and averages of power at these horizons could be used for prediction.

For more background information, Kalogirou (2001) is a detailed report of ANN applications in renewable energy systems, including but not limited to photovoltaic systems. It starts with a section on general neural network principles; the section "Network parameters selection" contains important advices for setting up the ANN based on the author's experience. When using a neural network for prediction, the following steps are crucial. First, a neural network needs to be built to model the behavior of the process. The values of the output are predicted on the basis of the model. Second, based on the neural network model obtained on the first phase, the output of the model is simulated using different scenarios. Third, the control variables are modified to control and optimize the output.

In back propagation networks, the number of hidden neurons determines how well a problem can be learned. This number was one difficult optimization task also mentioned in more recent articles. If too many are used, the network will tend to try to memorize the problem, and thus not generalize well later. If too few are used, the network will generalize well but may not have enough “power” to learn the patterns well. Kalogirou (2001) proposes an empirical relation for choosing the number of hidden layer neurons, equal to:

$$\frac{1}{2}(\text{inputs} + \text{outputs}) + \sqrt{\text{number of training patterns}}$$

Kalogirou (2001) also includes a review on significant studies performed during 1995–2001.

At the end of this section it is worth to mention the work of Da Silva Fonseca Jr. et al. (2011). The subject of their report is forecasting the power production of a PV plant using a methodology based on support vector machines (SVM) and on the use of several numerically predicted weather variables, including cloud amount. SVM is an advanced artificial intelligence technique based on statistical learning theory (Cristianini and Shawe-Taylor 2000; Smola and Scholkopf 2004). It is suitable for pattern recognition problems but can also be used for solving prediction problems.

The SVM model reported by Da Silva Fonseca Jr. et al. (2011) uses at input weather forecast (temperature, relative humidity, and total cloud cover amount at three levels) and calculated extraterrestrial solar irradiation at the hour for which the power production is forecasted. The power production data used in the study are recorded at the PV power plant located in Kitakyushu, Japan. The PV power plant has a nominal output of 1 MW.

The forecasts of power production were hourly, and they were carried out for a period of one year. The effect of using numerically predicted cloud amount on the quality of the forecasts was also investigated by authors. The forecast of power production obtained with the proposed methodology reaches a root mean square error of 0.0948 MWh and a mean absolute error of 0.058 MWh. Da Silva Fonseca Jr. et al. (2011) found that the forecasted and measured values of power production had a good level of correlation varying from 0.8 to 0.88 depending of the season of the year. The authors also pointed out that the use of numerically predicted cloud cover amount had an important role in the accuracy of the forecasts When cloud cover amount was not used, the root mean square error of the forecasts increased more than 32 %, and the mean absolute error increased more than 42 %. The authors concluded that the proposed forecasting method provides modestly accurate results of power production in partially clouded days and in days with sudden changes in the amount of solar irradiance reaching the power plant. Low accuracy in days with unstable weather conditions was also reported in Chap. 6 for ARIMA forecasting models and in Chap. 7 for fuzzy forecasting models. Results reported in Da Silva Fonseca Jr. et al. (2011) also clearly indicate that the use of a numerically predicted measure for the state of the sky at input is improving the accuracy of the power production forecast.

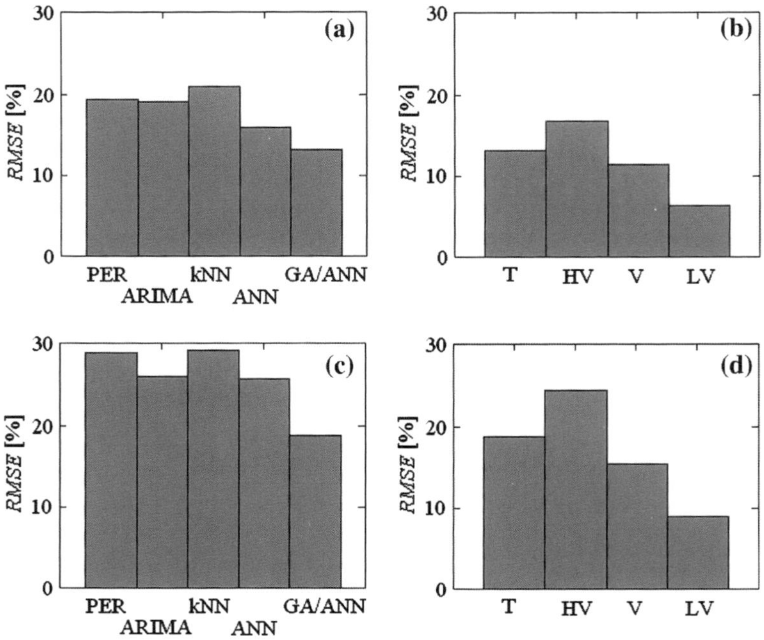

Fig. 10.7 RMSE of the five models tested in Ref. Pedro and Coimbra (2012): **a**, **c** 1 h time horizon. **b**, **d** 2 h time horizon. Notations on the graphs **b** and **d** are: *T*—entire validation data set; *HV*—High variability subset; *V*—Medium variability subset; *LV*—Low variability subset. Data from Pedro and Coimbra (2012) have been used to build the graphs

10.1.3 Comparison of Models Performance

The study Pedro and Coimbra (2012) is dedicated to the evaluation and comparison of five forecasting models for predicting the power output of a 1 MWp single-axis tracking PV plant operating in Merced, California. The models are selected to run without using exogenous data at input. Hourly average output power collected from November 3, 2009 (the first full day operating of the PV plant) to August 15, 2011 are used. Data prior to January, 1 2011 were used to train the forecasting models while data after January 1 have been used to test the model. The models studied in their work are: Persistent (PER) model, Auto-Regressive Integrated Moving Average (ARIMA), k-Nearest-Neighbors (kNN), ANN and ANNs optimized by Genetic Algorithms (GAs/ANN).

An exciting idea in this study is the clear sky decomposition of the power output data aiming to isolate the stochastic characteristic. More precise the output power P time series is decomposed as:

$$P = P_{\mathrm{CS}} + P_{\mathrm{ST}} \tag{10.17}$$

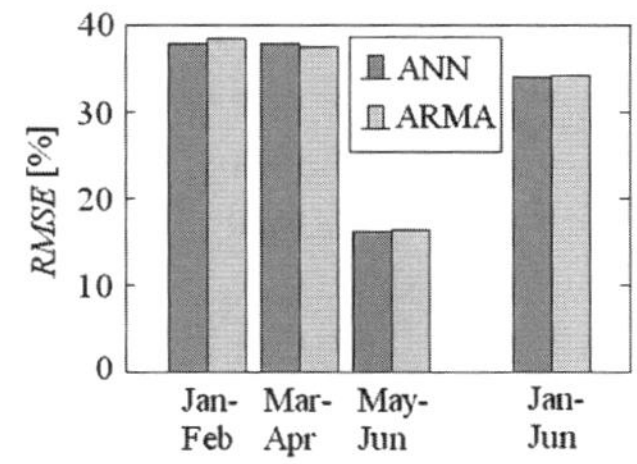

Fig. 10.8 Bi-monthly forecasting errors and 6 months forecasting errors of the electrical energy from PV plant for 1 day horizon. Data from Paoli et al. (2010) have been used to build the graph

where P_{CS} is the clear sky estimated output power and P_{ST} is the stochastic part. The clear sky output power P_{CS} is calculated by fitting the measured data with a smooth envelope surface. Excepting the persistent model, all four models forecast P_{ST}.

In terms of RMSE the performance of the five models are summarized in Fig. 10.7. For 1 h horizon of time (Fig. 10.7a) the best performance has been reached by the GA/ANN model (RMSE = 13.07 %). For 2 h horizon of time (Fig. 10.7c) the GA/ANN model was also found the best (RMSE = 18.71 %). Figures 10.7b, d show that the model performance increases with the decreasing variability of PV plant output power, i.e., with enhanced solar radiative regime stability.

The conclusions of Ref. Pedro and Coimbra (2012) are: (1) The ANN-based forecasting models perform better than the other forecasting techniques, (2) A substantial improvements can be achieved with a GA optimization of the ANN parameters; (3) The accuracy of all models depends strongly on seasonal characteristics of solar variability.

We return here to the Ref. Paoli et al. (2010) which basically deals with forecasting daily solar irradiation time series using ANN and ARIMA models. A brief summary of the paper is included in Sect. 7.1.1.2 of this book. Here, we refer only to the comparison of the ANN and ARMA(2,2) models performance as resulted from results reported in Paoli et al. (2010). The comparison is made for one day prediction of daily global solar irradiation on a surface tilted with 80°. This is the inclination of solar modules used further to validate the methods for PV out power prediction. Figure 10.8 gather the forecasting errors obtained for the best variant of ANN and ARMA(2,2) tested models during six month. Analyzing the results the author concludes: "ANN and ARMA models perform almost similar, denoting the stochastic nature of the time series and thus the impossibility to predict the cloud effect on solar radiation". Regarding the source of errors the authors emitted two valuable hypotheses: (1) High frequency noise series; it seems very unlikely that the ANN can predict "extra-ordinary days" at least if the previous day's cloud cover is ordinary. (2) ANN and ARMA models do not take risks and predict irradiation value centered on a mean value with a small standard deviation. The output is then an improved average that fits the history trend.

Finally, it is worth to acknowledge the paper of Picault et al. (2010) which reported a method for forecasting PV power in diverse environmental conditions. As a distinctive feature, reduction of mismatch losses by changing the interconnection wirings of the modules in PV arrays is addressed. The results are tested against field data, collected on a 2.2 kW_p PV system consisting of 20 Isofoton I-106 crystalline modules, part of the UNIVER project at Jaen University in Spain (Drif et al. 2007). In this experiment the PV generator was configured in four strings of five series-connected modules with a facility to rapidly change the module interconnection scheme. In addition to the most common topology for PV arrays in series-parallel (SP) (Fig. 9.4b) two other topologies, total-crossed tied (TCT) and bridge-link (BL), were studied.

The experimental procedure consisted in successively measuring the V–I characteristic of three topologies followed by the V–I characteristic recording of each module within the array. The procedure was carried out in both non-shaded and partially shaded scenarios. Static partial shade was performed by covering two modules with bubble wrap film, which decreased incoming irradiance by 40 % .

In order to model PV module performance, voltage-current curve in one diode model, conventionally described by Eq. (9.15), was expressed rather unusual as an exact analytical expression using Lambert *W*-function Picault et al. (2010):

$$I(V) = \frac{V_T}{R_s}\left[\frac{R_s(I_L + I_0)}{V_T} - W\left(\frac{I_0}{V_T}R_s \exp\left(\frac{V + R_s(I_L + I_0)}{V_T}\right)\right)\right] - \frac{V}{R_p} \tag{10.18}$$

where the notation are as for Eq. (9.15) and $R_s \ll R_p$ was assumed. The module shading circumstances are taken into account by applying a shade factor, with values taken between 0 (for totally shaded modules) and 1 (for non-shaded modules), to the irradiance received by the modules. Individual module shade factors are grouped into a matrix thus giving the shade scenario for the entire solar array.

Results of comparison of and field measurements reported by Picault et al. (2010) shows that in normal operating conditions (non-shading) all three topologies have similar power voltage characteristics. The mismatch losses are found between 1 and 2 %. In case of partially shaded conditions the difference between mismatch losses for the three interconnection scenarios became visible: TCT configuration experienced the smallest amount of mismatch losses (~3.8 %) while the SP configuration experienced the maximum losses (~7 %).

This paper demonstrates that in shadow conditions the performance of a PV plant can be enhanced by using alternative topologies of the PV generator. Real time control strategies for adaptive reconfiguration of solar PV arrays under partial shadow may be a future solution for maximizing the output power of a PV system under these circumstances. Research on this topic exists, e.g., Nguyen and Lehman (2008).

10.2 Smoothing PV Power Variability

For proper grid management, a forecast of the ensemble production of all PV systems contributing to a control area is necessary (Lorenz et al. 2009). Thus, regional PV power forecasts provide the basis for grid management and trading of

PV power on the energy market. On the local scale, smart grid applications define a sector with increasing need for PV power forecasting (Lorenz et al. 2011). Meanwhile, the geographic area of interest for forecasting can vary from a large area over which electricity supply and demand must be balanced to a much smaller region where grid congestion must be managed (Pelland 2011).

As shown in this book rapid changes in the output of PV plants are due to clouds. In comparison to the variability in solar irradiance measured in a point, the output power of large-scale plants exhibits a pronounced reduction in variability. Geographic diversity is another factor in smoothing power variations generated by an ensemble of distributed PV systems feeding into the same grid.

Smoothing PV power variability by aggregating spatially distributed solar systems in the same grid is the subject of this section. This is discussed in the light of several recent reports related to the matter.

Lorenz et al. (2009) report a move toward predicting regional PV power output based on weather forecast up to three days ahead provided by the European Center for Medium-Range Weather Forecasts (ECMWF 2012).

ECMWF provides forecasts of solar irradiance and cloud parameters with a temporal resolution of three hours and a spatial resolution of $0.25 \times 0.25°$. A resolution of three hours for expected solar power is too large for grid management. The authors investigated different approaches to refine the ECMWF global model irradiance forecasts, in order to derive optimized, site-specific, hourly forecasts. In brief these approaches refer to: (1) Spatial averaging and temporal interpolation, (2) Improved clear sky forecasts, and (3) Post processing with ground data. An optimum adjustment of the temporal resolution was achieved by combination with a clear sky model to consider the typical diurnal course of irradiance. Introducing an additional bias correction avoids systematic deviations for cloudy situations. In the study, solar data recorded at more than 200 meteorological stations in Germany have been used. According to the authors, irradiance forecasts one day ahead for single stations in Germany give a RMSE of 36 %. For regional forecasts, the accuracy is increasing with the size of the region. For the complete area of Germany, RMSE has been 13 %.

Lorenz et al. (2009) paper is mainly focused on the description and evaluation of the solar irradiance forecasting, as basis for PV power prediction. The authors evaluated PV power forecasts in a case study for an ensemble of 11 PV systems distributed over an area of 120×200 km in Southern Germany. The evaluation was performed for April and July 2006. These two months with different meteorological conditions were chosen in order to investigate the influence of weather conditions on the forecast accuracy. According to the results, for single systems a value of $RSME = 49$ % is reached in April, when cloudy situations were predominant. For July with mostly clear sky days in this region, a lower RMSE of 30 % was found.

Also, the authors state that the forecast errors are reduced to an RMSE of 39 % for April and 22 % for July when considering the power production of the complete ensemble of 11 systems. This corresponds to an error reduction factor of about 0.7 for the region of a size of 120×200 km.

A more recent study by the Lorenz's team (Lorenz et al. 2011) evaluates enhanced features of the regional PV power prediction system of the University of Oldenburg and Meteocontrol GmbH. As in Lorentz (2009) the study is based on forecasts of global solar irradiance and temperature provided by ECMWF, but this time 77 PV systems from Southern Germany are considered.

In the first part of the study, the performance of PV power forecasts for different spatial scales has been analyzed. The authors found that RMSE for a regional average (of size of 5 × 4.5°) is about half the RMSE of a single site. An evaluation of the irradiance forecasts in the same region revealed considerably smaller errors than for the PV power forecasts. In particular, during winter a strong increase of power forecast errors have been noted (over estimation of the power production). In order to improve forecast quality during winter, the authors have presented an empirical approach for enhanced PV power forecasting during periods of snow cover. Different criteria to identify snow cover on PV modules have been investigated. The parameters temperature and snow cover on the module on the previous day have been found to be robust indicators for snow-covered PV modules.

The authors compare forecasts of the proposed method with the operational forecast used by the German "50 Hz Transmission GmbH" grid operator for 1 year. Results show that RMSE of the forecasts could be reduced from 4.9 to 3.9 % for intra-day forecasts, and from 5.7 to 4.6 % for day-ahead forecasts by using the author's proposed method. For the winter period from 01 December 2009 to 28 February 2009, the RMSE of the regional forecast of PV power could be reduced to half compared with the errors when the snow detection algorithm was not used.

Marcos et al. (2011) reports a study regarding power output fluctuations in large scale PV plants. This study is based on one year data recorded at one second time interval at six PV plants in Spain. The power peak of the plants ranges from 1 to 9.5 MWp, totaling 18 MWp. In addition, data from two sections (48 and 143 kWp) of other PV plant have been used. The plants under analysis are scattered over roughly 1,000 km^2 area in the south of Navarra (Spain). The separation between the plants ranges from 6 to 60 km. All PV plants are equipped with vertical axis trackers and feed power into the 13.2 kV grid. Particular attention has been paid by authors to the analysis of the influence on the magnitude of power fluctuations coming from both the size of the PV plant and the sampling period. An analytical model to describe the daily frequency of encountering a power fluctuation of a certain magnitude has been presented.

The analysis of the data has revealed the smoothing effect of PV plant size on power fluctuations. The smoothing effect has been also found to be strongly dependent on the sampling-time considered Marcos et al. (2011).

Peland et al. (2011) proposed methods for hourly solar and PV power forecasts for horizons between 0 and 48 h ahead. The methods are based on post-processing of the outputs of the Canadian Global Environmental Multiscale (GEM) model. The solar and PV forecasts were compared with irradiance data from 10 North-American ground stations and with PV power data from three Canadian PV systems. A 1 year period was used to train the forecasts, and the following year was used for testing. Two post-processing methods were applied to the solar forecasts: spatial averaging and bias removal using a Kalman filter. On average, these two methods lead to a 43 % reduction in RMSE over a persistence forecast and to a 15 % reduction in RMSE over the Global Environmental Multiscale forecasts without post-processing. The authors noted that bias removal was primarily useful when considering a "regional" forecast for the average irradiance of the ten ground stations because bias was a more significant fraction of RMSE in this case.

The PV forecasting approach developed by Peland et al. (2011) is reasonably simple and requires only basic PV system information and historical output power data. The reported results of the tests placed RMSE in the range of 6.4–9.2 % for the three PV systems considered. About 76 % of the PV forecast errors were within ±5 % of the rated power for the individual systems, but the largest errors reached up to 57 % of rated power (Peland et al. 2011).

The issue of solar ramp occurrence and its smoothing is excellent summarized in the paper Mills et al. (2011). According to the authors, the apparent movement of the sun on the sky regularly leads to 10–13 % changes in PV output over a period of 15 min for single-axis tracking PV plants. Changes in solar irradiance at a point due to a passing cloud can exceed 60 % of the peak in a matter of seconds. The time it takes for a passing cloud to shade an entire PV system depends on several variables as the PV system size and cloud speed. For a PV plant with a peak power of 100 MW, it takes a time in the order of minutes rather than seconds to shade the system. Increasing the plant size decreases the output power ramp.

Another fine illustrated issue in Mills et al. (2011) is ramp smoothing when multiple power plants are aggregated in the same grid. The authors have analyzed a network of several time-synchronized solar irradiance measurements in the Great Plains region of the U.S. The measurement locations are in the place of six PV plants in the city of Las Vegas, four PV plants in Arizona and two PV plants in Colorado. The conclusions indicate that smoothing can occur on even longer time-scales between separate plants. The results presented indicate that the spatial separation between plants required for changes in output to be uncorrelated over time scales of 30 min is on the order of 50 km. The spatial separation required for output to be uncorrelated over time scales of 60 min is on the order of 150 km. The assumption that variability on a 15 min or shorter time-scale is uncorrelated between plants separated by 20 km or more is supported by data from at least one region of the U.S.

The authors conclude that when ramps over a particular time scale are uncorrelated between all N plants, the aggregate variability is expected to scale with $1/N$ relative to the variability of a single point. This diversity between multiple PV sites on all sub-hourly time scales needs to be accounted for in projections of variability that must be managed by system operators.

The conclusions of Mills et al. (2011) are general and can be applied everywhere. Integration issues are a major obstacle in increasing the share of solar generation in the energy mix. In this context, assessing the characteristics of aggregate PV output over large areas and correlation to load is critical. The variability observed by a point irradiance measurement is not the same with the variability exhibited by a spatially extended installation such a PV plant. A point measurement ignores sub minutes time scale smoothing that can occur within large-scale. Extrapolation to larger PV plants suggests that further smoothing is expected for short time-scale variability.

Both solar and wind energy have variable and uncertain output. The experience with managing wind variability will benefit solar integration efforts. Unified approaches for managing variable generation will ease renewable energy integration issues.

References

Almonacid F, Rus C, Perez-Higueras P, Hontoria L (2011) Calculation of the energy provided by a PV generator. Comparative study: conventional methods vs. artificial neural networks. Energy 36:375–384

Bacher P, Madsen H, Nielsen HA (2009) Online short-term solar power forecasting. Sol Energy 83:1772–1783

Brabec M, Pelikaan E. Krc P, Eben K, Musilek P (2010) Statistical modeling of energy production by photovoltaic farms. In: Proceedings of IEEE Electric Power and Energy Conference. doi: 10.1109/EPEC.2010.5697249

Brabec M, Pelikaan E. Krc P, Eben K, Maly M, Jurus P (2011) A coupled model for energy production forecasting from photovoltaic farms. Presented at COST ACTION ES 1002 Workshop, 22–23 Mar 2011. http://www.wire1002.ch/fileadmin/user_upload/Major_events/WS_Nice_2011/Spec._presentations/Brabec.pdf

Chen C, Duan S, Cai T, Liu B (2011) Online 24-h solar power forecasting based on weather type classification using artificial neural network. Sol Energy 85(11):2856–2870

Cristianini N, Shawe-Taylor J (2000) An introduction to support vector machines and other kernel-based learning methods. Cambridge University Press, Cambridge

Da Silva Fonseca Jr JG., Oozeki T, Takashima T, Koshimizu G, Uchida Y, Ogimoto K (2011) Use of support vector regression and numerically predicted cloudiness to forecast power output of a photovoltaic power plant in Kitakyushu, Japan. Prog Photovoltaics Res Appl. doi: 10.1002/pip.1152

Drif M, Perez PJ, Aguilera J, Almonacid G, Gomez P, De La Casa J, Aguilar JD (2007) Univer project. A grid connected photovoltaic system of 200 kWp at Jaen University. Overview and performance analysis. Sol Energy Mater Sol Cells 91:670–683

ECMWF (2012) European Centre for Medium-Range Weather Forecasts. http://www.ecmwf.int/

Green MA (1982) Solar cells: operating principles, technology and system application. Prentice-Hall, New Jersey

Izgi E, Oztopal A, Yerli B, Kaymak MK, Sahin AD (2012) Short-mid-term solar power prediction by using artificial neural networks. Sol Energy 86:725–733

Jang JS (1993) ANFIS: adaptative-network-based fuzzy inference system. IEEE Trans Semicond 83:378–406

Kalogirou SA (2001) Artificial neural networks in renewable energy systems applications: a review. Renew Sust Energy Rev 5:373–401

Lorenz E, Hurka J, Heinemann D, Beyer HG (2009) Irradiance forecasting for the power prediction of grid-connected photovoltaic systems. IEEE J Sel Top Earth Observations Remote Sens 2(1):

Lorenz E, Heinemann D, Kurz O (2011) Local and regional photovoltaic power prediction for large scale grid integration: Assessment of a new algorithm for snow detection. Prog Photovoltaics Res Appl. doi: 10:1002/pip:1224

Martin L, Zarzalejo LF, Polo J, Navarro A, Marchante R, Cony M (2010) Prediction of global solar irradiance based on time series analysis: Application to solar thermal power plants energy production planning. Sol Energy 84:1772–1781

Marcos J, Marroyo L, Lorenzo E, Alvira D, Izco E (2011) Power output fluctuations in large scales PV plants: one year observations with one second resolution and a derived analytic model. Prog Photovoltaics Res Appl 19:218–227

Mellit A, Pavan AM (2010) A 24-h forecast of solar irradiance using artificial neural network: Application for performance prediction of a grid-connected PV plant at Trieste. Italy Sol Energy 84(5):807–821

Mills A, Ahlstrom M, Brower M, Ellis A, George R, Hoff T, Kroposki B, Lenox C, Miller N, Milligan M, Stein J, Wan Y-h (2011) Understanding variability and uncertainty of photovoltaics for integration with the electric power system. IEEE Power Energ Mag 9(3):33–41

Ngyuen D, Lehman B (2008) An adaptive solar photovoltaic array using model-based reconfiguration algorithm. IEEE Trans Ind Electron 55(7):2644–2654

Osterwald CR (1986) Translation of device performance measurements to reference conditions. Sol Cells 18:269–279

Paoli C, Voyant C, Muselli M, Nivet M-L (2010) Forecasting of preprocessed daily solar radiation time series using neural networks. Sol Energy 84(12):2146–2160

Pedro HTC, Coimbra CFM (2012) Assessment of forecasting techniques for solar power production with no exogenous inputs. Sol Energy 86:2017–2028

Pelland S, Galanis G, Kallos G (2011) Solar and photovoltaic forecasting through post-processing of the Global Environmental Multiscale numerical weather prediction model. Prog. Photovoltaics: Res. Appl.. doi: 10:1002/pip:1180

Picault D, Raison B, Bacha S, De La Casa J, Aguilera J (2010) Forecasting photovoltaic array power production subject to mismatch losses. Sol Energy 84:1301–1309

Smola AJ, Scholkopf B (2004) A tutorial on support vector regression. Stat Comput 14(3):199–222

Zervas PL, Sarimveis H, Palyvos JA, Markatos NCG (2008) Prediction of daily global solar irradiance on horizontal surfaces based on neural-network techniques. Renew Energy 33:1796–1803

Chapter 11
Perspectives

A fact is definite: in the future renewable energies will be part of our life. Climate change concerns, high prices of fossil fuels, and increasing political support are driving renewable energy prospects. More and more clean electricity is generated by photovoltaic and solar-thermal systems for supporting our high-tech life. Solar power generators ranging from small standalone to large grid-connected systems are fitting well in the new energy paradigm and are put into operation on daily basis around the world. To date, there are two challenges standing against the growing share of photovoltaic systems in the energy mix.

The first challenge applies to all PV systems and refers to the price of solar generators which are still high in comparison to that of power plants based on fossil fuel. Major efforts are spent all over the world to reduce all costs associated to solar electricity production. Nevertheless, significant efforts are still to be done by researchers (to increase the module efficiency and reduce production costs) as well as by policy makers and governments (to improve legislation, incentives, commercialization), so that the day in which solar electricity will be fully competitive to come closer.

The second challenge stems from the intrinsic nature of solar energy which, although deterministic, is also stochastically fluctuating in time. Thus, the problem to be solved refers to technology developments and the integration of large solar power plants into the electricity grid. Solutions are searched on several levels: (1) The national energy policy related to sustainable development of power plants in order to maintain an optimal and safe energy mix in the grid; (2) Engineering efforts to reduce the response time of other power plants having the role to compensate the fluctuations of solar generators; and (3) Meteorology and atmospheric physics contributions to accurate forecasting of the solar plant output power for balanced power grid management.

Forecasting the output power of a PV plant requires forecasting the solar irradiance, translating the module V–I characteristics and evaluating the inverter efficiency in the anticipated meteorological condition. In many aspects,

M. Paulescu et al., *Weather Modeling and Forecasting of PV Systems Operation*, Green Energy and Technology, DOI: 10.1007/978-1-4471-4649-0_11,

the prediction of the PV plant output power can be identified with the forecasting of solar irradiance, which is by far the hardest task. The radiative regime is strongly correlated with the weather pattern. On time horizon of more than several hours, numerical weather prediction models offer the best solution. On shorter time horizon, forecasting solar irradiance is currently performed by extrapolating field measurements. At this time horizon, the solar ramp, i.e., the abrupt change in solar irradiance level, occurs when the sun is covered or uncovered by clouds. This brings the most important problem in inserting a major solar generator into the grid. Depending on its size, the output power of a PV plant follows almost instantaneous this change. Modeling and forecasting the sunshine number and sunshine stability number may be an adequate solution to anticipate the ramp moment. Moreover, including sunshine number in models of forecasting clearness index may also be a solution for increasing the accuracy of forecasting solar irradiance.

Studies reported in the literature do not come to a decision concerning the best performer between statistical or artificial intelligence models. The accuracy of both classes is comparable. The winner may be either an ARIMA, logistic, fuzzy, ANN or other model. Most likely the difference will be made by the ability of the model to accept at input, in addition to past values of the forecasted quantity, other variables measuring the state of the sky.

There is no doubt that forecasting is always accompanied by uncertainty. Overall results summarized in this book show that the level of accuracy in forecasting solar irradiance is still low. The performance is strongly linked to the stability of the radiative regime: the more stable it is, the more accurate the prediction. Therefore, efforts should be devoted for increasing forecasting accuracy in fast alternating state of the sky, which stress again a possible role for the sunshine number.

The models of estimating solar irradiance cannot be neglected in the forecasting task of solar irradiance. Weather models predict with high accuracy atmospheric meteorological parameters, which may be used as entries in the estimation models for forecasting solar irradiance under clear sky. The sunshine number may be subsequently used to adapt the result to the actual state of the sky.

To conclude, various models based on a multitude of methods exist to predict solar irradiance. This is the key undertaking in forecasting PV power output. The criterion of choosing a model is not only its performance but also the availability of data for the input parameters. Despite numerous results reported in the literature and analyzed in this book along with many examples, we abstain to recommend a specific model. We have endeavored to present a comparative study, leaving the reader an informed choice of the most suitable model for his application. On the other hand, the models quality presented here reflects the standing of June 2012 and is expected to improve incessantly in the years to come.

Appendix

A.1 Statistical Indicators

Several statistical indicators are shortly presented here. One denote by v_i, $i = 1 \ldots N$, a series of values. The first statistical moments of the evaluated and reference series, respectively, are the mean $\bar{v}$, the standard deviation σ, the skewness γ_3 and the kurtosis γ_4, defined as follows:

$$\bar{v} \equiv \frac{\sum_{i=1}^{N} v_i}{N} \tag{A.1}$$

$$\sigma \equiv \frac{\sum_{i=1}^{N} (v_i - \bar{v})^2}{N} \tag{A.2}$$

$$\gamma_3 \equiv \frac{\sum_{i=1}^{N} [(v_i - \bar{v})/\sigma]^3}{N} \tag{A.3}$$

$$\gamma_4 \equiv \frac{\sum_{i=1}^{N} [(v_i - \bar{v})/\sigma]^4}{N} - 3 \tag{A.4}$$

For convenience one reminds some basic facts (Neter et al. 1979; Griffith and Amrhein 1991). The mean is sensitive to extreme values. In normal or symmetrical distributions these extremes balance out. In skewed distributions there is one long tail which is not balanced by the values in the other tail. In these cases the mean is not a good summary statistic. The standard deviation is a measure of data spreading given in the same units as the actual values. The standard deviation is a good unbiased estimate for normal distribution but can become a highly unreliable estimate if skewness exists in the data. Skewness measures deviations from symmetry. It will take a value of zero when the distribution is a symmetric bell

M. Paulescu et al., *Weather Modeling and Forecasting of PV Systems Operation*, Green Energy and Technology, DOI: 10.1007/978-1-4471-4649-0,

shaped curve. A positive value indicates the observations are clustered more to the left of the mean with most of the extreme values to the right. Kurtosis is a measure of the relative peakness of the curve defined by the distribution of the observations. A normal distribution will have a kurtosis of zero while a positive kurtosis indicates the distribution is more peaked than a normal distribution.

One denote by $v_{\text{eval},I}$, $i = 1 \ldots N$ a series of values to be *evaluated* against another *reference* series of values $v_{\text{ref},I}$, $i = 1 \ldots N$. Here the index i is associated to a set of states which is similar for both series. Three usual statistical indicators of accuracy are the mean bias error (MBE), the mean absolute error (MAE) and the root mean square error (RMSE), defined by:

$$\text{MBE} \equiv \frac{\sum_{i=1}^{N} \left(v_{\text{eval},i} - v_{\text{ref},i} \right)}{N} \tag{A.5}$$

$$\text{MAE} \equiv \frac{\sum_{i=1}^{N} \left| v_{\text{eval},i} - v_{\text{ref},i} \right|}{N} \tag{A.6}$$

$$\text{RMSE} \equiv \left(\frac{\sum_{i=1}^{N} \left(v_{\text{eval},i} - v_{\text{ref},i} \right)^2}{N} \right)^{1/2} \tag{A.7}$$

MBE is a measure of systematic errors (or bias) whereas RMSE is mostly a measure of random errors. MAE is a quantity used to measure how close forecasts or predictions are to the final outcomes.

Another indicator is the second centered moment of the error distribution s_D, defined by:

$$\text{SD} = \frac{\sum_{i=1}^{N} \left(v_{\text{eval},i} - v_{\text{ref},i} - \text{MBE} \right)^2}{N - 1} \tag{A.8}$$

The units for the indicators MBE, MAE, RMSE and s_D are those of the quantities in the time series. Sometimes, dimensionless indicators MBE, MAE, RMSE and s_D are obtained by dividing Eqs. (A.5–A.8) through $\bar{v}_{\text{ref}}$.

A dimensionless indicator is the index of agreement d_2, defined by Willmott et al. (1985) as:

$$d_2 = 1 - \frac{\sum_{i=1}^{N} \left(v_{\text{eval},i} - v_{ref,i} \right)^2}{\sum_{i=1}^{N} \left(\left| v_{\text{eval},i} - \bar{v}_{\text{ref}} \right| + \left| v_{\text{ref},i} - \bar{v}_{\text{ref}} \right| \right)^2} \tag{A.9}$$

The index of agreement varies between 0.0 and 1.0 where a value of 1.0 expresses perfect agreement between the series v_{eval} and v_{ref} while 0.0 describes complete disagreement.

A measure of how well future outcomes are likely to be predicted by the model is the coefficient of determination r^2:

$$r^2 = 1 - \frac{\sum_{i=1}^{N} \left(v_{\text{eval},i} - v_{\text{ref},i}\right)^2}{\sum_{i=1}^{N} \left(v_{\text{ref},i} - \bar{v}_{\text{ref},i}\right)^2} \tag{A.10}$$

The coefficient of determination r^2 stands for the proportion of variability in a data set that is accounted for by the statistical model.

References

Griffith DA, Amrhein CG (1991) Statistical analysis for geographers. Prentice-Hall, New Jersey
Neter J, Wasserman W, Whitmore G (1979) Applied Statistics. Allyn and Bacon, Boston
Willmott CJ, Ackleson SG, Davis RE, Feddema JJ, Klink KM, Legates DR, O'Donnell J, Rowe CM (1985) Statistics for evaluation and comparison of models. J Geophys Res 90(C5):8999–9005

Index

M. Paulescu et al., *Weather Modeling and Forecasting of PV Systems Operation*,
Green Energy and Technology, DOI: 10.1007/978-1-4471-4649-0,

Printed by Printforce, the Netherlands